The Texas 36th Division

A HISTORY

Bruce L. Brager

EAKIN PRESS Fort Worth, Texas
www.EakinPress.com

Dedication

*To Private Bruce Kouser (1925–44), uncle and namesake,
replacement rifleman 142nd Regiment, 36th Division.
To his comrades in the 36th Division,
its ancestors, and its descendants.*

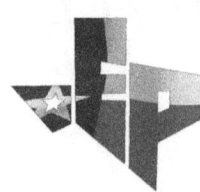

Copyright © 2002
By Bruce L. Brager
Published By Eakin Press
An Imprint of Wild Horse Media Group
P.O. Box 331779
Fort Worth, Texas 76163
1-817-344-7036
www.EakinPress.com
ALL RIGHTS RESERVED
1 2 3 4 5 6 7 8 9
Paperback ISBN 978-1-68179-321-4
Hardback ISBN 978-1-68179-323-8
eBook ISBN 978-1-68179-325-2

Contents

Introduction		v
Chapter 1	Prelude	1
Chapter 2	Reconstruction and Beyond	9
Chapter 3	Getting Ready for War	27
Chapter 4	Combat at St. Etienne	49
Chapter 5	Forest Farm	67
Chapter 6	Between the Wars	78
Chapter 7	Getting Ready for War Again	96
Chapter 8	Salerno	121
Chapter 9	San Pietro	144
Chapter 10	The Rapido River	158
Chapter 11	Velletri	178
Chapter 12	Anvil/Dragoon	195
Chapter 13	The Battle of Montelimar	208
Chapter 14	Routine Combat	222
Chapter 15	Lost Battalion	231
Chapter 16	Winter Warfare	241
Chapter 17	The End of the War	256
Postlude and Meaning		265
Appendix: The Cost		271
Endnotes		273
Bibliography		305
Index		319

Acknowledgments

The author wishes to thank:
* Pathfinders for the story of the 36th Division—Vincent M. Lockhart (who also reviewed part of this work), Robert L. Wagner, Lonnie J. White.
* The late Edwin Eakin and the staff of Eakin Press, particularly Virginia Messer and Angela Buckley.
* Staff of research depositories, including:
    Daughters of the Republic of Texas Library, San Antonio, Texas
    Texas Military Forces Museum, Camp Mabry, Austin, Texas
    The Center for American History, The University of Texas at Austin, Texas
    The Confederate Research Center, Hill College, Hillsboro, Texas
    Archives/Museum, The Citadel, Charleston, South Carolina
    Main Branch, Arlington Country Library, Arlington, Virginia
    The Institute of Texan Cultures, San Antonio, Texas
    The Library of Congress, Washington, D.C.
    The National Archives, Washington, D.C.
    The George Washington University Library, Washington, D.C.
    National Guard Association Library, Washington, D.C.
    The Texas Library and Archives Commission, Austin, Texas
* The collectors of the "36th Division Collection" and other material at the Texas State Library, including the Library and

Archives Commission, Robert L. Wagner, and the late Governor John Connally of Texas.
* The authors, editors, and gatherers of the multitude of documents and books checked, or read in full, for this project.
* The 36th Division Association, including the staff of the *Fighting 36th Historical Quarterly*.
* Dell Computers, Austin, Texas, for the 333P computer. Obsolete and rickety by today's standards, running MS-DOS, since retired, but more than sufficient.
* All my history and social studies teachers and college professors. Something got through.
* Assorted family and friends, particularly Laurie Brager (sister) and Dan Laurence (cousin and Texas tour guide).
* My Mother, the late Beverly Brager, whose estate financed this work and who named me after a veteran of the 36th Division. Hope it was worth it.
* The others who helped, whose names I forgot to write down, or never knew.

Introduction

"Texans! The troops of other states have their reputations to gain; the sons of the defenders of the Alamo have theirs to maintain! I am assured you will be faithful to the trust!"
—Jefferson Davis, Confederate President, 1861[1]

"Let me say that on every hand I heard that Texas is wanted here if but for the moral effect of her fearful name."
—John Marshall, Virginia citizen, 1862[2]

August 26, 1944, southern France. The only offensive action the 36th Division planned was an attack by a temporary task force, Task Force Butler (commanded by Brig. Gen. Frederick B. Butler, VI Corps deputy commander), through the Condilac Pass to restore a roadblock near the French city of Montelimar. The American roadblock had been blocking German withdrawal north up Route N-7, the main north-south artery along the Rhone River.

The American attack began early in the afternoon. Butler sent two rifle companies, from the 3rd Battalion, 143rd Infantry Regiment (36th Division), over the northern slope of Hill 300. As they neared the highway, the American troops were reinforced by a platoon of medium tanks and some tank destroyers coming directly out of Condilac Pass. Unfortunately, in addition to the Germans holding the highway, Task Force Butler quickly found itself facing additional German forces moving down from the north and heading up from the south. In the face of the heavy German opposition, the American attack failed. The infantry held part of Hill 300, with the armor withdrawing back into the pass.

The 3rd Battalion's commander remembered: "We didn't have much trouble taking the high ground, but we had trouble keeping it ... Fortunately, we had good artillery communications, and they were blasting the valley. As the Germans attacked over our positions, we called the artillery down on our own location. There was nothing else we could do."[3]

The Presidential Unit Citation the battalion won described the rest of the battle:

> The following day, when attacked by an enemy force of battalion strength, units of the battalion fought valiantly to repel the attackers and inflict upon them an estimated 30 percent dead, while other units of the battalion courageously beat off successive enemy tank and infantry attacks from the north.
>
> The members of the battalion directed thousands of rounds of mortar fire into the enemy, blocking the highway with the debris of destroyed vehicles and trucks.
>
> On 29 August 1944, the enemy in overwhelming numbers desperately attacked the battalion and succeeded in infiltrating through and dividing it into small units. Although completely isolated from other units and faced with possible annihilation, the members of the 3rd Battalion fought furiously to hold their positions and by mid-morning had completely beaten the hostile forces, who suffered tremendous losses in personnel and equipment.
>
> During this action, the 3rd Battalion captured more than 600 prisoners, including the Commanding General of the German 198th Infantry Division.[4]

The formal history of the 36th Division began twenty-seven years before the fighting over N-7, with the formation of the division just after the United States entered World War I. The 36th Division's documented ancestors in the Texas National Guard, the Texas Volunteer Guard, and the Texas militia trace back to the 1870s. The tradition in which the 36th played so great a part stretches back even before the 1836 defense of the Alamo. The tradition continues today.

Whatever lessons he may present, this writer set out to produce a book to be read for entertainment, and hopes to be judged according to whether he has produced a "good read." He does not claim to have produced the comprehensive history of a military unit with a story more than a century and a half long. This writer never intended to include all the details, to trace all the trails to their end, or to answer all the questions one might ask about the various phases and aspects of the history.

This is not *the* history of the 36th Division. However, this is *a* history of those men, giving the flavor of several fascinating periods. Tracing this history shows how a story which began as Texas history became American history.

CHAPTER 1

Prelude

"The war we are going into will be only a breakfast spell and we are in an awful hurry to get to the front to have a part in whipping the Yankees. It is a fact that one Southern man can whip ten Yankees."[1]

—Capt. William K. Martin,
Henderson Guards,
Fincastle, Texas, May 1861

Texans had a hard time getting to the Civil War. As was the case in most of the United States, there was little effective militia structure to expand to meet wartime needs.

The American militia had stumbled along with the rest of the American military system until 1861. The militia, like the regular army, performed well on occasions, badly on other occasions. Outside of wartime, militia organizations were social clubs far more than military training organizations. "Muster day, in the period before the Civil War, was primarily a social occasion."[2] Militia had played some role in earlier wars. Most, but not all, militia units panicked at the Battle of Bladensburg, August 1814, and allowed British units to enter—and burn—Washington, D.C. Other American militia units, however, took part in the smashing American victory over the British at New Orleans in January 1815.

"Citizen soldier," the idea that every able-bodied man has the responsibility to defend his homeland and his monarch, goes back to ancient times. The Anglo-American tradition has been traced

1

back to Anglo-Saxon England, well before the Norman invasion of 1066.[3] As the system evolved, all men, within certain age limits, owed an obligation to serve when needed in defense of the realm. The Normans adopted a similar system, eventually adding the variation of "trained bands," volunteer and drafted men who received better training, predecessor to the "ready" militia. Even with these variations, the militia was a home service force, never approaching the effectiveness of knights or of a regular army.

Colonial Americans inherited this tradition and structure. Every man within a range of ages was a member of the ordinary militia. The better-trained volunteer militia, the first to be called out in case of crisis, is considered an ancestor of the modern National Guard. The closest equivalent to the ordinary militia would be the liability of all men to the draft.

One very significant difference emerged between the American and British militia systems. The American militia was not considered a tool of the central government, but a means of protection *from* the central government. "Democratic" militia versus "tyrannical" standing army, versus emergency volunteers, would be as significant a part of the debate over military structure as would the question of effectiveness.

George Washington was not happy with the way American militia performed in the Revolutionary War. Modern historians have both agreed and disagreed. One historian has even offered an interesting analysis of one overlooked role the militia played in the Revolution:

> Repeatedly it was the militia which met the critical emergency or, in less formal operations, kept control of the country, cut off foragers, captured British agents, intimidated the war-weary and disaffected . . . While the regular armies marched and fought more or less ineffectually, it was the militia which presented the greatest single impediment to Britain's only practicable weapon, that of counter-revolution. The militias were often much less than ideal combat troops . . . But their true military and political significance may have been underrated.[4]

The United States Constitution compromised between centralized operations and quality control and "more democratic" control of the states. Article I, Section 8, gives Congress the power:

To provide for calling forth the Militia to execute the Laws of the Union, suppress Insurrection and repel Invasions;

To provide for organizing, arming, and disciplining the Militia, and for governing such Part of them as may be employed in the Service of the United States, reserving to the States respectively, the Appointment of Officers, and the Authority of training the Militia according to the discipline prescribed by Congress.[5]

The government compromised by neglecting the regular army. By 1784 the official complement of the army was eighty privates and just a few officers.[6] Growing problems with Indian tribes, and unsettled issues with the British and the French, led to expansion of the regular army and the 1792 efforts to regularize the militia (the last such laws until 1903). Both forces were put to the test in the War of 1812. The war itself was settled by treaty, not by force of arms. Militia troops reached their low point when the Maryland militia fled the field at Bladensburg, allowing the British to enter and burn Washington, D.C. Tennessee militia, however, played a major role in the overwhelming American victory at New Orleans, January 8, 1815, after the war had formally ended. Little had changed by 1835, when the militia was supplementing the small regular army in meeting the limited military needs of the United States.

The militia formally began in Texas almost at the same time as English-speaking settlers arrived. On February 18, 1823, the emperor of the newly independent Mexico authorized Stephen Austin "to organize the colonists into a body of the national militia, to preserve tranquility."[7] Soon after, Austin's militia unit was authorized to "make war on Indian tribes, who were hostile and molested the settlement."[8] Austin's militia soon battled raiding parties of Karankawa Indians, finally forcing them to cease raiding. According to a study of the Texas militia, however, "Despite this success Austin's militia remained small and imperfect, relying on mostly small units recruited for the duration of an emergency."[9]

This was the period in which a better-known competitor to the Texas militia was formed. Volunteer "ranging" cavalry companies were established as separate organizations from militia. They quickly became known as "Texas Rangers," though the name did not become formal until after the Civil War. Stephen Austin formed the first temporary ranger company in 1823, to protect his colony from Indian (primarily Comanche) attack. Further companies were raised, for limited terms of service, during the 1835–1836 Texas Revolution from Mexico. Rangers fought in the Mexican War and handled Texas frontier defense during the Civil War. However, the early rangers were best known for their role in active campaigning against the Comanches.

The Texas militia did little. The Texas War of Independence depended on Sam Houston and the regular Texas army, and informally organized volunteers, most famously at the Alamo. Eight months after the War of Independence ended at San Jacinto, on December 6, 1836, the Texas Congress passed a law organizing a militia.[10] However, the Congress refused to appropriate any money to carry out this law.

The Texas Revolution seems to have been another example of the very human trait of failing to learn proper lessons from experience, in this case making the very easy mistake of learning the wrong lessons from success. A military study of the Texas Revolution warns us that:

> [The Texas Revolution] left a legacy of valor that has inspired Texan soldiers on battlefields all over the world. While remembering the courage, however, one should also recall the disorganization, the pettiness and the lust for power that required so much needless sacrifice. Moreover, it should be acknowledged that the sacrifice was not all on one side.[11]

Texas settlers, mostly from the Old South, had brought the American stereotype that only a few dedicated and independent-minded men were needed to accomplish a lot, even in war. The military study continues, "The American colonists of Mexican Texas were no strangers to war: they were born to it. Most descended from America's first revolutionaries, and many had fought with Andrew Jackson in 1815 at New Orleans."[12] Some Texans may have learned from New Orleans, and Jackson's expert choice of ground and placement of artillery.

Others thought elan and spirit were sufficient. Just look at the American Revolution. (Benjamin Franklin's expert diplomacy, Washington's military intelligence, and Baron von Steuban's training programs were forgotten.) The romantic adventurism of Walter Scott's novels is credited with adding a mood, making war romantic. The antebellum southern character, so strong with the Texas settlers, included a taste for quick action without regard for consequences. These "Texas" characteristics, good and bad, were only strengthened by their almost accidental victory at San Jacinto. They would appear again.

In 1846, just before the start of the Mexican War, Gen. Zachary Taylor arrived in Texas with regular United States army troops. Expecting war, Taylor asked the governor for militia troops to supplement his army. Gov. J. P. Henderson replied that the militia was not sufficiently well organized, and issued a call for volunteers. Taylor praised the Texans at the Battle of Monterrey. However, he also wrote, "I regret to report that some shameful atrocities have been perpetuated by them" since the town was captured.[13]

When the Mexican War ended, so, effectively, did the Texas militia. The regular United States Army took over protection of the frontier. Ranger companies provided a second line of defense. Texans were not happy with the quality of Federal defense of the frontier. By 1860 many Texans in the frontier areas felt that the federal government had "displayed a cold blooded indifference to our condition."[14] This apparent government failure "was one of several factors that moved the frontier counties to endorse secession."[15]

In 1856 the sole remaining formal member of the militia, the adjutant general, was suspended from office. A new militia law was passed on February 14, 1860, at the urging of Gov. Sam Houston. Houston would appoint a commanding adjutant general, A. B. Norton, and several brigadier generals. Though some units continued to exist, the Texas militia itself was never organized. This likely resulted from a combination of inertia and Houston's reluctance to help recruit troops that might be used to fight the federal government. Events of the next several months proved Houston's fear to be warranted.

The military needs of the country, and of Texas, changed drastically in 1861, from repelling foreign invasion to engaging in civil war. The two "national" governments faced similar problems: how to meet massive military needs with portions of the tiny regular army. The enlisted soldiers of the army, unable to resign at will, remained Federal. About one-third of the officer corps "went south" to fight for the Confederacy. The total size of the army in 1860 was about 16,000 men. For the North, doubling in size would be substantially insufficient to meet its needs. The South had to recruit fresh virtually all its enlisted personnel. Both sides eventually adopted conscription, the first national draft laws in United States history, but such a drastic measure was not feasible in 1861. A historian of the Civil War military command structures summarizes the problems facing both armies:

> Both the Union and the Confederacy followed the same procedures in establishing their military commands and recruiting the huge armies they perceived as necessary for attack and defense. Both relied on new volunteer forces, rather than on the ill-trained militia units, to provide the framework for mobilization. Both central governments depended on the individual states to play a crucial part in the creation and mobilization of the mass armies. This was a natural, and indeed an essential, approach in view of the available machinery of government... In their turn the states depended on a good deal of local and individual entrepreneurship... Notwithstanding the military imperative and the issues at stake, politics had much to do with the raising of the armies. This was natural in an era when people took their politics very seriously...[16]

This was also natural in a civil war, that most political of conflicts. State power and influence, even in the far more nationalistic North, made it politically necessary to pay close attention to the wishes of the states. States' rights was the foundation concept for the Confederacy. However hard this made things for Jefferson Davis and the central Confederate government in Richmond, this was as real a part of the South's war as any military necessity. The same was true for Abraham Lincoln. Northern state political leaders were more responsive to the idea that some things had to done by a central government. Lincoln, however, still had to take state wishes into account when creating Federal policy.

The political realities of the Civil War would often be ignored in some later analysis of the conduct of the war. The best-known of the post–Civil War military reformers, it was later said, "never understood the necessities under which Lincoln operated, which forced him to appeal to the governors for aid and to give them something in return."[17] Lincoln and Davis could not afford the luxury of ignoring politics.

★ ★ ★ ★ ★

Early recruitment of Texas forces in the secession crisis was hampered by lack of support from Governor Sam Houston. The strongly unionist Houston would not let his state recruit men to fight his country.

By the time war began, however, Houston was out of office, removed after Texas voted for secession. Houston had refused to take an oath of allegiance to the Confederacy. His fellow Texans preferred to ignore his advice about the danger of secession, about the Yankees who, "when they begin to move in a given direction, where great interests are involved . . . move with the steady momentum and perseverance of a mighty avalanche . . ."[18]

Opinions with less foresight prevailed, and Texas geared up to join the Confederate war effort. After the firing on Fort Sumter, Texas began full-scale recruiting. Eventually, Texas would send approximately 90,000 men into Confederate service,[19] though estimates vary. About 4,000 of these enlisted in the thirty-two companies,[20] from southeast and central Texas (including one directly recruited in Texas in 1862 by the Confederate government[21]), that became Hood's Brigade, claimed as an ancestor by the 36th Division. The Confederate companies received cursory training in Texas, and a wide variety of equipment and uniforms. Sharing Captain Martin's obvious enthusiasm, and with little real preparation for war, the companies traveled to Richmond, Virginia.

A veteran of one Texas company described part of the journey:

> On about the eighth day out we reached Alexandria [Louisiana], with sore feet, and worn out. We had marched something like two hundred miles over bad roads. This was a character of exercise that very few men in the company knew anything about, but during the next three years they were destined to know more about it . . .[22]

Three years later, Company M of Hood's Texas Brigade was still walking. Early on the morning of April 9, 1865, the Texas Brigade was assigned a position one mile from Appomattox Court House:

> We fought and marched seven days, reaching Appomattox Court House on April 9th, 1865 . . . The Federals in great numbers surrounded us and surrender was inevitable. Only six

of the old company remained to surrender at Appomattox . . . A few of the boys had returned home during the war maimed and crippled and a few were in Federal prisons, but the great majority had been laid to rest in soldiers' graves.[23]

The April 14 surrender ceremonies marked the formal end of Hood's Texas Brigade, and the Army of Northern Virginia. The brigade became legend. The men still had to get home.

The trip began with a twenty-mile walk the next day[24]—the old habit of long marches was hard to break.

"During the war the Texans fighting in Hood's Brigade earned a reputation as prolific foragers."[25] The brigade would also gain a reputation as one of the best brigades, if not the best, in the Confederate Army. The reputation was earned the hard way. However, the end of Hood's Texas Brigade was followed by the beginning of the 36th Division.

CHAPTER 2

Reconstruction and Beyond

"The condition of civil affairs in Texas is anomalous, singular, and unsatisfactory."[1]
—Maj. Gen. Philip H. Sheridan, 1866

The largest Confederate army west of the Mississippi, under Gen. Edmund Kirby Smith, surrendered on June 2, 1865. Jefferson Davis and the Confederate government were still fleeing south, and Federal general in chief Ulysses S. Grant could not be sure that fighting had ended. French troops occupying Mexico might also prove a problem. Maj. Gen. Philip Sheridan, one of Grant's best subordinate commanders, was sent to the Southwest. Texas was occupied, though Federal troops did not arrive until June 19. Many Texas slaves first learned they had been free for two years. In the interim, however, in the three weeks between the sudden collapse of Federal authority and the arrival of Federal troops, "The state was swept by chaos and anarchy."[2] Surrendered Texas troops roamed the countryside, looting almost at will. Some Confederate and state arsenals were looted, in violation of the surrender terms. Few people, however, were injured.

Texas was in an unusual situation. As a member of the Confederacy, it had lost the Civil War. However, unlike the other Confederate states, Texas had not been conquered by the Federals.

Much the contrary. What was probably the last battle of the Civil War, the Battle of Brownsville, was fought in Texas on May 13, 1865. The Confederates won this engagement, not the first time they had defeated Yankee efforts to conquer Texas. The Texans who stayed home, though certainly not those who had left the state to fight Grant and Sherman, had so far seen no conquering blue uniforms to bring the reality of defeat literally and figuratively home. They knew, however, that the war was lost. According to a Texan who lived through the period:

> At the beginning of the Civil War, the people of the South were enthusiastic, hopeful and had scarcely a doubt of triumph and ultimately securing their independence from the North. This hopeful feeling weakened as time passed . . . While the early feeling of hope and confidence in ultimate success had in great measure been destroyed, yet no true Southern man had entirely despaired of success . . . he still had a hope, until the surrender of Lee at Appomattox, that in some unknown and mysterious way the cause of the South would succeed . . .
> When this news was received . . . It fell with stunning effect on the whole community. It seemed to denumb the mind, still the heart, and freeze the blood in the veins."[3]

Maj. Gen. Gordon Granger arrived in Galveston on June 19, 1865, to take command of all Federal troops in Texas. He immediately issued orders abolishing the Texas militia, and all other Texas military units. Two days before Granger arrived, A. J. Hamilton had been appointed provisional governor of Texas by President Andrew Johnson. A former member of Congress, in 1861 Hamilton was one of the few Texas politicians to join Sam Houston in opposing secession and joining the Confederacy. Hamilton fled Texas in 1862 to avoid arrest. Eventually becoming a Federal brigadier-general of volunteers, in 1863 Hamilton was appointed military governor of Texas. He never actually served in that position.

Hamilton's Reconstruction governorship, lasting about one year, was not untroubled. Particular problems were general lawlessness and Indian attacks on the western frontier. One cause of secession, at least in the frontier counties, had been the feeling that the United States government was not sufficiently defending Texas against the Indians—though the "Indian problem" was probably overstated.[4] Texans had the same problems with the Confederate government. A historian who studied this period in Texas has written that "Part of the difficulty, indeed, lay in the belief of most non-Texan Confederate authorities that the Indian menace did not really require the military might or expenditure that Texans demanded."⁵ The Texans handled frontier defense themselves during the Civil War. The "Indian menace" seemed to worsen after the war, with first the delay in getting federal troops into position, and then their concentration on civil rights matters. Governors were not given the benefit of the doubt, though their policies were not that much different than during the war. However, "outside" governors were seen as impositions, even less acceptable than Confederate generals.

Texans were slow to accept Reconstruction, though there was at least one notable exception. John H. Reagan, former Confederate postmaster general, in his famous "Fort Warren" letter of August 11, 1865 (written while in prison at that fort), urged acceptance of the results of the war. This included recognizing the supreme authority of the federal government, its right to prevent secession, the end of slavery, the right of freedmen to protection of the law, and freedmen being allowed to vote—though with literacy and other tests. Reagan saw Texas as being in the position of a conquered nation. Acceptance of the conditions was necessary to get back real state government.⁶

> Your condition as a people is one of novelty and experiment, involving the necessity of political, social and industrial reconstruction after a sweeping and thorough revolution in all these respects, and this is to be accomplished in opposition to your education, traditional policy, and prejudices.
>
> I do not propose to discuss either what belongs in the past, or the policy of what is now required of you, but to accept the present condition of things, as the result of the war, and of inevitable necessity, and from this, as a starting point, to inquire what policy our people should adopt for the future.
>
> You must, in the first place, *recognize the necessity of making the most you can of your present condition* [underlined in typescript], without the hope of doing all you might desire. This is required both by reason and necessity.
>
> The state occupies the position of a conquered nation. State government and state sovereignty are in abeyance, and will be so held until you adopt a government and policy acceptable to

the conquerors. A refusal to accede to these conditions would only result in a prolongation of the time during which you will be deprived of a civil government of your own choice.[7]

Reagan was released from Fort Monroe in October of 1865. According to Reagan, his letter "met the approval of many of the ablest men in both the North and the South," but the people of Texas "misunderstood the . . . purpose . . . they were not in condition to reason on the subject."[8]

Former 1861 unionists were a minority of the Texas constitutional convention elected on January 8, 1866. However, out-and-out secessionists were also a minority. "A considerable element in the convention, the group which really held the balance of power, should be classed as merely conservative. They were likely to vote against the unionists out of opposition to radicalism rather than because of hostility to the United States government."[9]

Slavery was declared at an end in Texas, but the convention refused to ratify the Thirteenth Amendment. The federal government was asked to remove the Freedmen's Bureau, the organization trying to protect the rights of former slaves, from the state. The convention claimed that the bureau was "not only unnecessary to the protection, but absolutely prejudicial to the interests of the black race."[10] Blacks were granted the right to property, basic legal protection, the right to sue and be sued, and the right to testify in court cases involving blacks. The wartime state debt was repudiated. Secession was also repudiated, but as a result of the war, not as a rejection of its legality. Texas laws not in conflict with the U.S. Constitution, provisional government decrees, or the results of the war, were accepted.[11]

On July 6, 1866, running as a conservative, convention president James W. Throckmorton was overwhelmingly elected governor of Texas. The amended state constitution was also approved, but by a far smaller margin. About six weeks later, though somewhat reluctantly, President Johnson approved the results of the state election.

Despite a reputation as an "unreconstructed rebel," Throckmorton had been an 1861 unionist, and expected to cooperate with the federal government and the army. However, his legislature was strongly former secessionist. Its first task was to elect two United States senators. One of these, David G. Burnet, was considered an "outspoken rebel," though one biographer claims he was actually unionist in 1861.[13] There was no doubt about the second, Oran M. Roberts, president of the 1861 Texas secession convention. The United States Senate refused to seat both of them.

The new legislature again asked that the Freedmen's Bureau be withdrawn from Texas. Texans particularly resented the local agents and their alleged lack of objectivity in black/white legal quarrels. Texas was also following the example of other southern states and passing a series of measures designed to regulate black labor, the "black codes." These laws specifically limited the right of blacks to change employment, established sharecropping (and liens on crops) as a preferred method to normal wages, and enacted strict vagrancy laws to get blacks into the system. Another of the laws called for entire families to contract for work through the head of the family.

The black codes may have been "An honest attempt by the legislature, blinded as it was by racial prejudice, to make what it thought was a workable system of free labor."[14] Others, particularly in Congress (where it counted most), saw these codes—in Texas and elsewhere in the South—as a major part of a pattern of a new form of slavery. In March of 1867, over President Andrew Johnson's veto, the first two Reconstruction Acts were passed.

The South was divided into five military districts, each to be commanded by a general appointed by the president. Existing southern governments were subordinated to army rule, with a July 1867 act giving generals the power to remove governors. This power was used by Gen. Philip Sheridan, commanding the military district which included Texas, to replace Governor Throckmorton with Elisha M. Pease, the loser in the 1866 election. Black codes were eliminated. The generals were ordered to protect the civil and property rights of all persons; to suppress insurrection, violence, and disorder growing throughout the South; and to ensure that all criminal actions were punished, if necessary by military courts. Violence and mistreatment aimed at freedmen was targeted, with the activities of groups such as the Ku Klux Klan one reason for the imposition of congressional Reconstruction. Texas was no exception.

> White duplicity and maltreatment in dealing with Negroes and their supporters is a constant theme in [Freedmen's] Bureau and Army reports to Washington... The statistics on murder

committed in Texas after the Civil War . . . substantiate the charges of violent white hostility to Negroes and to Unionists.[15]

Revived southern militias, after the 1865 restoration of militia rights, had been part of the problem Congress was trying to correct. According to a militia/National Guard historian, "Members of the Southern Organized Militia units immediately brought out their Confederate gray uniforms and spent most of their drill periods in terroristic activities calculated to keep the . . . Freedmen in a quiescent frame of mind."[16] The Army Appropriations Bill, passed the same day as the Reconstruction Acts, removed militia rights from all southern states.

The spirit of reconciliation is fine, an American strength. However, immediately returning militia rights to southern governments seemingly more interested in restoring the status quo antebellum with minimum acceptance of the new realities from the war had been a mistake. Abuses of these powers had led to the rights being removed in 1867. This, in turn, led to increased law enforcement problems for the congressional Reconstruction governors. They asked their supporters in Congress to restore militia rights to the South.

Early in 1868, Texas Governor Pease reported to Congress on the law enforcement situation in Texas. His message claimed that outlaws had committed 411 assaults and murders, mostly unsolved, in the past year. Pease was able to list names and dates for many of the cases. Crimes ranged from a shootout between two former Confederate officers to the lynching of a black man accused of stealing a knife. The crimes even included a former Confederate colonel shot to death by a unionist revenue collector. Victims were both unionist and former Confederate. Twelve crimes were black on white, forty black on black, to use modern terms. Nearly half the crimes, however, involved black victims and white perpetrators.[17]

Throughout 1868, Bvt. Maj. Gen. Joseph J. Reynolds, commander of federal troops in Texas, reported widespread lawlessness. "Secret organizations were in action during the summer. Unionist and negro murders were common and U.S. officers and soldiers were attacked,"[18] as a historian later put things. Troops were redeployed to keep order, but murders of blacks had become too frequent to allow an accurate count.[19]

Under the supervision of the army, a Texas constitutional convention met in 1868. On June 15, 1868, crime was declared out of control in parts of the state. The convention resolved to ask the United States Congress to restore Texan militia authority. The convention, citing figures similar to those in Governor Pease's report, called for sending two representatives to Washington to personally explain the situation in Texas regarding crime and the threat of Indian attack.

March 2, 1869, two days before Andrew Johnson left office, and exactly two years after militia rights were lost, the bill restoring southern militia rights went into effect. "Reconstruction" had to be completed under the terms of the new law. As a result, Texas, Virginia, and Mississippi would have to wait until 1870 to formally get back their militias.

When the Texas legislature met on April 26, 1869, Gov. Edmund Davis submitted legislation calling for the formation of a militia—an organized State Guard and a reserve militia—and a state police force. Men between eighteen and forty-five, unless exempted by federal or state law (including paying a commutation fee) would be members of one of the two forces. Effectively, though, service in the State Guard was volunteer. Two months of intensive debate followed, including the arrest of some opponents in the legislature, who had fled to break the quorum, and their forced return to the capitol. The act was finally approved and signed on June 24, 1870, technically before militia rights were restored to Texas. By July 22, 1870, the first company was organized. James Davidson, who later fled the state after embezzling $38,000,[20] was appointed adjutant general and chief of State Police.

The act provided that:

> In time of war, rebellion, insurrection, invasion, resistance of civil process, breach of the peace or imminent danger thereof, the Governor shall have full power to order into active service the military forces of this state . . . It shall be the duty of the Governor, and he is hereby authorized, whenever in his opinion the enforcement of the law of this State is obstructed, within any county or counties, by combinations of lawless men too strong for the control of the civil authorities, to declare such county or counties under martial law, and to suspend the laws therein until the Legislature shall convene, and to take such action as may be deemed necessary.[21]

Thirty-nine companies of State Guard were rapidly enrolled. A state police force was also recruited. A considerable number of members of both organizations were black. This was more a result of circumstances, less of planning. One recruiter reported that "In raising volunteers for my command I found only eight or nine white citizens who showed a willingness to offer their services, consequently colored men were selected for the duty. Many hundreds more than were required offered their services."[22] Texan resentment of the heavily black militia still smoldered several generations later. A 1940 history of the Texas guard states that "a number of the sergeants and privates were negroes, who when armed and clothed with power were arrogant, domineering and vindictive."[23]

The years of the joint existence of the State Guard and State Police would be eventful. The State Police, with substantial black membership, was the focus of much of the trouble until its abolishment in 1875. An academic study of the State Police feels that it did not deserve its reputation.

> The Texas State Police have traditionally been regarded as one of the more unsavory aspects of Radical rule . . . Reexamination of the sources suggests that these judgements were too harsh. While not above criticism, the State Police was a worthwhile agency created for a legitimate purpose. During its brief career it accomplished much that was laudatory, and its premature disbandment by the conservatives for political purposes left a void which was not soon filled.[24]

The militia was supposed to back up local law enforcement, and the State Police. When law enforcement was "beyond the control of the civil authorities, militia units were placed on a war footing and sent in to enforce gubernatorial declarations of martial law."[25]

One of several serious incidents occurred in January 1871, in Walker County. L. H. McNelly, later a prominent Texas Ranger,[26] was commanding a State Police unit sent in to investigate the slaying of a freedman, Sam Jenkins. McNelly and his men arrested four suspects. Three were convicted in their trial. As soon as the verdict was announced, the three began shooting guns smuggled them by friends. McNelly himself was wounded in the incident, which sent the other court officials fleeing for their safety. The prisoners, aided by a sizable portion of the town, escaped. Only two people answered a subsequent call for a posse. Governor Davis, when he learned of the incident, declared martial law, sending in Adjutant Davidson and a militia unit. Though twenty people were tried by Davidson's military commission, only one of the escaped prisoners, the interestingly named Ned Outlaw, was convicted. He received five years' imprisonment.[27]

Davis reviewed the case, throwing out the conviction. Outlaw filled suit for false imprisonment and was eventually awarded $20,000 in damages.[28]

Davis used the State Guard for political purposes as well as law enforcement purposes. In August 1871, just before an election, Davis issued a proclamation giving several rules for conduct by prospective voters. State Police, militia, and other law enforcement personnel at the polls were supposed to enforce these rules. Whatever his motives, Davis should not have been surprised that it appeared he was trying to intimidate voters—and that protests followed.[29]

Elections in November 1872 restored the Democrats to power, though by small majorities, in both houses of the legislature. When the new legislature opened in January, the members quickly set to work to abolish the State Police and remove the governor's power to declare martial law.

December of 1873 saw Davis losing his bid for reelection to the Democratic candidate, Judge Richard Coke, by a considerable margin. Davis tried to have the results thrown out, claiming the legislature passed an illegal law to govern the election. The Texas Constitution called for elections to be held at county seats, and for the polls to be open four days. These provisions were separated by a semicolon. The March 1873 law—which Davis had signed—changed both provisions. Davis admitted that the legislature could hold elections in each precinct, as the possibility of such a change was specifically allowed. David claimed, however, that the semicolon made it illegal to change the number of days the polls were kept open.[30]

On January 5, 1874, the Texas Supreme Court, with judges appointed in 1867 by Gen. Philip Sheridan, supported Davis's position—giving legal meaning to what appears to have been a typographical error.[31] The court earned itself, and its entire tenure, the nickname "semicolon court," and a reputation that "no court accepts as authoritative precedents the opinions of the semicolon court."[32]

Davis was determined not to yield his office, and called out mostly black units of the militia. They took over the lower floor of the capitol. The newly elected legislature, and its militia supporters, controlled the upper floor. On January 15, after finally getting the election records, Coke was declared the new governor and inaugurated.

On January 16, after seeing at least some of "his" militia units join the Coke supporters,[33] Davis requested help from President Grant in Washington. Grant refused. Davis gave up, and Coke become governor.

★ ★ ★ ★ ★

The documented history of the 36th Division begins with the formation of its oldest recognized ancestor, the Houston Light Guard, in 1873. Efforts at forming the unit began in October 1872, when it become apparent the Democrats would gain a majority in the November elections. The editor of the *Houston Daily Telegraph* played a key role in encouraging participation in the new militia company. Allen Charles Gray purchased the paper in late 1873. He was the brother of Edwin Fairfax Gray, the key organizer of the company. An 1874 editorial in the *Daily Telegraph* declared:

> There are young men enough in the city with ample means and time to go into the military companies and organize a handsome battalion of infantry, and there certainly ought to be enough willing to form one good company, especially as the trouble of obtaining a charter and arms has already been gone through with.[34]

The unit was Confederate in sympathy. The company had a charter insuring that the unit would have to approve by vote any attempt by the governor to use it in establishing martial law. At the beginning of the Civil War, Confederate volunteer units had elected their own officers. No unit presumed to have veto power over its use by higher commanders.

By 1877, some members of the militia from the Galveston area thought that the organization of the militia should expand beyond the company level. They looked into holding a convention, inviting representatives from other units. Adjutant General William Steele favored militia reform and had discussed the convention idea with Governor Oran Roberts. Steele, however, wanted the convention postponed. He was concerned about its possible effects on the North, presumably not wanting anything to interfere with the ongoing Hayes-Tilden presidential election dispute and the pending end of Reconstruction.

The convention met anyway, on February 12, 1877, and organized a provisional brigade with two regiments and a battalion. Governor Roberts, also initially opposed to the convention, confirmed its decisions.[35] Two years later, the adjutant general's report showed that the militia had expanded to include a third regiment and twenty-two unattached companies.[36] Three of these companies were all-black, the others all-white. Some militia companies had been integrated under the Davis administration.

A new militia law was passed in 1879, formalizing the expanded structure. With amendments ten years later, this law would govern the Texas militia until the time of the 1903 Dick Act.[37] A similar active/reserve structure was maintained, with the more formal militia now called the Texas Volunteer Guard. Universal military obligation was "rendered meaningless"[38] by the failure to provide any means of identifying those so obligated. However, procedures were established for forming militia companies, and for organizing these companies into regiments and brigades. Company officers were to be elected, regimental officers appointed by the governor on the recommendation of line officers of the regiment. Brigade and division officers would be appointed by the governor. Drill and encampment requirements were specified, though political exigencies would cause their alteration over the years. Military offenses were listed and made enforceable by civilian authorities. Procedures for use of troops in civil disturbances were also specified. Guard uniform laws were still a bit flexible. Except at times specifically prohibited, units could adopt different uniforms.[39]

The new militia law was printed and distributed in the late summer of 1879. It was September of 1880 before forty-six companies had taken an new oath of enlistment and reorganized themselves according to the new law. On April 6, 1880, the earliest recognized regimental ancestors of the 36th Division, the 1st and 2nd regiments of Infantry, Texas Volunteer Guard, were "Constituted ... and organized from existing companies in southern Texas..."[40] Company A of the First Regiment, Texas Volunteer Guard, was the Houston Light Guard.[41]

"In a history of the Houston Light Guard, written for publication, one had to decide what good things to reject and what will be of most interest to the readers."[42] The author of the early-1900s brochure from which the quotation comes thought drilling was the only thing of interest to the readers. The Houston Light Guard was good at drilling, though. At one time they were actually barred from interstate competition.[43] Yet, during their early history they would do more to "show the people of the United States that they were of the stuff that soldiers are made; that the martial flames burned so brightly in the sunny Southland that her youths have few lessons to learn in war's school!"[44] One example occurred in September 1887 when the Houston Light Guard cooled down a potential race riot in Matagorda County.[45]

September 19, 1887, saw an attempt to serve an arrest warrant on John Nuckols, a white resident of Matagorda County.[46] Jerry Messina, a black constable, left Columbia, Texas, to serve the warrant. He asked directions from a man named Stafford, a white, who offered to guide the constable. Messina disappeared. Blacks in the county claimed he had been murdered, an opinion shared by the county sheriff. Whites in the county claimed Messina's mule was found in a pasture, that Messina had stolen a horse in place of the mule and fled the county to avoid indictment for stealing a cow.

Blacks were a heavy majority of Matagorda County. Black farmers had a particular reputation for independence, being less dependant on whites than blacks who lived in town. Oliver Shepard and Burton Hawkins were known as leaders of the rural blacks. Messina had vanished on a Monday. By Friday of that week, September 23, Sheriff Wadsworth still had no definite clues as to what had really happened. Rumors were starting of blacks organizing, and impending violence. Sheriff Wadsworth sent for aid from the sheriffs of Wharton and Brazoria counties. Each came to Wadsworth's assistance with posses of fifty to sixty men.

When they arrived, things seemed peaceful, so Wadsworth sent them home. The Brazoria posse, back home on Sunday the 25th, received a message from Wadsworth asking their return and stating that instead of law and order being restored, as he thought, riot reigned supreme . . . Negroes were flocking from south and north, east and west and from Brazoria and Wharton Counties, as well as Matagorda . . . he must have help.[47]

Shepard and Hawkins organized their own posse, described by Olson as "an armed vigilante-type force of forty-seven men."[48] They rode to Nuckols' house, but found only his wife and daughter. After threatening to kill Nuckols and Stafford if Messina had been harmed, the black posse headed to Cedar Lake, meeting a party of ten armed whites concealed in the woods. After some tense moments, the parties broke contact with no firing.

On September 25, a Sunday, things did not end so well. A group of whites, including Stafford, exchanged shots with some blacks, including Hawkins. Hawkins and another black were killed. The twenty whites then rode on to Shepard's cabin. On the way, they killed a black man hiding in a cotton patch. The victim may or may not have been armed.

Shepard was in front of his cabin when the whites arrived, and they immediately opened fire. Shepard shot back but wisely retreated inside his cabin. He held the attacks off until nightfall. The next morning, with heavy shooting, the whites attacked—but Shepard had escaped. The whites found his trail but lost it after ten miles.

Wadsworth telegraphed the governor that "A fight occurred tonight between negroes and whites in the southeast corner of Matagorda County, near the Brazoria line. The negroes are collecting from adjoining counties and there are indications of a general uprising."[49] Though seeking to protect the safety of all parties, the authorities' biggest fear seemed to be a black uprising,[50] a traditional southern fear, not just a breakdown in law enforcement.

The commander of the Light Guard at this time, Capt. Frank A. Reichardt, was a Houston wholesale grocer. On Monday, September 26, 1887, he received a telegram from Governor Ross ordering him to take twenty men, and sufficient ammunition, to Columbia "as soon as possible." Reichardt was told to report to Sheriff Hickey to suppress "expected trouble between whites and colored." The rest of the company was to be ready to follow, if necessary. A second company was also to be put on alert.⁵¹ Reichardt immediately sent word to his men and requested a special train.

Problems with the train delayed departure until the next morning, but a total of twenty-eight men and officers left on the train. Cigars were passed out, but all alcohol was confiscated. They arrived at Columbia at 1:30 in the afternoon, to be ferried across the Brazos River and then marched to a local residence, the Harris house, being used as a gathering point. Members of another company, the Pearson Guards, were also at the house. That afternoon they left by ferry for Brazoria, where they received a cheering welcome. After a "grand ball" that night, the next day the unit returned to Columbia.

On arriving at Columbia, Captain Reichardt received a telegram from the governor. The telegram ordered them to stay at Columbia until Wednesday, if requested by Sheriff Hickey, but no longer. By that evening, the unit was back in Houston. Regarding this operation, which may have seemed "pointless and unnecessary,"⁵² on September 28 the *Houston Daily Post*⁵³ commented:

> Their very presence has benefited [Brazoria County] no little. The average negro being extremely ignorant, and knowing how much his race outnumbers the whites in this county, could not be made to believe except by ocular demonstration, that forces sufficient to outnumber him could be brought.

The Houston Light Guard, though located 100 miles from the scene of trouble, was a trusted unit which had done well in an earlier deployment. Its use might have represented a show of the ability to move force, as well as a pure show of force. But basically it was a show of force, which successfully cooled down a dangerous situation.

★ ★ ★ ★ ★

Describing the Texas Volunteer Guard just before the start of the Spanish-American War, one historian states that:

> it is evident that the Texas Volunteer Guard had now come of age. Given a minimum of state support it revealed widespread soldierly talents which require only proper training and incentive to produce a sound military organization as distinguished

from a collection of clubs having merely the outward appearance of some military skills.[54]

Another historian is not so generous. "Put simply the Texas militia from 1835 to 1903 was superfluous to the survival and prosperity of the state and its citizens."[55] He probably is ungenerous in slighting the law enforcement role of the militia. However, the record shows that his general conclusion is probably right. San Jacinto was won by regular Texas Army troops and emergency volunteers. The Texas Rangers fought the Comanches and other Indian tribes. Texas participation in the Mexican War, excellent combat service and near-barbarous behavior after the battles, was through rangers and volunteers. Texas Civil War troops came from the same source as most of the troops the first two years of that war—volunteers.

The Texas militia shared a lot in common with militia of this period in the rest of the United States, however. When the Civil War ended, the United States had a substantial number of trained and experienced soldiers, Federal and Confederate. Some militia activity continued after the war. In the South, however, this quickly took on an anti-freedmen tone, resulting in its banning in the 1867 Reconstruction Acts. When southern militia rights were restored a few years later, it was to fill a law enforcement need, not to function as a reserve for the regular military. The federal armies declined drastically in size during the same period. Just over 1,000,000 men were serving in the United States Army (regular and volunteer) in 1866. Just over 57,000 served in 1866, roughly 37,000 in 1869. In 1870 total strength dropped below 30,000, staying below that level until 1898.[56] The declining strength of the army showed the decline in interest in things military among the general public, and among the Congress. "The former citizen soldiers came back home again to become citizens, and after four years of bitter war, to devote little time to soldiering."[57]

Even the volunteer militia units, the antebellum "cream" of the citizen militia, declined. One historian has noted that "In the decade or so after the Civil War, the militia was at its lowest ebb in our history. Only a few Northern states attempted to maintain a militia organization; and during Reconstruction, Southern states could not do so."[58]

Things began to change in 1877. In the South, Reconstruction came to an end as part of the political deal leading to certification of the election of Rutherford B. Hayes. Nationally, a series of railroad strikes began in West Virginia but spread to ten states. "The railroad strike of 1877, and the industrial warfare it induced, was the stimulus that set off the development of the modern National Guard,"[59] another historian states.

In some cases, the militia was not able to handle labor violence, and the regular army had to be summoned. Use of the regular army to settle domestic disputes did not sit well with the general American dislike of standing armies. In other cases, such as in Pittsburgh in 1877, the militia (which most states were already starting to call the National Guard) was willing to open fire on demonstrators. The guard was willing to do what the government, and business leaders, defined as its job. Whatever the motivation, law enforcement, concern about standing armies, or expert lobbying, interest in the militia, and appropriations, began to rise. Labor violence continued periodically into the 1890s, culminating in the strike against the Pullman Sleeping Car Company. The guard was a major choice for law enforcement—often really just strikebreaking—efforts. This led guard leaders from around the country to seek to reform the guard, to refocus the guard on its primary military function, and to formalize its position in the federal military structure.

Many states began to increase funding for guard units, enabling these units to improve their quality and frequency of training. Some regular army technical assistance was sought and received. The bulk of the costs of training, however, was still borne by the states and the individuals involved.

Then, on April 15, 1898, the American battleship *Maine* blew up in Havana harbor. Historians now blame an accident, but the explosion and loss of life led to a United States declaration of war against Spain ten days later.

At the start of the war with Spain, the United States Army had an authorized strength of just under 28,000—though of relatively good quality. This was proportionally smaller, relative to national population, than the 16,000 officers and men at the start of the Civil War. The year before, at a banquet of the Grand Army of the Republic (Union Civil War veterans) Secretary of War Russell A. Alger had expansively toasted the quality and fighting power of the

United States Army. A few days later, an editorial in the *New York Times* pointed out that Secretary Alger knew that "we could offer no resistance on either coast to a first-class or second-class naval power, and that two army corps could traverse the country as far as their commanders choose to take them without meeting any effectual opposition."[60] Fortunately, by this time Spain had deteriorated into a third-class naval power, distracted by the additional need to fight Cuban and Filipino rebels.

A prescient 1897 editorial on possible use of the guard, in the *Army Navy Journal*, stated,

> If the United States should be involved now with Spain or any other power of consequence, the first move would be to place the regular Army into the field, and immediately afterwards the National Guard would be mobilized for active service. Right here would ensue . . . a confused tangle, made all the more desperate by the patriotic enthusiasms of delays, bickerings, rivalries and dangerous disputes, all permitted to thrive because a general war organization was not previously laid down . . .[61]

The war ended up being fought by regular army troops, and volunteers raised for the war. (Actually, President William McKinley issued the call for volunteers two days before war was declared.) Guard units could not be accepted into federal service. Individual members could volunteer, and reform their units in the federal military. A sizable number volunteered as individuals. Volunteer units ended up with varying records when the treaty of peace with Spain was signed on December 10, 1898. Some participated in putting down the rebellion in the Philippines, following the islands' transfer to United States control. "As it has done in the Civil War and would do again in World Wars I and II, the War Department in 1898 responded to crisis with improvisation, trial, and error, followed by a growing mastery of the circumstances that confronted it."[62]

Two immediate results followed from the war. Casualty figures from the war showed a remarkable pattern. Battle deaths totaled 379. Disease, food poisoning, and other causes killed 5,083.[63] These figures, and the outcry that followed, are credited with inspiring efforts to better guarantee food safety. Further motivation from the construction of the Panama Canal eventually led to successful efforts began to eradicate yellow fever. The United States was now a world power and needed to have the military of a world power.

Texas militiamen saw no formal military action between the end of the Civil War and the 1898 start of the Spanish-American War. Texas militiamen responded to President William McKinley's April 23, 1898, call for volunteers to fight Spain. Militia members eventually served in all five of the Texas regiments raised for the brief conflict. Only one of these regiments, however, the 1st Texas Infantry, even left the United States. They almost did not get overseas. Disputes over the proper role of militia/guard volunteers hindered recruitment. Units from all over the country, not just Texas, questioned their proper place in the federal military structure and whether they would be able to keep their own state identities.

Three months of occupation duty in Cuba began on December 26, 1898, ending when the regiment returned to Galveston, to be mustered out, on March 25, 1899. Some of its members had died of disease, including the regiment's commander, Col. Woodford H. Mabry. (Camp Mabry in Austin, Texas, still the headquarters for the Texas National Guard, is named after Colonel Mabry.) No Texas regiment, however, saw any combat. The pessimistic historian of the Texas Guard, quoted earlier (page 23) summarizes, "No Texas militiaman fired a single shot in anger at the Spanish during the entire war and it seemed time had passed the Texas militia by."[64]

CHAPTER 3

Getting Ready for War

"They are always inspecting us. I feel like a piece of prize beef."[1]
—Pvt. Leo Muller
36th Division, September 1918

American reaction to the 1912 Mexican revolution, and the raids of its best known protagonist—Francisco "Pancho" Villa—into American territory, is an interesting story, but one in which the 36th Division's ancestors played little part. As one history of Villa put it, "President Wilson took the unprecedented step, on June 18 [1916], of ordering the entire National Guard into federal service [on] the border."[2] Virtually none of the guard saw combat, doing little more than camp on the border with Mexico.

Deployment had been somewhat clumsy. One troop train at San Antonio was delayed by a broken rail. The train commander formed his men into defensive positions to prepare for a Mexican attack. There was no attack.

The law at the time said the guard could not be required to serve outside the United States. The guard was of uncertain combat value. However, the guard's presence on the border gave the people a feeling of security and freed the regular army for pursuit into Mexico.

At the end of the Villa crisis, the 2nd Texas Infantry spent an entire week in 1917 out of federal service. Budgetary considerations led the War Department to try to get the guard back into civilian life as soon as possible. However, late in March, rapidly worsening relations with Germany made it necessary to put a halt to demobilization and start to remobilize and expand the guard. By the end of the first week of April 1917, most Texas and Oklahoma guard units had started remobilizing, most Texas units returning to border guard duty via Fort Sam Houston, in San Antonio.

They were assigned to protect the Mexican border from possible saboteurs entering the United States, and to watch for trouble from Mexico. The intercepted and published "Zimmermann telegram" had raised this possibility, as it proposed an alliance between Germany and Mexico if the United States entered World War I.[3] As to the men being returned to border services, "This time there were no reports of undue grumbling, for the American entry into World War I on April 6 made it almost inevitable that sooner or later the Guardsmen would get all the action they wanted and more."[4]

As unprepared as it was, the army that entered World War I benefited from twenty years of efforts at reorganization, fifty years of theorizing as to the best balance for the American military among professionals, draftees, militia, and emergency volunteers. Much of the debate was politically motivated. John Logan, a Civil War "political general" who turned out to be an excellent commander, was an outspoken critic of West Point and the professional military establishment. In 1887 Logan wrote:

> Away off in the wilds of America a soldier had been found totally different from any that ever walked a battlefield. Upon one day he was a citizen, quietly following the plow; upon the next he became a soldier, knowing no fear and carrying a whole destroying battery in his trusted rifle. He was a soldier from conviction to principle, from loyalty to his country, from duty to his family. He moved with the discipline of an educated soldier but he fought with the desperation of a lion at bay.[5]

The best-known military theorist of the later nineteenth century was Emory Upton, an 1861 West Point graduate and a Civil War "boy general" with a distinguished and innovative combat record. In 1875, after some work revising a manual on infantry tactics, Upton was sent on a world tour to study the military structure of other nations.

In 1878, after his return from Europe, Upton published *The Armies of Asia and Europe.* Upton recommended that the United States adopt a modified version of the German military system—general staff, professional conscript army, and a trained reserve system. Upton was particularly impressed by the German army's apparent freedom from civilian control. The public, and the Congress, however, were not impressed. America had always won its wars by using a small professional army augmented by mass, state-controlled, civilian volunteers in wartime. With the congressional desire for economy, and the public desire to forget about things military, Upton's plans got nowhere.

Americans did not care what other countries' armies were doing. Upton's colleagues, therefore, urged him to make his points by writing a history of the United States military experience. Focusing on the Civil War, the major American military experience to that point, Upton started writing what become *The Military Policy of the United States.*

Upton made specific suggestions as to how the army should be organized in the future. He called for a system based on an expandable "cadre" army, a structure able to incorporate additional troops into a wartime force. "National volunteers," based on trained conscript reserves, would provide much of the emergency army. The militia would be the last line of defense, only to be used in the United States itself.

These, however, were just details. Control of the military was the real issue. Upton chose to blame civilian control of the military, and resulting politicization of command appointments, for Northern delay in winning the Civil War. Upton wrote:

> In time of war the civilian as much as the soldier is responsible for defeat and disaster. Battles are not lost alone on the field; they may be lost beneath the Dome of the Capitol, they may be lost in the Cabinet, or they may be lost in the private office of the Secretary of War. Wherever they may be lost, it is the people who suffer and the soldiers who die, with the knowledge and conviction that our military policy is a crime against life, and crime against property, and a crime against liberty.[6]
>
> In every country save our own, the inability of unprofessional men to command armies would be accepted as a self-evident proposition ... The disasters [of the North during the

American Civil War] ... must therefore be credited to the defective laws which allowed the President to dispense with an actual General in Chief and substitute in his stead a civil officer supported by military advisors, disqualified by their tenure of office and occupations from giving free and enlightened opinions.[7]

Upton's book was incomplete in 1884, when unbearable headaches drove him to suicide. The book was finally published in 1904. Upton's ideas, however, were known within the army.

Like other prominent American commanders, including Philip Sheridan, Upton was quite impressed with the German command structure. An autonomous general staff was apparently able to rely solely on its professional judgment in using a well-trained mass conscript army. Upton failed to notice the great degree of effective civilian political control the highly skilled Chancellor Otto von Bismarck exercised over the German military. During the Franco-Prussian War, the generals constantly complained about Bismarck's control.

Upton's expandable "cadre" army evolved into the system this country used in both world wars. However, Upton's lack of political realism in creating his ideas may have killed any immediate impact they might have had on United States military reform. Noted military historian Russell Weigley feels that:

> Emory Upton did lasting harm in setting the main current of American military thought not to the task of shaping military institutions that would serve both military and national purposes, but to the futile task of demanding that the national institutions be adjusted to purely military expediency.[8]

The Spanish-American War proved the impetus to begin military reform. Secretary of War Russell Alger became an easy target for criticism of the conduct of the military during the Spanish-American War. He resigned in 1899, to be replaced by Elihu Root. Several years younger than Alger, Root was a successful attorney from New York state. Root had no military experience when he took the job, but was considered able to administer the new territories acquired as a result of the war with Spain. Root found a War Department that was "a mess. The war had demonstrated its inefficiency and corruption. Its red tape was proverbial. Personal jealousies and spite crippled the efficiency of the personnel."[9]

Root saw that properly administering the new territories, his specific assignment from President McKinley, required an effective military system. War Department problems during the Spanish-American War could not be repeated. The revolt in the Philippines provided the impetus for Root's first measure, increasing the size of the regular army. Expiration of the acts establishing the military force for the war would have caused the army to shrink back to pre-war size. However, Congress had already increased the authorized strength of the military, until July 1901, to 65,000 regulars, 35,000 volunteers.

In his first annual report, Root addressed what he saw as the fundamental problem, American fear of a large standing army. He criticized the practice of solving military problems singly. Root then called for first determining the purpose of the military, and then seeing how current structures met this purpose:

> Two propositions seem to me fundamental in the consideration of the subject:
> First. That the real object of having an army is to provide for war.
> Second. That the regular establishment in the United States will probably never be by itself the whole machine with which any war will be fought.[10]

Root pointed out advantages to immediate military reform—the lessons of the war were still fresh, and troops were needed to fight the Philippine rebellion. The size of the army was increased in 1901 and never again slipped below 60,000.

Root added qualitative changes to the army and the War Department. He created the post of army chief of staff, abolishing the old command structure with a general-in-chief. The specific powers, duties, and responsibilities of the new position would be worked out over the next twenty years. Root created the Army War College. Partially influenced by Emory Upton, though he had not yet read (or arranged to have published) Upton's *Military Policy of the United States*, Root included a planning division. His designers hid the planning staff inside the War College to avoid legislators' fear of permanent standing armies.

Most significantly, Root saw the need for the United States to have a workable reserve system to back up the regular military. This view was reinforced in 1900. Col. William Sanger, inspector general of the New York National Guard (later assistant secretary of war), returned from a trip to Europe studying the militia systems of Great Britain and Switzerland. Sanger's report concluded that "We have always been and always shall be largely dependent on our citizen soldiery to fight our battles . . . an effective militia is a force of the greatest value."[11]

Root's familiarity with Emory Upton's ideas proved most influential in structural reforms within the regular army. Root thought that volunteers would still have to play a major role in any future conflict. The regular army would only be a part, and a small part, of the American military in a war. The best way to obtain qualified volunteers would be to train and support the militia, to have a pool of experienced individuals who could volunteer. The guard would be the training school for any future emergency volunteer army. Guard officers would be trained at the War College to better serve in their positions. What Root was proposing was a system similar to that used during the Spanish-American War, but better organized and with increased professionalism of both the regular army and the guard-based pool of volunteers. "In the event of war, it would be the guard to which the War Department would turn to take care of the volunteer army, to train it, and make it into an effective force."[12] Root did not seem to be calling for the National Guard, in that form, to play a major role in the wartime army. However, the guard would be valuable. As Root stated:

> National Guard organizations are the great school of the volunteer to which the country must look, in order that its young men, when they go out to fight the battles of their country, shall find officers competent to lead them, to organize them, to transport them, to equip them, to feed them, to keep them in health, and lead them against the enemy.[13]

Root proposed efforts to bring the guard up to federal standards. Full federal control, as Upton desired, was not politically feasible.

The actual Militia Act of 1903, the Dick Act, was a joint effort of the War Department and the National Guard Association. Congressman Charles Dick, the chief sponsor of the legislation in Congress, apparently worked with the War Department during its initial drafting by the NGA. Despite Dick's hedging, the opinion of at least one historian (and retired National Guard general) is that Dick's committee had a War Department draft among the papers, drafts, and suggestions it was considering.[14]

The association convention approved a draft bill, which was introduced in the House of Representatives on January 23, 1902. The House and Senate made further changes. "An Act to promote the efficiency of the militia and for other purposes"[15] was signed into law on January 21, 1903. The organized militia, now formally called the National Guard, was a recognized part of the United States military. Procedures existed for calling it into federal service, for giving militia members equal treatment with regular army troops, and for training the National Guard to federal standards. Officers would have access to the same military education given regular officers. Federal financing would pay for training, and cost of the guard in federal service. Important changes in 1908 would eliminate a nine-month limitation in service, allow the guard to be used outside of the country, and specify that it was to be called into service before any volunteer forces. Legislation in 1916 further refined the 1903 and 1908 acts. They did not end the debate over the role of the guard, or the quest of the guard to be both a federal and state force. On July 1, 1903, provisions of the Dick Act were adopted by the Texas legislature for the Texas National Guard.

★ ★ ★ ★ ★

Though the exact nature and degree of American participation in the Allied cause was being worked out, some token contribution to the Allied World War I effort was needed as soon as possible. Four infantry regiments of the regular army were collected, the 16th, 18th, 26th, and 28th infantries, formed into the 1st Division—the "Big Red One," still in existence as of the 1991 Gulf War—and sent overseas. The regiments still had to be expanded, and sailed to Europe composed of about two-thirds new recruits.

One battalion of the 16th Infantry paraded through Paris on July 4, 1917. They were still outfitted with cavalry hats similar to cowboy hats, which seems to have quite fit the French image of Americans. After the parade, however, the division had to undertake three months of intense training behind the lines until it could be deployed in a quiet sector.[16]

American officials quickly realized that recruiting volunteers would not be sufficient to expand the small regular army of less than 200,000 men, and the National Guard, to a force sufficient to actively contribute to the war in Europe. Conscription had been proposed by the War Department two months before the United States declared war on Germany. After overcoming resistance to the idea of a national draft, a bill was signed by President Wilson on May 18, 1917. The bill was a substantial improvement on past experience, eliminating many of the administrative problems in the Civil War Federal and Confederate draft systems. "The most significant difference in the draft of 1917-18 ... was that local civilians rather than army officers administered it,"[17] as one historian summarizes things. General policy, and quotas, would be set from Washington. However, eligible men would be drafted by boards composed of their neighbors.

During the course of the war, the army would expand from 213,000 men in federal service on April 1, 1917 (including some remobilized guard) to 3,685,000 men in the "National Army" at the time of the armistice, November 11, 1918. Nearly two million Americans were in France when the war ended.[18] The mobilization of manpower and economics was massive, surpassing anything this country had done before. Supplies arrived where needed. Food, unlike during the Spanish-American War, was not more deadly than enemy bullets.

Even in medical care, the improvement over past American wars was significant. The Medical Department of the army worked closely with the Red Cross to prepare and operate needed facilities. World War I was the first war, at least among American troops in Europe, in which enemy bullets (and other weaponry) killed more Americans than did disease.[19] For every thousand soldiers who fought the American Civil War, sixty-five died per year. Fifteen per thousand per year died in World War I.[20]

An early step in organizing the "National Army" was to designate the divisions. The government settled on a system of reserving certain numbers for regular divisions, certain numbers for National Guard divisions, and certain numbers for "National Army" divisions, primarily draftees, organized especially for the conflict. State names, however, were eliminated from unit names. It would not be the "36th Texas Division," nor the "141st Texas Infantry

Regiment," nor "Smith's Texas Division," after its combat commander. The War Department wanted to give the army a more national and less regional identity.

The divisions would be "square" divisions. Each had two brigades of two large regiments each. The total strength of a full-size American infantry division, with its attached artillery, would approach 30,000 men. This is twice the size of a World War II division, larger than Lee's entire army at Appomattox. Each brigade had a regiment of light artillery. The division had its own heavy artillery regiment. Further combat elements included three machine gun battalions, a trench mortar battalion, and engineer troops. Normal supply and support units were attached.

The division was large and hard to move, as this was before the day of the mechanized army. However, the huge divisions "had frontal attack, breakthrough power unknown to any army until the advent of full motorization of divisions and the creation of large units of fast, armored vehicles, self-propelled guns and tanks."[21] A theory is that the size of the divisions was partly related to a perceived shortage of officers qualified for high rank.[22] The huge World War I division had a major general in command and three brigadier generals. Splitting it into two World War II–size divisions would have required a total of two major generals and four brigadier generals.

The Texas and Oklahoma National Guard units were scheduled to be formally federalized on August 5, 1917,[23] though most were already back in federal service by this time. The units would have to be expanded, through recruiting, to four times their size at the August federalization. It would be several months before the 36th Division existed as a practical military unit. Individual units would be recognized as meeting federal standards for guard units at different times.[24] However, August 5, 1917, can be considered the day the 36th Division was born.

Brig. Gen. John A. Hulen, of Houston, had commanded a Texas brigade on the Mexican border. Hulen was given a presidential appointment as brigadier general in the national army and placed in command of the first Texas National Guard brigade. The 2nd, 3rd, and 4th Texas infantry regiments were part of this brigade. The 1st, 5th, and 6th Texas infantry regiments were in the 2nd Brigade, under Brig. Gen. Henry Hutchings. Also formed, but not yet brigaded, were the 7th Texas Infantry Regiment, the 1st Texas Cavalry, and the 1st and 2nd Texas field artillery regiments. Commander of the last regiment was Claude V. Birkhead of San Antonio. Birkhead would play a role in division history until 1946.

General Hulen set up headquarters in Houston, where he was in charge of recruiting men and selecting officers. Several thousand written applications for commissions were given Hulen by Texas governor James E. Ferguson and the state adjutant general. Of the total, 417 were designated for commissions at Hulen's recommendations. All officers became guard recruiting officers. General Hulen's first general order specified that "All commissioned officers of the National Guard of Texas are authorized to act as recruiting officers in all units of the National Guard of Texas being organized under the provisions of the National Defense Act."[25] Each officer was supplied written instructions, five typed pages long, on how to exercise his authority.[26]

As a whole, the instructions were logical. Some items, such as the slimmer accepted weights for specific heights over 1990s standards, show the differences between the periods. The officers were also told that:

> Upon receipt of these papers, you should proceed to your home station for your recruiting, taking with you the papers and blanks which have been given you. You should take immediate steps to provide, at your own expense, an office or place in which to carry on organization work. You should see that the newspapers give the fact that you have been authorized to organize a unit for the National Guard of Texas, the widest possible publicity. You should procure the cooperation of the leading business men of your locality. And you should take all other steps which may occur to you which are proper towards securing recruits.[27]

This was before "be all that you can be" television commercials. The officers had to rely on extensive press coverage to aid the recruitment effort, but only seventy-three companies and batteries were filled by the end of June. The governor of Texas had to issue two public appeals. The governor said it was the duty of the men of Texas to fight, and they were likely to be drafted anyway. So why not do their duty in a way which would enable them to serve with their neighbors? The appeals worked, as did other, less sanctioned methods. General Hulen's second general order complained that

> It has been noted that in advertising and speeches used in recruiting for the National Guard of Texas, reference, in some instances, has been made to 'Conscription' and 'Conscripts' in a way that suggests an odium attaching to persons who shall be selected for service under the selective draft system. This is improper and no persons engaged in recruiting for the National Guard of Texas will make use of such terms, or similar expressions, in any way or manner tending to reflect upon either the selective draft system or persons selected for service thereunder.[28]

By the time recruiting was stopped, the Texas guard had risen in size from roughly 5,200 to just under 20,000 men. Oklahoma, with its much smaller population, had roughly doubled the size of its guard.

General Hulen complained about the problems "getting suitable and trained officers for the increased guard."[29] However, Texas and Oklahoma guard officers were probably comparable to militia officers earlier in past wars.[30] Two of the four brigadier generals in the division had seen combat in the Spanish-American War, a third had extensive guard experience. Edward St. John Greble, appointed 36th Division commander, was a brigadier general in the regular army, a major general in the national army. Greble was a West Point graduate, age fifty-eight at the time of his appointment.

Texas was not the only source of troops for the 36th Division. Before 1890, Oklahoma was known as "Indian Territory," land set aside for resettlement of Indian tribes from other parts of the country. Use of the territory began to change after the American Civil War. Indians in what became Oklahoma had tended to support the Confederacy during the Civil War. Motivated by this, and by a desire to resettle plains tribes, the postwar federal government persuaded and pressured the tribes to give land to the plains tribes. In 1899 land was open to settlement by non-Indians. White settlers quickly petitioned the government in Washington to grant territorial status to the western part of the territory. The request was granted in May 1890. A territorial legislature was conveyed in August of the same year.

One of the first acts of the legislature was to establish the

Oklahoma Territorial Militia, the direct ancestor of the Oklahoma National Guard. The territorial governor was authorized to organize "such military companies, battalions and regiments . . . as he may deem proper."[31] An upper size limit of two regiments of infantry, two battalions of cavalry, and one artillery battery was placed on the Oklahoma militia.

The militia followed standard practice and was subject to the call of the governor when he felt the territory's safety was in danger. The territorial legislature followed Texas pre–Civil War practice in one particular. A historian of the Oklahoma Guard has written that "Although eager to establish Oklahoma's first true citizen-soldier units, the First Territorial Legislature neglected to appropriate any funds to equip and maintain the men."[32] The men trained with obsolete and donated weapons, and wore castoff uniforms. Despite these handicaps, one regiment of infantry was successfully organized. Under terms of an act passed by the Third Territorial Legislature, in 1895, improvements were made in the Oklahoma militia. The name was also changed, to the Oklahoma National Guard.

President McKinley's first call for volunteers, two days before war was actually declared on Spain in 1898, yielded two cavalry troops, which became part of the 1st Regiment of United States Volunteer Cavalry, or the "Rough Riders." Five companies joined from Oklahoma and Indian Territory, in response to McKinley's second call a month later. The Oklahoma men in the Rough Riders saw action in the famous July 1, 1898, attack on San Juan Hill, part of the defenses of Santiago, Cuba. The other Oklahoma and Indian Territory units saw no action.

In 1907, when both territories were combined into the state of Oklahoma, the Oklahoma Territory National Guard structure was expanded to include what used to be Indian Territory. A 1908 act formally organized the new militia.

Units of the militia were deployed in 1909 to help quell disturbances resulting from Indian resistance to incorporation into the new state. In 1916 Oklahoma units were deployed to the Texas-Mexico border. The only action they saw was the night of October 16-17, when a patrol chased Mexican raiders into Mexico. Oklahoma guardsmen were mustered out of federal service March 1 and 2, 1917. A month later, with war against Germany pending, the Oklahoma units were back in federal service. On August 30, 1917, strengthened by recruitment, the units bordered trains for Camp Bowie, Texas, to join the 36th Division.

★ ★ ★ ★ ★

National Guardsmen—345,000 of them—throughout the nation had to be trained. Locating the planned thirty-two new training camps was a task assigned by Secretary of War Newton Baker to the six army department commanders. The 36th Division would train at a site two and a half miles outside of Fort Worth, Texas, to be called Camp Bowie. Contracts to build the camps were given to reputable local construction firms, as the military assumed there was no time to go through normal competitive bidding procedures. By late August and early September, Camp Bowie was considered advanced enough to start bringing in the men. Construction of Camp Bowie continued well into 1918. In five months, the National Guard was called up and the national army created. The 36th Division was created and brought up to strength. A camp was built for the training.[33]

The first units to arrive at Camp Bowie, on July 26, 1917, were four troops of the 1st Texas Cavalry. They were assigned to guard the camp during construction. Ten additional companies arrived during August. Major General Greble arrived August 23, officially taking command of the camp the next day. The three brigadier generals assigned to the division arrived within a few days of Greble. Only one held a regular army commission, while "The initial staff of the 36th was composed largely of Regular Army officers."[34] A historian of the period has written,

> As a general rule, Regular Army and Guard officers did not mix well in World War I; basically their animosity for each other, where it existed, was one of full-time professionals versus part-time citizen soldiers. Regulars tended to be rigid, formal, well-trained, perhaps rank-hungry career officers while Guard officers were likely to be somewhat loose, informal, moderately-trained non-professionals.[35]

Soon after the camp opened, the two infantry brigades, and their four component regiments, were organized. The 36th's organization, though planned by General Greble, was actually supervised by George Blakley, the senior brigadier general serving as acting division commander. All new division commanders, including General Greble, were sent to Europe for a firsthand look at the war.[36]

The 71st Brigade, 141st and 142nd regiments, and 132nd Machine Gun Battalion would be formed from the 1st, 2nd, and 7th regiments of Texan infantry, and the 1st Oklahoma Infantry. The 72nd Brigade would have the 143rd and 144th regiments and the 133rd Machine Gun Battalion. Its sources would be the 3rd through 6th Texas infantry regiments and the 1st Texas Cavalry. The 61st Field Artillery Brigade would be made up of the 131st, 132nd, and 133rd regiments and the 111th Trench Mortar Battery, and be formed from the 1st and 2nd regiments of Texas Field Artillery and the 2nd Texas Cavalry regiment. The division would also have engineers, signal men, supply and hospital personnel, military police, and headquarters personnel.

The commanders sought to mix trained guardsmen with untrained recruits, to benefit the latter and produce the most efficient division possible. The Oklahoma Guard leaders were not happy that their men were losing state identity, not even making up an entire regiment. The 1st Oklahoma Infantry became part of the 142nd Infantry. Troop A, 1st Oklahoma Cavalry, became part of the 111th Ammunition Train. The company of engineers was absorbed into the 111th Engineers. The 1st Oklahoma Field Hospital became part of the 111th Sanitary Train. The Machine Gun Company of the 1st Oklahoma Infantry and part of Company M were merged into the 131st Machine Gun Battalion, the divisional battalion.[37] Ironically, the Oklahoma Guard units were better equipped than some of the Texas units, probably due to the fact that the Oklahoma units, though smaller, had a smaller proportion of recruits.[38]

The World War I army was segregated. One individual, George Gonsalves, half black, was apparently the only soldier with any black blood in the division.[39] There were other minorities, particularly Mexican Americans and Indians. One of the few Mexican-American officers, Capt. Augustine de Zavala, of the 143rd Infantry, was the grandson of Lorenzo de Zavala, an important Hispanic leader in the Texas Revolution and the first vice president of the Republic of Texas.

Chapter 3: Getting Ready for War 41

Some of the German Americans in the unit would be discharged at their own request. A few officers would be left home when the division went to France, including one colonel.

Over 600 American Indians, many from Oklahoma (some only part Indian), served with the 36th Division. The Indian guard members seemed fairly well educated, but this seems not to have been true of draftees. Some Indian leaders in Oklahoma had wanted all-Indian units, but it was decided to mix them in with other ethnic groups. This was not integrationist liberality. The real motivation, and the racial attitudes of the time, are reflected in a statement by the commissioner of Indian affairs. Discovering one company almost entirely Native American, Commissioner Cato Sells thought the Indians should be "mixed indiscriminately among the whites, elbow to elbow, so that they will absorb the English, habits and civilization of their white brothers."[40] Though some Indians were dispersed, this particular company retained its character throughout World War I.[41]

One source, from the 1980s, evaluated the racial situation by stating,

> There are indications in the sources to suggest a measure of prejudice on the part of the majority toward some of the minority, but there is no evidence of difficulty between the various racial and national groups or of blatant discrimination in the 36th.[42]

The American troops would have to be trained far better than ever before:

> Against an enemy like the Germans, training rigorously in peace and conditioned by years on the Western Front, it was hardly enough to induct men and equip them. All soldiers agreed that, for the first time, the American Army would have to complete a systematic and thorough training program before committing its troops to battle.[43]

How long would this training take? Common opinion in the army was that, starting from scratch with a new inductee, it would take at least a year to produce an effective soldier. The experience of the 1st Division seemed to show this to be accurate. The 1st was a new division, but composed of what were supposed to be trained and experienced regiments of the regular army. Pershing still took three months before he felt the division was ready to take over a quiet sector of the front. New divisions were bound to take longer. The War Department was trying to create an army from the ground up. The men had to be trained as individuals and then as part of increasingly larger units up to division. There was no substitute for actual combat experience, but the army would do the best it could.

When it came to equipment, guard-based divisions were better off than totally new national army divisions. Some equipment had been sent to the units, and some equipment came with the guard components. However, the 36th Division was handicapped, at least at the start, by equipment shortages. There were not enough basic weapons to go around, let alone artillery and machine guns. Many of the Camp Bowie facilities were not completed when the division first arrived. A historian has pointed out that

> The raw state of camp construction at the time of the arrival of the Guard units, the mobilization and organization of the division, the almost constant flow of transfers, the temporary absence of General Greble, the many inexperienced officers and non-coms, and the sometimes inclement weather hampered the progress of training at Camp Bowie. More serious impediments were the shortages of equipment and ordnance, the lack of adequate training facilities, and illness. The majority of problems were encountered and alleviated during the first four months of training. The slow start of the 36th was not unique, for similar problems plagued the instruction of other divisions.[44]

The 111th Engineers, though short of tools, ended up receiving what may have been the best training during this period. They helped build the camp.[45]

Initial War Department plans were to use the regular army to train the national army and the National Guard. The department first estimated that it could provide 961 enlisted men to each national army division,[46] a small percentage in a division of close to 30,000 men. Even this figure proved to be higher than could be managed. Further problems with using experienced men in new divisions arose from the need to send a token force to Europe as a symbol of commitment to the French and the British.⁴⁷

There was a shortage of officers and noncommissioned officers. The officer problem would eventually be solved by special training schools, producing many of what came to be known as "ninety-day wonders," a term that remained famous in World War II. Officers' training schools produced 80,568 officers during World War I.⁴⁸ They become lieutenants. Captains also tended to have little experience. Higher ranks, unlike at the start of the Civil War, overwhelmingly tended to be filled on the basis of merit, primarily with West Point graduates.

The initial officer shortage was exacerbated in the divisions by sending experienced regular and guard officers to train new recruits. The 36th Division had such problems. Some division leaders were afraid it might become a depot division, for training and providing replacements to other divisions. Other divisions served this purpose. Experienced men were also extracted from combat-intended divisions to be sent to divisions higher on the list for dispatch to Europe. However, it was not until March 1918, when the War Department announced a new draft law to fill understrength divisions, that the 36th's ranking officers knew it would stay together as an active unit.

British and French officers and noncoms arrived during the fall of 1917 to help with training. An artillery range was obtained, and used with such frequency that ranchers began to complain that the noise was affecting their stock. A training trench system was established, and the men could be trained in the type of warfare that awaited them—until the American army could take the offensive.

Camp Bowie, though a substantial improvement from Civil War campsites, was a relatively undeveloped facility when the men arrived. The first real cold spell, in late September, found the men without winter clothing and sufficient blankets. Half the buildings still lacked heating by the time of a second cold spell on October 8. The men now had winter uniforms, but still lacked overcoats and extra blankets. The camp soon suffered from outbreaks of measles, pneumonia, meningitis, and similar respiratory diseases. Measles was the most common, but pneumonia the biggest killer.

On November 28, General Blakely imposed a two-week quarantine, keeping his men in camp and allowing only relatives of the men to visit. Similar problems were occurring at other camps. One cause may have been that many of the men were raised in isolated rural areas and had not been exposed to common childhood diseases. The army soon established a two-week observation camp for newcomers. The medical epidemics, while they lasted, proved obstacles to effective training.

Additional medical problems were caused by what at first were only limited restrictions on access to prostitutes and bootleggers in Fort Worth. Cooperation between town officials and the 36th Division eventually eased the twin problems of excess drinking and venereal disease. Approved diversions for the soldiers were created in camp, including sports teams and the World War I equivalent of USO facilities.

Training itself could be dangerous. On the afternoon of May 8, 1919, the headquarters companies of the 141st and 142nd infantry regiments were engaged in firing trench mortars. After an hour of slow fire, the crews were ordered to fire rapidly. This was apparently going well until one shell exploded while still in the mortar. Investigation later showed that one gunner had apparently loosened a second safety pin on a shell, arming the shell when it was put in the barrel, rather than in flight. The explosion killed an officer and five men immediately. Five others died of their injuries in the hospital. A further four were badly injured.[49]

Some divisional officers, especially in the higher grades, would end up being transferred or retired after a harsh inspection by the War Department inspector general. General Blakely was among these officers being transferred out of the division. General Greble would not take the 36th Division to Europe. He learned, in March 1918, that he had been found physically unfit for combat in Europe. He believed that his age, fifty-eight, was responsible. His relief appears to have been more a direct decision by General Pershing. Greble was one of seventeen division commanders who visited Europe in 1917. Five were judged unsatisfactory by Pershing, who preferred younger commanders. Greble also badly hurt his chances of going to Europe with the division by ignoring Pershing's wish that the touring generals spend some time with British units.[50]

★   ★   ★   ★   ★

Chapter 3: Getting Ready for War

On May 16, 1918, General Greble received warning orders that the 36th Division (now officially nicknamed the Panther Division) would be leaving Camp Bowie soon afterward. On July 2, the division received orders to leave camp and head to New York City. The next evening, a small advance party left Camp Bowie for New York, from where the division would sail for Europe.

Units of the 36th Division (which did not travel as a complete division) took about four days to reach New York City from Camp Bowie. Most of the division went to Camp Mills, a processing camp on Long Island, to prepare for the voyage to Europe. Average stay at such a camp would be twenty-four hours. Individual supply problems were corrected, records complied and checked, citizenship confirmed for each soldier, and other problems rectified. Some leave was permitted, with passes into New York City, but those given leave were cautioned about security.

General Greble's successor joined the division at Camp Mills. Maj. Gen. William R. Smith was fifty, a regular army colonel also holding a national army commission, a West Point graduate, and a native of Tennessee. Smith had been confirmed as major general in the national army on July 6, the same day he was appointed to command the 36th. Smith had been in New York waiting to ship out with his previous unit when he learned of his promotion and transfer. He had to go to Washington to find out where the unit was, and then immediately return to New York.

At the same time, Col. Jules S. Muchert was relieved as commander of the 144th Infantry. "The army was much concerned, perhaps unduly so, with the possibility of enemy spies in the military."[51] Muchert had served in the Prussian army before moving to Texas. His patriotism, loyalty, and devotion seem not to have been questioned.* However, in the words of the unpublished division history prepared by staff in 1919,

> While Colonel Muchert had been connected with the National Guard of Texas for many years, and had shown himself to be an able and efficient officer, he spoke with a German accent, and it was known that he was born in Germany of German parents, and that he had at one time served in the Prussian army;

---

* This writer remembers talking with his grandfather about the problems the grandfather's German-sounding last name had created in World War I.

and, although neither his patriotism not his loyalty was in the least questioned, his view point was hardly American, and it was felt that, in justice to him and the troops of his command, he should not be taken to France.[52]

One division officer, serving with the 111th Field Signal Battalion, wrote about his travel experience of getting to France. (His spelling is unchanged.)

> [After a few days in Camp Mills] We received our travel orders on the 17th and left the next morning for Hoboken were we boarded our ship . . . At 2 o'clock on the afternoon of July 18, 1918, whistles blew, bells rang, and off we shoved toward Berlin. It was a wonderful clear day and the harbor was beautiful in all its activity. We glided out from our dock and into the main channel smoothly going seaward, picking up here and there a sister ship. The Statue of Liberty was far behind before all our convoy was assembled . . .
> We were issued life belts and initiated in the arts of "abandon ship" drill . . . at the first try it took nearly an hour to unload the ship. But we were soon able to empty the ship in 18 minutes. A ship will float at least 30 minutes after being hit, so Uncle Sam had it figured out pretty well.
> [On the third day after leaving] I was in the smoker, which was aft, very much interested in a thrilling adventure at the front written by a wounded soldier when all at once we had an earthquake on our port side, which proved to be the unloading of one of our 6-inch guns. It was followed by our other three guns, after which nearly every ship in the convoy opened up. It was exciting and nearly deafening. Some of the ship's officers declared it was a submarine . . . It was every interesting to see the destroyers maneuvering around like so many little fox terriers . . .[53]

Most of the division arrived in the port of Brest, in France, on July 30 and August 12, after twelve-day voyages. For the last few days, considered the most dangerous, they were escorted by destroyers which came out to meet them. Despite bad food, seasickness, and one U-boat attack on their convey, the division arrived in good order, disembarked, and went into camp near the port.

> The trip from Brest to Arrentiers was full of interest . . . We passed through some pretty places, saw much of the farm and

peasant life of France, and enjoyed it very much. It was funny when we unloaded at Bar-sur-Aube. Of course we did not know where we were going, how far we were from the front or anything about the country, as that is all handled by the officers, so when we got off the train we could near the big guns ... and thought we were right near the front. It was soon discovered, however, that a field artillery company was practicing only a short distance away.

By the end of August 1918, the division had arrived in a training area about 120 miles southeast of Paris, with divisional headquarters set up at the town of Bar-sur-Aube. Life in the field seems to have had its challenges totally apart from the Germans. A member of an unidentified unit wrote of other opponents:

I declared war on the rats last night, they are almost as much trouble as the "huns" only they don't grow quite so big. I don't think they get much larger than a cat, but they will just as soon sleep with you as not because I've had a dozen of the devils to sleep with me every night. Every time I kill one it seems like about forty comes to his funeral ...
They [apparently referring to officers, not rats] are always inspecting us. I feel like a piece of prize beef. They never inspect a man all the way through ... Just to show you how technical the army is, the other day the sergeant says, inspection arms! I laid down my gun and rolled up my sleeves and he didn't want to see my arms, but my gun.[55]

The division started a program to prepare it for the hard combat of World War I. American training camps had focused on trench warfare, as ordered by the War Department. Pershing, however, was determined to undertake offensive warfare to end the trench stalemate, and offensive techniques were to be taught the Americans. In the words of a War Department report,

Great stress is laid upon the subject of training for the reason that it is believed that the satisfactory results obtained on the Allied fronts are chiefly due to the constant training of the officers and men in every phase of operations. It is considered that it would be reckless to neglect this subject knowing that we are pitted against an enemy who has been engaged in such training for a period of more than fifty years.[56]

Specific orders to the 36th called for training which "must contemplate the assumption of a vigorous offensive," able and energetic leadership, and discipline.[57] As the soldier who battled the rats put it, "the principle of the thing is to get the other fellow and not let him get you."[58] The division also underwent another extensive shift of officers, including some in higher positions, and some shifting of men as men were transferred to other divisions headed immediately for the front.

In late August and early September, the Allied high command planned offensives along the entire front. The 36th Division was nearing the end of its advanced training. Contingencies of war made it impossible to follow Pershing's original plans for all his units and give the division some time in a quiet sector of the front. However, a member of Pershing's staff later wrote a biography of Pershing, in which he states that "The 36th had no trench experience, but it could charge and keep on charging in the full flight of American initiative which is not at its worst in their home country of Texas."[59] On September 23, 1918, General Smith was notified that the division would be moved, assigned as a reserve division to the French Group Armies of the Center, the GAC. On October 4, the division was formally assigned to the French 4th Army, and the 71st Brigade moved out.

Within a week, the division would see its first World War I combat.

CHAPTER 4

Combat at St. Etienne

"A German major said he had been in the war ever since it began but never had seen such fighting."[1]
—A. PAGE, member of
144th Infantry Regiment, c. 1918

Elements of the 36th Division (the 71st Brigade) first saw action on October 8, 1918. World War I had been underway for four years when division members had their first direct contact with the German enemy. The United States had been at war eighteen months. The nature of this war had greatly changed at least twice, as had ideas for how American troops might best be used as part of the Allied war effort.

Most people's image of World War I is that of trench warfare. They think of shell-ravaged ground, devoid of any vegetation, as desolate as the moon. People think of opposing trench systems slicing their way across the landscape. The image comes to mind of soldiers going "over the top" of the trenches, accompanied by massive cannon fire, in frontal assaults on enemy positions. Finally, the popular image is of enemy machine guns slaughtering attacking troops by the thousands. This is a grim picture, and was true of much of World War I on the "western front" of France and Belgium. Opposing Allied and German trenches, with the moonlike landscapes of no man's land in between, stretched from the

Swiss border in the south to the North Sea. There was little room for a war of broad maneuver, even if the commanders had been able to conceive of such moves.

World War I did not start as static trench warfare. On August 3, 1914, the Germans began a broad rightward sweep into Belgium and France. The relatively rapid German advance caught the French army, and the few British troops in France at the time, off guard. The Allies were forced back for a month, until the early-September Battle of the Marne stopped the Germans' advance and sent them into retreat. The Germans managed to limit this retreat. Each side then tried to outflank the other to the north. The marines dug in, and trench warfare began.

Neither side considered a major attack outside of a central corridor, though the French made some halfhearted advances into Alsace. According to one historian, writing in his study of World War I tactical innovations,

> The chief and most unrealistic error committed by both sides was continuing to emphasize the corridor stretching from Verdun to the Rhine: The traditional route of access to Germany by a French army or to France by a German army. Because neither side could forget the history of this corridor, neither could really gamble profitably on an alternative.[2]

Prewar military planning had looked to a war of mobility. Though recognizing the power of defense, neither side was able to react quickly when the war began to stalemate. Troops would dig in for protection where an engagement had ended. Short-term entrenchment is a valuable military tactic, one still used by modern armies. The World War I armies, however, stayed where they entrenched in September and October 1914. "Little did we think when we were digging those trenches that we were digging our future homes,"[3] is the way one British soldier described the almost casual start of trench warfare. A German wrote that "the war has got stuck into a gigantic siege on both sides."[4]

Trench warfare began in all its horrors. "Quite early in the war the opposing high commands became aware of the stalemate when they read reports showing a sudden increase in casualties coupled with inconsequential gains of ground."[5]

The armies still thought offensively, though. The French, for example, refused even minor withdrawals to more defensible positions. The French general staff preferred to advance. Acting quickly and with audacity, bringing French "elan" into play, was the way to overcome problems. Fine-tuning attack procedures was the first thing the generals tried. With both sides looking to martial ardor to combine with massive frontal assaults, results such as the Battle of Verdun, fought on and off from February through April 1916, could be expected. Virtually no ground changed hands permanently in this German attack. The Germans, however, suffered 300,000 casualties. The French total was somewhat greater.

Even the slowest of military minds saw this slaughter could not continue. In France and Britain, politicians pointed out the dangers to popular support of the war. By August 1915, the Germans introduced the first special-assault units, or *Sturmtruppen*. In December 1915, special training started to expand the number of these units. These troops did not play a major role in the Battle of Verdun. A year later, however, in the spring of 1917, things began to change. Erich Ludendorff, in operational command of German forces in the West, pushed for greater and more effective use of storm troops in attacks.

The World War I German army was surprisingly open to tactical innovations. Tactical command was decentralized, within the limits of operational assignments. Decentralization evolved from the unusual nature of the World War I German army—a loose combination of the armies of individual German states, particularly Prussia and Bavaria. Tactical innovation could move up from field units, as well as down to these units. "The German Army was a highly decentralized, mission-oriented organization that placed a great deal of trust in its officers."[6]

Greater centralization in the British and French armies made it substantially harder for innovation to be adopted. Changes had to be proposed to the high command, and then sent down through the chain of command. German military tradition encouraged independent tactical thinking and made it easier to try out their ideas. Inconspicuous small units of well-armed and highly trained men, able to lead an attack by infiltrating through enemy lines to wipe out enemy strong points, was such an idea. Despite the German openness to innovation, it still took well over a year for the *sturmtruppen* to become fully effective.

The Allied incorporation of tanks as full partners in their attacks was similarly slow. Armored land-fighting vehicles had been conceived as early as 1855. Later in the nineteenth century, H. G. Wells called for a "land ironclad" to destroy enemy defenses. Just before World War I broke out, Hugh F. Marriott, a British mining engineer, thought that a vehicle known as a Holt tractor could be adapted for military purposes. Marriott started the process of creating the tank when, in July 1914, he told a British staff officer friend of his, E. D. Swinton, about the idea. Swinton forwarded the idea to the War Office, but nothing happened. There still was no progress a few months later, despite Swinton's further efforts and firsthand reports from the front. Swinton later complained that "we ought to have realized that our established methods were useless in warfare of such a nature."[7]

The first tank prototypes were not demonstrated until January of 1916. Tanks were not used in actual battle until September of the same year, at the Battle of Flers. By late summer of 1917, the role of tanks in battle was still uncertain. Tanks could be highly effective when properly used. Terrain, however, could make it almost impossible to use tanks successfully. This included terrain pockmarked by too many artillery shellholes. Tanks could also be mechanically unreliable. Tanks were a weapon with great potential, but tactical and mechanical refinements were necessary before they could be put to maximum use.

★ ★ ★ ★ ★

The United States Congress declared war on Germany on April 6, 1917. The next day, Secretary of War Newton D. Baker told the House Military Affairs Committee that "the plans of our military cooperation are in the making rather than having been made."[8] British and French delegations arrived in Washington later that month to plead for immediate assistance. The War College Division of the General Staff (the primary United States Army planning bureau at the time) recommended that all trained officers, regular army and National Guard, stay in the States to train recruits and draftees.

---

*Not published until 1948.

Part of the token first American troop contribution (the 1st Division) paraded through Paris on July 4, 1917. After the parade, however, the division had to undertake three months of intense training behind the lines until it could be put into a quiet sector.[9]

The German U-boats were being defeated, making travel to Europe much safer. The British had found enough shipping to get American units to France. There was still the question of how they would be used once they got there. The British had initially wanted American troops to be incorporated into British forces. However, by 1918 the Germans started what one official telegram described as "what may well be prove to be the decisive battle of the war . . . the situation is undoubtedly critical and if America delays now, she may be too late."[10]

The British and French were forced into flexibility by their increasing desperation as the German spring advance continued.[11] Pershing got the other Allies to agree that the Americans would fight first in one, then a second, independent American army—after some time training with the British and French. Such an arrangement had been slow to work out, with the Allies arguing—with much logic on their side—that American forces were substantially less experienced in combat than British and French. They felt United States troops would be most effective if used to bolster experienced units.[12] The Americans would get experience and training, and Allied provision of supplies would be easier. Allied units would be restored to functional strength.

John J. Pershing had been assigned to command in Europe in an order dated May 26, 1917. The order specifically stated that Pershing was to cooperate with the Allies, "but in doing so, the underlying idea must be kept in view that the forces of the United States are a separate and distinct component of the combined forces, the identity of which must be preserved."[13] The reasoning behind this order, and Pershing's full support of the need for independence, was both practical and emotional. There would be distinct problems integrating American troops, even in company- or battalion-size units, into Allied armies. Culture was different from the British. Language problems would create difficulties working with the French.

American desire to undertake offensive warfare as soon as possible was an issue. Though considering infantry as the key to victory in battle, Pershing did not want his men used up in the trench-warfare meatgrinder. In the words of Pershing's 1919 final report, "Trench warfare naturally gives prominence to the defensive as opposed to the offensive. To guard against this, the basis of instruction should be essentially the offensive both in spirit and in practice. The defensive is accepted only to prepare for future offensive."[14] The offensive, and the new *Sturmtruppen* assault tactics, had began to restore mobility to the war. A memo on training noted that "All operations reports emphasize the importance of the ability of troops to maneuver. This applies especially to the infantry."[15]

National pride was also a factor in Pershing's reasoning. American troops could not be made to feel they were not trusted to fight on their own. "High officers of the Allies have often dropped derogatory remarks about our poorly trained staff and high commanders . . ."[16] Extensive association with war-weary and depressed Allied troops, aside from that needed for training, might rub off on the Americans. Without independent American armies, it would be harder to have a real say in any peace discussions after the war. The Allies were never fully reconciled to this approach. The emergency would sometimes require American troops to be used wherever and however needed. However, on July 24, 1918, Pershing issued orders creating, effective August 10, the American 1st Army.

American units would still receive field training with the British or the French, with somewhat different systems depending on which of the two administered the training. For those to work with the British, according to the official government history of the American military in World War I,

> Under the Six-Division Plan, six American Divisions were to be moved from the United States to France in British ships and trained by the British . . . Immediately upon debarking, each division was equipped by the British and dispatched to its previously selected training area.
>
> There, under the supervision of veteran British units and experienced British training cadres, intensive training was begun. By agreement with the British, American units in training were to be used in emergency to man the rear defenses. As the training progressed, some units participated in front line operations with the British units to which they were attached; but from the American viewpoint these operations were incidental to training.[17]

Chapter 4: Combat at St. Etienne

For those units training with the French, such as the 36th, the procedure was somewhat different:

> For infantry units—to billet a French and an American unit in the same locality, the French unit to assist in instructing the American unit. After a period of careful, deliberate instruction, small units of the American division would be sent into the lines where they would serve for a few days side by side with French units. This method was to be continued until all units of the American division had short service in the front line... When this preliminary training... was completed, entire divisions were to be concentrated at key points... for divisional training.[18]

Pershing wanted American troops fully trained before they saw combat. He saw the role of the American troops as winning the war in 1919. Pershing must also have foreseen the morale effect on the Germans of suddenly having to face enormous numbers of fresh American troops. Ludendorff also foresaw their effect. The German March 1918 offensive was timed to take advantage of a "window of opportunity," to use a modern term. The Russian surrender in the east let the Germans bring reinforcements to France. With the slow buildup of United States forces, American troops had not yet arrived in anything like their expected numbers.

The near success of the German offensives made it necessary for Pershing to modify his policy. While they were underway, according to a 1944 government history of the 36th Division in World War I, the German offensives

> so depleted the Allied reserves that the Allies faced almost certain defeat unless they received immediate support. In this crisis [Pershing] postponed for the time being the concentration of American divisions for the formation of an American army, and made all American combatant forces available for service with the British and French armies.[19]

American units helped blunt the German attacks. Ludendorff's offensive eventually ran out of steam when the Germans lacked the logistical capability to exploit breakthroughs. American participation in the defense, including the successful counterattack and recapture of the village of Cantigny, provided, in the later words of a senior American general, "The first cold foreboding to the German that this was not, as he had hoped, a rabble of amateurs approaching."[20]

Allied forces began, on July 18, 1918, a counteroffensive in the Marne salient. A July 24 meeting of the Allied commanders, undertaken while the Marne offensive was still in progress, agreed on a further offensive with a role for an American army. "The immediate purpose of this plan was to reduce the salient which interfered with railroad communications which were essential to further offensives operations."[21]

The next step was a series of major converging offensives. An American attack was to be launched on September 26 between the Argonne Forest and the Meuse River. The French 4th Army would be on the left of the American attack. The attack was to be directed against a German railroad, their principal supply line, running between Carignan, Sedan, and Mezieres. At Sedan this lay fifty-three kilometers from the front. If the Allies could cut the supply line, the Germans would not be able to hold their positions to the west and northwest of Sedan. This was a vital line of communications, certain to be fiercely defended by the Germans. "The enemy is strongly resisting the advance of this corps,"[22] a September 30, 1918, dispatch would read. A 36th Division staff officer later wrote, "While it is undoubtedly true that the morale of the German troops had deteriorated since the . . . termination of the spring offensive, such was the force of Prussian discipline that there was little indication of weakened spirit . . ."[23]

"East of the Argonne Forest, the American First Army was to make an initial advance of 16 kilometers and penetrate the hostile third position on its front,"[24] the staff officer continued. This was the Kriemhild Stellung, part of what was popularly called the "Hindenburg line." As the Americans were advancing to the east of the Argonne, the French 4th Army would advance to the west. By October 4, eight days after the attack started, the Americans had advanced twelve kilometers and gained a foothold in the advanced positions of the Hindenburg line. The French 4th Army had initially been stopped in front of Blanc Mont, a "sloping chalk-limestone ridge, which rose about 250 feet at its highest elevation and extended several miles east to west,"[25] as later described by a historian.

The 36th Division, and the 2nd Division, were assigned to the

French army group controlling its 4th Army. When the French attack bogged down, 4th Army commander Gen. Henri Gouraud obtained the services of the 2nd Division. The 2nd Division was known as a "marine" division. Though commanded by a marine, Maj. Gen. John A. Lejeune, the division actually had only one marine brigade. On September 29, the 2nd Division began its assignment to the 4th Army. The night of October 1–2, the division went into line, and attacked October 3. The division, and cooperating French troops, suffered heavy casualties but succeeded in capturing the ridge and advancing several miles further.

On October 5, Lejeune felt it necessary to request that his division be relieved. Gouraud had anticipated the need to relieve a part of the 2nd Division, and the 36th was available. General Smith, commander of the 36th, had already been asked to send one brigade of infantry and the 111th Field Service Battalion to the area the night of October 4–5. Lejeune was given command of the 71st Brigade, which included the 141st and 142nd regiments, the afternoon of October 5. At a conference that evening, Lejeune told the brigade and infantry regimental commanders that the next night the 71st would replace the 3rd and 4th brigades of the 2nd Division. Lejeune promised to make up ammunition shortages and have his brigade leave their heavy weapons until the 71st Brigade's weapons could be brought up. The 71st was ordered to move out early the next morning.

The brigade had to stay off the main roads, so the march to the line took most of the day. With the exception of shelling, the march was relatively uneventful. However, the men were particularly impressed by the remains of the German defensive positions of the Hindenburg line. The many bodies seen by the men of the 72nd Brigade led one officer to later remark that "the sun was sitting [sic] on our day of posing in one of Uncle Sam's uniforms and it was up to us to make good on what we have told the folks at home we were going to do."[26]

Looking north, the direction in which they were advancing, the men of the 36th could see smoke clouds, airplanes, and observation balloons. As the shelling "increased in the early evening the wandering troops frequently found it necessary to break up into detachments and hug the places of shelter to avoid casualties."[27] The brigade had halted for dinner, and to await the return of their commanders, away getting specific instructions from General Lejeune. Some shells were coming very close. A division member wrote,

> There's absolutely nothing so uncanny as to hear a shell approach. It is not comfortable in broad daylight, but at night it is positively bloodcurdling.
> ... It does not take a man long to differentiate between a shell or "G.I. Can"* coming his way, and one going the other way—at the foe. It took me only about four seconds, more or less. A shell coming in your direction . . . warns you of its approach when thousands of feet away by a chortling whistle, like that of a distant locomotive. This whistle, more of a longdrawn moan than a whistle, grows louder and hoarser as it approaches. A swish-swishing sound is added as it whirls nearer until you have a reproduction of the sound wind causes when blowing into an empty bottle. Just before it strikes, the sound is like that of a wild cat whistle and then comes the all-shaking sharp "BOOM." [Capitalized in original typescript.] To hear one of these "G.I. Cans" approach from an utter darkness, knowing not where it will land . . . is enough to cause your knees to wobble and a chill to run down your spine. And then to have one strike near, and the next one nearer . . . and then after a pause to hear another one coming, with the full knowledge that if its progress forward were equal to that of the preceding, brings you mighty "near to God" as one man said . . .[28]

Guides from the 2nd Division were supposed to meet the 71st Brigade and escort its components into proper places in line. The maps given out proved to be almost useless. The front was not marked, nor were the locations the brigade was to take. A further problem came from the 71st's sector being on the corner of four maps, difficult to read even if pasted together.[29]

The guides took shelter during an evening shelling and did not reappear until dark. It turned out, also, they did not fully know the area near the front. According to the unpublished official division history, "the original guides furnished the [141st Infantry, 2nd Battalion] led them around all night under shell and machine gun

---

*The division staff officer preparing the authoritative official history, Captain Spence, describes the "G.I. Can" as being shells from a 155mm cannon. They were nicknamed after United States government–issue garbage cans, an early use of the term G.I. The other main artillery piece was the 88mm. The G.I. Can made a whistling noise when incoming; the 88 was silent. Spence, footnote on page 84.

fire, and only when new guides found them were they able to get into place."³⁰ One lieutenant in the 141st later said that his battalion's guide become lost when they were almost at the front. After this, "we . . . wandered around in the woods lost while the artillery was playing on us and machine gun snipers were very active."³¹ Fortunately for the 71st Brigade, there was little movement on the front on October 7.

Sixteen members of the brigade, however, were killed or wounded by shelling in the evening. Entrenching tools become well appreciated. As one soldier put it, "I have two weapons of defense, one a rifle the other a small shovel which I carry on my belt. It may sound foolish but both are of equal importance and I would not part with my latter friend which I use solely for digging."³²

Another member of the 144th Infantry wrote a young woman, asking her to

> Tell your dad and Mr. White that if they want any pools dug, just to wait until Argie and I get back and all they will have to do is get something that sounds like a machine gun or a cannon and turn it loose and either one of us can move more dirt laying flat on the ground with a shovel than they could both move with a team and scraper.³³

Second Division commanders, thinking they might be called back into action, refused to turn over maps and related data to the 36th. Commanders from the 71st Brigade were hampered from the resulting lack of information.

German artillery fire was coming from the hills to the north and northeast of the village of St. Etienne, between the lines. German planes were out in force, strafing, bombing, collecting information and, apparently effectively, spotting for the artillery. French planes were not effective at preventing German efforts.

German and American lines at places were little more than a hundred yards apart. The 71st Brigade's machine gun battalion, the 132nd, was divided between the two regiments. French units were on both sides of the Americans. There were also gaps in the line, inherited from the 2nd Division. The gaps made communication difficult within the brigade.

The area near St. Etienne was far less wooded than at Blanc Mont. The unpublished division history prepared by staff just after the war notes, "The nature of the terrain and the character of the

German defenses were well adapted to a stubborn rearguard defense such as the enemy was making."[34] An Oklahoma officer with the 36th later wrote that "The northern slopes of Blanc Mont are covered with growths of pine and underbrush which thin out as the village of St. Etienne and the Arnes are approached. Beyond the edge of these trees and underbrush the country is open and for anyone to venture out of the cover was to draw fire immediately in the day time."[35] The Germans hid machine gun nests in wooded areas. The posts were arranged in depth, and arranged so that if one was captured it could be fired on by another position. The system worked, causing casualties in the ensuing American attack.

Several small hills were visible, as were observation posts and deep dugouts. Barbed wire was used to complete the defensive positions. The Oklahoma officer remembered that "To a naturally defensive position the Germans had added strength by stringing barbed wire from tree to tree in the woods and placing strands of entanglements in the low places . . ."[36] The forces were at worst relatively evenly matched in this sector, or the Americans might have had an advantage—German force strength is uncertain. However, the Germans had the tactical advantage of defending positions on slightly higher ground. The particular topography of the area pushed the 141st regimental line, to the brigade right, 600 meters in back of the line of the 142nd.

While the 71st Brigade was arriving and digging in, the morning of October 7, French 21st Corps commander Stanislas Naulin was conferring with General Lejeune. General Naulin told Lejeune that "a general attack would take place at daylight on the 8th, and that he anticipated that the fresh brigade would achieve a success equal to that gained by the 2nd Division on October 3rd."[37] Lejeune protested that the French general "was expecting the impossible of untried troops," urging that they be allowed a few days of training in the line under fire—basically Pershing's technique of the year before—before being used in an all-out attack. Naulin did not accept the suggestion and left declaring that "Tomorrow will be another great day for the 21st Corps!"[38]

General Gourard had actually ordered the three-corps attack for October 7. The attack was postponed to give the 71st Brigade time to get into line. "The proven aggressiveness of the Americans was why the French had asked Pershing for the services of two

American divisions for the Champagne drive in the first place,"[39] as a recent history notes. Possibly the failure to let the 71st adjust to the line justified Pershing's original concern of the treatment of Americans fighting under foreign commanders. However, the one 71st Brigade had been assigned by Lejeune to relieve two brigades of the 2nd Division. Lejeune issued verbal orders to his brigade commanders that morning, subject to written orders to be sent later. General Whitworth, commanding the 71st, relayed the orders to his units.

The 71st was to be supported in its assigned advance of about two kilometers by the 2nd Division units on its flanks, two battalions of French tanks and artillery. The tanks were supposed to help the infantry wipe out German machine gun nests. A short preliminary barrage was to be followed by the advance of the 71st behind a rolling barrage—a valuable, but complex, artillery technique of having shells targeted to stay in front of the advancing unit.

Warning orders were issued the day before. However, Brigadier General Whitworth did not get Lejeune's written orders for the 5:15 A.M. attack until after midnight. It took his staff three hours to get copies and his own orders out to his subordinate commanders. Some company commanders were not actually informed until H-hour had arrived.

Field telephones had not been used to transmit orders, for security reasons. "It was well understood," a division memo noted, "that the German was a past master in the art of 'listening in.' "[40] Orders had to be hand delivered. Some elements never received their orders, "at least one of the runners having been killed by shellfire,"[41] and ended up attacking at the "sound of the guns." However, despite precautions, the Germans seemed to have guessed that something was up. A German officer present at the battle later wrote that the Germans knew something was planned and were ready, due to observing "strong traffic along the enemy rear lines of communications moving chiefly in the direction of St. Etienne, and apparently also a number of tanks."[42]

The German artillery started firing the same time as the Allied, but more effectively, including gas shells. The Germans were slightly overshooting, inflicting heavy casualties on the support and reserve battalions. One veteran of the battle remembered that "whenever a barrage comes over every one has to dig a hole in the earth and get in it for protection. A great burlesque among the fellows is 'these dugouts are bomb proof until a shell hits one.'"[43] German planes and balloons directed the fire without any opposition from Allied planes.

The Allied 2nd Field Artillery overshot more seriously, leaving the German entrenchments and machine gunners virtually untouched. German machine gunners were quite effective at long range, but apparently surrendered easily when the Americans came close.[44]

Waiting could seem more nerve-racking than actual combat, remembered one division veteran who took part in the attack. "There is no fear in those thrilling moments when you are in the fight but, ah! it is before the fight, in those still, awesome, soul-trying moments and hours when you realize that you are soon to grapple with the Hun, and as you contemplate the dreadful ordeal."[45]

The rolling barrage, designed as covering fire just ahead of the advance, was moving away before some of the companies went "over the top." The advance, due to lack of written orders being received in time, lost coordination a few minutes after it started, when the troops reached the first barbed wire. German machine gun fire was heavy as the Americans tried to cut their way through barbed wire.

The normal confusion of combat grew worse as the attack continued. Many officers had been killed or wounded shortly after starting out. An example was the commander of the 1st Battalion, 141st Infantry. The unpublished division history noted that "He had only advanced some three hundred yards when the burst of a shell immediately in front of the advancing lines scattered the Battalion Headquarters detachment, knocked the Battalion Adjutant unconscious, and instantly killed the Major."[46]

The men in the support and reserve battalions pressed so close to the assault line that switching position, and dispersing casualties among the units, was harder. The 141st and 142nd regiments lost contact with each other. The attack soon degenerated into a series of small-units actions. "In a short space of time all liaison between units of the assaulting companies had been broken up . . . Platoons broke up into small groups and then became intermingled as the men fought for themselves,"[47] the Oklahoma officer, quoted earlier, later wrote in his history of the 36th Division.

Both regiments were hurt by machine gun fire from their right flanks. The starting gap between the 142nd and 141st was never rectified during the attack. The 141st was also unable to advance as far as expected, at least partly due to heavy losses among battalion and company commanders, leaving inexperienced combat troops with insufficient leadership.

Supporting French tanks come into action about fifteen minutes late. Their commander was one of those who never actually received an attack order. They provided little assistance and drew heavy enemy fire. Worse, they mistook the 3rd Battalion headquarters detachment, and nearby Company B, for the enemy and twice fired into the detachment. The tanks were eventually withdrawn. A battle participant from the 144th Infantry remembered,

> We got so mixed up the Germans could not use their guns. Our men used their bayonets and rifles for clubs, some of them had pick handles and trench knives. A German major said he had been in the war ever since it began but never had seen such fighting.
> . . . By sun up we had killed or captured (mostly killed) every Boche in our front and had left the French on both sides of us about a mile to the rear . . . while the Germans were retreating from the French, they tried to pass us, but we cut them all to pieces with our machine guns. In the early part of the fight there were such little firing, the French thought we were not charging. We were using clubs and bayonets. The French general commanding the fourth army corps sent a message to us asking why we were not attacking. The Messenger went back and reported that he could not find the Americans. We had gone too far for him. The French tanks had come down. They didn't know we had gone so far, and started to shoot us up.[48]

Portions of the 2nd Division were able to provide some assistance, partially compensating for the lack of any real progress by the French 73rd Division. The 141st Regiment, and the 2nd Division elements, were pushed back from their maximum advance. At 5:30 P.M. they repulsed a counterattack, ending the action for the day on the brigade's right.

Members of the 1st and 2nd battalions of the 142nd Infantry received no advance notification at all of the attack. An official report sent to division headquarters after the attack noted that "As far as can be found out, no regimental orders were issued for the attack of October 8th."[49] Their first "word" was the start of shelling by both sides. On the right of the 142nd's sector, troops began their advance from a hilltop position, running immediately into German machine gun fire, directly ahead and on their right flank. The troops used a sunken road as cover, rushing forward short distances and taking shelter, then repeating the process, until the frontal German position was flanked. The captain of the leading company was hit almost immediately. An American doctor was also killed early in the attack, while giving aid to some wounded men.

One veteran of the 142nd infantry described what he saw of the attack. The men around him

> crawled out in a little bunch of shrubbery and the Boches were in two good sized woods one in front of us and the other on the right of it with an open space of about one hundred yards between. To get to either we had to cross an open space of about seventy five or a hundred yards and of course the enemy were shooting at us from both woods with machine guns and snipers located in trees or any place where they could get a view of us, but protected from our view. We did not stay in the shrubbery long and then we started into the weeds . . .
>
> The first Germans I saw were coming out of a dugout yelling "Kamerad" at every breath, so I picked up a few German hand grenades, which we call potato mashers, and when I came to a dug-out would pull the string and throw a couple in. If any one was at home, they had a hard day.[50]

The 142nd was taking its worst fire from the left front, causing the regiment to veer to the right toward a wooded knoll called Hill 160. "In going forward, the assault battalion was to skirt the eastern edge of St. Etienne . . . These instructions were given because the Marines were thought to be in possession of St. Etienne, as they had reported."[51] The knoll was about 1,000 yards southeast of St. Etienne and would provide some protection from the unexpected heavy fire coming from the town. The marine brigade to the left of the 142nd was supposed to have occupied St. Etienne during the night. However, a patrol found it empty and the marines did not move in. The Germans did. Particularly effective was a machine gun in the steeple of a church. The Oklahoma officer remembered that "This machine gun had been able to cover the entire area of open ground over which the support and reserve battalions had been compelled to advance . . ."[52] The position was finally destroyed by an American artillery shell.

The French tanks were little more helpful than those "assisting" the 141st.

The 142nd managed to end German resistance on Hill 160 and then head toward the strong enemy positions in the cemetery. Some of the attacking American troops came in through the village, at the flank of the Germans. Hard hand-to-hand fighting finally resulted in the capture of prisoners and the evacuation of the cemetery. The Germans evacuated the town before noon, as it was empty when the marines entered. The marines were unable to advance north of the town. The 142nd kept going toward the northeast, reaching the St. Etienne-Semide road. German resistance stiffened. Without flank support, the 142nd was taken heavy machine gun fire on the right. German artillery was effective. Allied artillery was not effective.

The 142nd was forced to pull back from its most advanced position to meet an expected heavy counterattack. Some members of the regiment were captured when they failed to get the word to withdraw, another example of major communications problems. Most of the maximum advance of the 142nd, however, was held and, aided by artillery, the German attack was repulsed.

The regiment held this position throughout the night and the next day, following orders from the 2nd Division to "rectify your line, organizing it in depth. Gain and maintain contact with the enemy along your front and close liaison with allied troops on your right and left." The brigade was also ordered to scout for a possible attack and to take over the 2nd Division lines when that unit pulled out.

War is confusing. Perhaps this is an even greater characteristic than the "hell" William Sherman applied to war. The regiments had to be prepared for the continuing advance, and their remaining strength evaluated. This was not easy. Officer casualties were heavy in the attack—frequently an ironic sign of a well-led unit. Units were disorganized and only provisionally sorted out the next day.

Message runners suffered particularly heavy casualties, as they made good targets. Reports were often inaccurate when sent back to headquarters, creating major problems for the commanders. One historian writes that "At first the messages were optimistic; later, they were the opposite. Casualties were grossly overestimated. Determining the exact locations of the various organizations was impossible presumably because many officers did not have maps and those who did found them inadequate. The uncertainty explains at least to some extent the poor placement of the artillery."[54]

Estimates are that the 71st Brigade suffered more than 1,600 casualties in the attack of October 8 and 9. Thirteen hundred of these known casualties occurred on the 8th. Two hundred ninety-eight officers and men were killed. Seventy-four men were missing. The rest were wounded, including some by gas. An unknown number of other men were evacuated for "shell shock" (battle fatigue) and exhaustion. Two battalion commanders were killed, a third wounded, a fourth evacuated for shell shock, out of a total of six. Fighting had also taken a heavy toll of company commanders. The surgeon for the 3rd Battalion, 142nd Regiment, was killed in action.

The attack's execution had problems. One German observer later described the attack as a "failure." He explained the problems, somewhat sympathetically, as due to the 36th being under fire for the first time, and the tactics of the advance—battalions in line—leading to confusion if the lead battalion got held up or stalled at all.[55] However, the German position was strong, and the German soldiers' reputation as tough, skillful fighters well deserved. The 2nd Division, the French, and the artillery did not supply such good support as might have been expected. The lack of Allied effort in the immediate air was vividly apparent even during the battle. Casualties were heavy, but World War I was characterized by heavy casualties, brave soldiers trying to make up for a frequent lack of skill in the upper levels of command. The frontal-attack tactics were more reminiscent of the Civil War—when they caused even heavier casualties among American troops—than those used in World War II. Heavy casualties also resulted from the enthusiasm with which the Texas and Oklahoma soldiers pressed the attack.

Finally, it had been fifty-three years since the 36th Division, or its ancestors, had seen real war. Even so, the attack was not a failure. The Allied advance continued.

CHAPTER 5

Forest Farm

"The town was still burning..."[1]
—GEORGE MCCALL, 144th Infantry, 1918

Attacks on the Germans continued on October 10. "Field orders of the 36th Division issued at 2:00 p.m. [October 10, 1918] had directed the 72d Infantry Brigade to execute a passage of lines through the 71st Infantry Brigade and attack at 5:00 p.m. No objective was set, but the 72d Infantry Brigade was ordered to keep abreast of the French divisions to its flanks." The 71st had been sufficiently battered by its previous efforts that the division commander, General Smith, felt it could not executive a "delicate side-slipping maneuver."[3]

Smith wanted to keep the pressure on the Germans, who appeared (accurately) to be retreating. He also had been ordered to stay in line with the French divisions on his flanks.[2] One of these divisions was already moving forward. The pass-through was ordered, and the 71st Brigade's commander told to maintain close contact with the enemy until the 72nd could get into line. The 142nd Infantry Regiment, of the 71st, was already in close contact, being kept in place by heavy machine gun fire. The Germans did not seem to be withdrawing as quickly on this part of the front.

The 141st, on the other hand, the 71st's other regiment, found only a German rear guard. A 500-yard advance was ordered and was successful in spite of intense artillery and poison gas. Machine gun and rifle fire were light.

Late on the afternoon of October 10, the 72nd Brigade began to advance. The 144th Infantry, on the left, was ordered to relieve the 142nd. The 143rd would relieve the 141st. The French corps commander in charge of the 36th Division, General Naulin, ordered that both brigades be placed alongside each other as soon as practical. The entire Champagne-area advance would shift slightly to the right, northeastward. The Allies could now aim their advance directly at Germany itself.

Shellfire on the 72nd Brigade's left flank was sufficiently intense to prevent the 144th from completing its advance that day. The regiment took position to be able to cover gaps in the 142nd's front. The 143rd also encountered heavy fire but was able to keep going until it came within range of German machine guns. Part of this regiment ended up passing through the 141st. The 2nd Battalion headquarters detachment actually crossed into German lines. Fortunately, only the battalion adjutant was captured when the Americans withdrew in the darkness. About dawn the next morning, October 11, the Germans withdrew from the immediate division front. The pass-through maneuver was completed that morning.

The advance continued after the 143rd and 144th regained full communications with each other and with the adjacent French units. Little opposition was encountered until the 144th crossed a ridge about a mile north of St. Etienne. The leading elements of the 2nd Battalion, on the right of the regiment, encountered heavy machine gun fire from nests on either side of a German supply railroad. Capt. Ben Chastaine, an Oklahoma officer with the 36th, takes up the story:

> Immediately these positions were uncovered the task of reducing them was begun. Flankers soon brought fire to bear that enabled the troops in front to advance. During this skirmishing, which lasted more than an hour, enemy observation was secured by an airplane which appeared overhead, and almost immediately this was followed by artillery fire that brought numerous casualties within the ranks of the attacking force. During the artillery fire two small ammunition dumps, abandoned by the enemy, were exploded in the vicinity, causing the impression generally that the ground was mined...
>
> During the artillery fire some of the German machine gunners apparently withdrew... while others were shot or bayonetted at their posts, being unable to get away before their positions were reached...[4]

That afternoon, elements of both regiments reached the town of Machault. Both encountered heavy machine gun fire from the town. The right flank of the 143rd, to the east, began taking fire from machine guns in a wooded area in front of the French 73rd Division.

Another German artillery-spotter airplane flew over. Soon after, the enemy artillery again opened fire—on the Americans, on the French, and on the town of Mauchault itself. Captain Chastaine later wrote that this fire on the town saved the American troops

> any real effort directed to dislodge the enemy. The German's own artillery suddenly opened up on the town and caused the "boche" machine gunners to scurry away to the north in order to save their lives. Due to this kindly aid from the Hun artillery Machault was evacuated before an assault could be launched against it. Almost immediately afterwards the enemy ceased altogether to direct his artillery on the positions in the vicinity, apparently being uncertain of the location of his own lines and being afraid to fire at random.[5]

German troops were observed evacuating positions in front of the 144th Infantry. Advanced detachments of the 144th opened fire. Some officers urged that a further advance would put demoralizing pressure on the German withdrawal. However, the 144th was not ordered to advance. Coordination was felt to be insufficient with the 143rd infantry to its right, and with French troops to its left.

Supply lines to the rear had also been hard to maintain during an advance. That night it proved impossible to get hot meals to the front-lines troops, a frequent problem.[6] The next day, as the rapid advance resumed, "The main problems of the day were routine for rapidly advancing troops . . . the various trains and miscellaneous units were pressed to keep up with the infantry."[7]

Another sign of the growing Allied success was the great amount of ammunition and supplies captured by the 36th and the nearby French units. Attigny, a main German supply center, was captured on October 13. "The town was still burning"[8] when the Americans entered. Another participant later wrote that the "enemy had maintained an ammunition dump of gigantic proportions."[9] Trucks were found abandoned, standing loaded and "ready to continue the journey"[10] to the front.

Some Germans later described their retreat as "deliberate and easy."[11] However, the retreat was so rapid that massive quantities of materiel could not be removed.

American and French patrols pushed forward to the Ardennes Canal, paralleling the Aisne River. The Germans had destroyed all major bridges. One American patrol, however, found a footbridge. The American officer commanding this patrol reported that

> This is an improvised bridge and simply to accommodate the crossing of small patrols. To the right of this foot bridge about 300 meters is the remains of what was once a strong stone bridge crossing both canal and river, but has been mined. About 10 yards to the right of these remains is a temporary plank bridge crossing the river, same was constructed (presumably) by the enemy ...
> On one of these planks was laying a spool of slick wire and four objects that resembled large bologna sausages, as they were stuffed with a material of unknown composition and tied at either end with a string. Judging from these two items and the fact that the Boche which was killed [by the patrol] was on the south side of the river bank, I should say that they were laying a mine to blow up the footbridge across the river whenever any of our patrols attempted to cross ...
> The north bank of the river at this point is thickly covered with underbrush and trees. The houses that run parallel with the road on the north side of the river are believed to be occupied by the enemy ...[12]

A large portion of the countryside had been wrecked. The official history of the division reports,

> In making this retreat, the Boche of course blew up or burned as many of his ammunition dumps as was possible, but the vigor of the pursuit prevented anything like the complete destruction of the immense quantities of ammunition and supplies that had accumulated during the four years of German occupation.[13]

Fresh water could not always be found. The war, however, had developed movement, contrasting strongly to the previous years of bloody trench warfare.

On the night of October 12, 1918, the commander of the French corps to which the 36th belonged directed all his division commanders to establish bridgeheads across the Aisne. This was the first step in pursuing the enemy to the north. General Smith was not able to immediately carry out these instructions, as he was busy getting both his brigades back into line and taking over a section of the front to his right. Smith did place division artillery at the disposal of the French division on his left, as support for that division's planned surprise crossing of the Aisne.

The 71st and 72nd brigades spent most of Sunday, October 13, getting into position. The 72nd Brigade had finished by noon and spent the rest of the day clearing out remaining Germans south of the Aisne. The 71st Brigade had farther to travel, and was not in position until that night.

The French army commander, General Gouraud, decided to delay the attack across the Aisne for a few days. His army had outrun its supply bases. As a historian of the 36th wrote, "During the advance from St. Etienne to the river, the supply department had labored under the greatest difficulties."[14]

The American 1st Army, to the right, was advancing more slowly than the French. Unfortunately, a "strategically located pocket"[15] of Germans remained south of the Aisne. With the pause in the advance, the 36th spent a few days strengthening its positions, being resupplied, patrolling in the area, and planning for the upcoming continuation of the advance. Patrols were sent across the river to gather information and prisoners for questioning. Sometimes these patrols succeeded. Sometimes the men had to hurry back across the river after being discovered. The division took some shelling, including gas. Gas, however, was proving less of a hazard than the men expected.

Several messages of congratulations came in regarding the St. Etienne fighting, though one historian feels they were "a little magnified."[16] Their French corps commander, General Naulin, in a general order, included two paragraphs on the 36th:

> The 36th Division, U.S., recently organized, and still not fully equipped, received during the night of 6th-7th October, the order to relieve under conditions particularly delicate, the Second Division, to drive out the enemy from the heights to the north of St. Etienne-a-Arnes, and to push him back to the Aisne.
>
> Although being under fire for the first time, the young soldiers of General Smith, rivalling, in push and tenacity with the older and valiant regiments of General Lejeune, accomplished their mission fully. All can be proud of the work done. To all,

the General commanding the Army Corps, is happy to express his cordial appreciation, gratitude, and best wishes for future success. The past is an assurance of the future.[17]

The *New York Times* and the Associated Press were similarly complimentary of the 36th and its conduct in combat. They might also have complimented how well Americans usually treated civilians. One division member wrote that

> In every town where the Americans are you will always find all the children are our friends. One great Frenchman has said it was one of the greatest tributes he could pay the American soldiers was that they fought like the devil and were never too busy to be good to the children . . . If you are kind to the French people they cannot do enough for you.[18]

Combat for the 36th Division was not over. On October 18, the division was reassigned to serve as the right flank of the 11th French Corps, under General Prax. Soon after its reassignment the 36th Division took over, to its right, the section of the front formerly held by the French 7th Division. The 36th now faced part of an odd-looking "loop" in the Aisne River. The river made a ninety-degree turn to flow westward. In making the turn, it swung around high ground before turning west. A U-shaped salient was formed, with the opening facing Allied controlled territory. The Germans still held this salient.

The French thought the loop could create problems if it remained in German hands. It could be used to launch attacks or raids into the French army. Bypassing the area would expose the French, and the 36th, to flank fire. At the very least, it gave the Germans a useful vantage point for observing Allied activity in the area. The Germans seemed to agree with the French analysis of the bend's importance, and the bend was well fortified. The Operations Report of the 36th Division states,

> The Boche had fortified the mouth of the loop with a series of strong points, defended by approximately sixty machine guns, and in the 141st Infantry sector, consisting of 1400 meters frontage extending along the West half of this loop, six prepared machine gun emplacements of concrete were found, two consolidated trenches, one minewerfer emplacement, and twenty dug-outs, showing indication of machine gun occupancy.[19]

Chapter 5: Forest Farm

There was rolling high ground within the loop. The German defense line was at the mouth, the southern end. Any approach would be blocked by barbed wire, taking full advantage of the terrain and natural cover. Artillery and machine guns enjoyed clear fields of fire, and, in some cases, excellent cover. The officer preparing the official history of the division later wrote that "Field of fire was the paramount object of the German dispositions, regard for cover being sacrificed to attain it."[20] Even with cover a secondary consideration, one machine gun emplacement was totally concealed by natural brush. On the higher terrain north of the Aisne River, near the town of Voncq, was a concentration of heavy artillery. According to the same unpublished official history, "The mouth of the loop was particularly well fortified by means of a series of strong points and organized trench systems."[21]

The position was officially referred to as Forest Farm, after a nearby abandoned farmhouse. German defensive emplacements had already proved sufficient to beat off French attacks the nights of October 16-17 and 17-18, though the French were able to defeat a German counterattack after the second French assault.

The French army commander was being urged by Pershing to join in a major attack all along the front. The French were not able to do so on October 14. After a successful three-day attack, Pershing's army was at the northern tip of the Argonne Forest. Pershing now created a second American army and assumed command of the American "group of armies." Pershing felt that the Germans were clearly in retreat, that a well-organized and coordinated offensive would achieve their complete defeat.[22] October 28 was tentatively set as the date for the coordinated attack, but at the request of the French on October 27 this was postponed to the first of November.

Corps Commander Prax, on October 24, ordered General Smith and the 36th Division to take Forest Farm before October 27, probably meaning before the end of that day. The same day, Smith was notified that the 36th would be pulled out of line, starting on the 26th, through early morning of the 28th. The 36th and 2nd divisions had only been lent to Marshall Foch and the French for the emergency a few weeks before in the Champagne. The emergency was over, and Pershing wanted his divisions returned. Confusion arose, in that the planned 36th Division assault battalions were due for relief the second night. The 71st Brigade, designated to make the attack, could not be ready before the 27th. Smith protested the orders to Prax, pointing out the problems in replacing assaulting troops practically in the middle of an attack, with a counterattack expected. Withdrawing these troops, according to Smith, as paraphrased by Captain Chastaine,

> would be extremely inadvisable because of the confusion that would result. Also, if the troops in the support and reserve positions were relieved that night it would place new troops, unfamiliar with the ground, in a position where they would not be able to render the desired assistance in the event of an attack by the enemy in force.[23]

General Smith might not have agreed with the French about the importance of the salient. The French apparently gave Smith and his staff different reasons for the need to clear out the loop. "It is well to consider, at this juncture, that no satisfactory reason for the attack was ever received,"[24] the official history pointed out.

Smith also wondered if the French could be counted on in case of problems. General Prax appears to have been harder to work with than General Naulin, commander of the 36th's former corps. Smith also seems to have suspected that the French were using the 36th to take casualties the French wanted to avoid.[25]

Prax overruled Smith's objections, but reassured him on several points. The battle-experienced French 22nd Division would relieve the assault battalions after the objective was secured, and one full day after the attack. The French would not leave the Americans in trouble. More experienced artillery would remain to help with the attack, and additional artillery would be assigned. Liaison would be improved. Finally, the relieved troops would be trucked, not marched, to their new assignments.[26]

For a few days before the attack, German artillery bombardment increased in intensity and accuracy. Shelling followed every troop movement. General Smith suspected that the Germans might be listening in to phone communications, as a result of wires left when they withdrew. He gave some false orders over the phone, identifying some troops as being on a particular hill. The hill was heavily shelled. Coding messages would take too long, as would runners—who could always be killed. The solution was to look to Indian troops in the 142nd, particularly Choctaws. After a successful test, having the Indians transmit orders in their own language, Smith decided on this first documented[27] example of Indian "code talking" that would become famous in World War II.[28]

The attack on Forest Farm was set for 4:30 P.M. on October 27. The timing was set so that near-darkness would lessen the value of German observation posts. The initial attacking units would be the 2nd Battalion of the 141st Infantry (with an attack company from the 1st Battalion) and the 3rd Battalion of the 142nd Infantry. Both battalions were depleted due to the St. Etienne fighting of a few weeks before, and to the flu, but were considered dependable. The rest of the 1st Battalion would support the 141st's attack. The second battalion of the 142nd would support its attack. Three machine gun companies would cover the attack. An officer with the division wrote that "A high state of morale existed in all ranks. The men had gained confidence in their ability, that had come with experience under fire."[29]

Before the attack, and the preparatory artillery barrage, the attacking troops were withdrawn several hundred yards. This was a margin of safety during the barrage, to lesson American casualties if any shells fell short. "Due to the fact that our farthest advanced units were within sixty meters of the enemy positions, and within one hundred and ten meters of the enemy wire, it was necessary to withdraw our troops from their advanced positions,"[30] stated the 71st Brigade's operations report.

Another measure was taken to avoid the confusion of the St. Etienne attack. All officers and men were fully briefed on the attack. For security reasons, the men were not given this briefing until just before the attack started. The buildup had been undertaken as carefully as possible, in what appears to have been a successful effort to fool the Germans. Increasing mist helped.

After twenty minutes of intense bombardment, ending exactly on schedule at 4:30 P.M., the artillery began a rolling bombardment. Commanders blew whistles, and it was time for the infantry to go "over the top." "Slowly and steadily they pressed onward up the slope toward the two strong points, one at [Forest Farm] and the other in the little wood to the east. They were following the barrage so closely that they were almost 'leaning against it.'"[31] At St. Etienne, the rolling barrage had been too far ahead of the attacking troops, giving the defenders time to recover. This error, and the opposite mistake of shelling one's own troops, was being avoided. Though the Germans were able to start firing full-scale barrages in only about a minute, they were shooting in back of the attacking troops. The only Allied mistake was when an Allied battery fired at a company of the 142nd Infantry. Combined with a diversionary shelling across the Aisne a few miles away, the artillery was helping. A participant in the attack remembers,

> The enemy were about 1,000 yards away, and so we entrenched. We were in support a few days, and then the order came for our battalion, the Second Battalion of the One Hundred and Forty-first Infantry, to take the German stronghold at Forrest Farm, on another part of the canal. The line was behind and the French were unable to bring it up, though they had tried three times. At that time, our battalion was at low ebb. There should have been a thousand men and officers in it, and there were only 179. We had to go up against 3,000 or 4,000 Boches, well entrenched behind barb wire entanglements.[32]

The two attack waves advanced about a hundred yards behind the rolling barrage. The operations report of the 71st Brigade states that

> The firing schedules of our artillery started and proceeded in the most excellent manner, exactly on scheduled time. A personal reconnaissance after the attack showed that the chief enemy strong points had been demolished. The rolling barrage preceding our Infantry was perfect.
> The paths cut in the wire by the Trench Mortar Batteries were excellently created for the purpose of our advancing Infantry, being sufficient to meet the situation and ample enough to allow our Infantry to proceed through the wire before the enemy could offer strong resistance.[33]

Some men were able to just step over the barbed wire. Others could walk through where artillery had cut the wire. Other wire could be cut with little interference from the Germans, trapped in their shelters by the artillery barrages. Before the Germans could react, American troops were in their trenches and dugouts. The trench lines were quickly captured. The advancing troops paused at the north end of the trench system to dig in against any counterattack. Artillery shelled the rest of the loop.

The capture of the trench system took less than one hour. It took little longer for patrols to clean out the rest of the loop. The highest of several different 36th Division casualty reports for the whole attack gave 14 killed, 36 wounded, 63 gassed, 5 missing.[34] German casualties were calculated as 49 dead, 200 prisoners, and an unknown number wounded. Unlike what was so often the case in World War I, the attack had gone every bit as well as had been intended. According to a participant,

> Our battalion of 179 men went over the top . . . and took the trenches thirty minutes later. We captured 200 Boche prisoners. Our casualties were few compared to the difficulties we had to overcome. We had to advance about 1,500 yards in the open, down hill, to where they were entrenched behind a fringe of trees, as is always the case.[35]

On October 29, as promised, the French took over from the assault battalions. Also as promised, the battalions were taken by truck to rejoin the rest of the 36th Division, though the 141st Assault Battalion did not arrive in time and had to walk. For about a day and a half after they left, "as if to retaliate for the clean cut manner in which the Americans had wiped out the German battalion within the loop,"[36] the Germans heavily shelled the loop area. The loop proved to be as useful against the Germans as the French had feared it would be against them.

A modern historian writes that "The 36th left the front with a substantial record and an excellent reputation. It performed well enough under the worst of circumstances at St. Etienne and enjoyed remarkable success under the more favorable conditions at Forest Farm."[37] After a day's rest at Camp Montepelier, the division was assigned as reserve to the 1st Corps, 1st American Army, and dispatched to a rest area about twenty miles from Verdun. It arrived on November 3 and started retraining on November 5. Replacements and supplies arrived. On November 10, Sunday, special memorial services were held within the division.

A few days before this, Pershing had created a third American army. The former 2nd army was renamed the 3rd, with the new 2nd being the new army. The 36th was requested by the commanding general, to serve initially in reserve for a November 14 attack. The 36th was scheduled to move toward the front on November 11.

CHAPTER 6

Between the Wars

"I can tell you that the silence that night shouted to us louder than all the bells and whistles and noise makers . . ."[1]
—BRUCE SIDES, 111th Engineers,
January 1919

Mr. Sides describes how his unit, attached to the 36th Division, learned of the events of November 11, 1918:

So on the morning of November 11 we shouldered our packs and started on the long hike out. The guns were booming as usual. It did not sound like peace. Up in the day we began to get word from passing couriers that the armistice was signed. We laughed at them, for the guns still roared away. But at 11 a.m. came the silence and we heard no more guns. We halted for our noon meal and then resumed the march.

From time to time during the afternoon we got news . . . but still nothing official. Finally we halted for supper and prepared to camp for the night . . . In the gathering dusk we lined up for chow. It was then that the Colonel came along and said "Boys, you may build fires and have all the light you want—it is finished." We were almost too tired to cheer. A celebration was in order. You can't celebrate much in rain and mud, nor spread very much joy with corned beef, hardtack and coffee.

But we did celebrate—we built fires. Can you realize what it meant to us? For sixty one days to live in the dark; never to light even a match or cigarette in the open after dark . . .

So we gathered scraps of plank and broken timbers, splinters of torn trees and fences, and each group built a fire. Then we sat around those fires and talked, and talked, and talked, with just the safe stars overhead. And I can tell you that the silence that night shouted to us louder than all the bells and whistles and noise makers of christendom. We just built fires, that was our celebration.[2]

No complaints were recorded, but "The Armistice obviously necessitated a change of plans."[3] Newly arrived divisions were sent to serve in the 3rd Army for occupation duty in Germany. The 36th, like other combat divisions, would be withdrawn to better training areas. Before leaving, the 36th Division members received an official division patch, to be worn on the left sleeve at the shoulder, a change in the World War I practice of not permitting these patches. The Civil War "Kearny" patches, visual symbols to designate divisions and corps, were being reintroduced. The original 36th Division design, a now unknown variation on a star, was rejected as being too close to the star adopted by the 2nd Division. The new design was a khaki "T" on a cobalt blue downward-pointing arrowhead, on a khaki disc.[4]

One history of the 36th Division in World War I entitles a chapter "Marking Time at Tonnere."[5] Training eased up, but started again in December 1918, to the disgruntlement of the men. "Social disease" increased. Flu also increased, though not even close to the catastrophic extent elsewhere—the Spanish flu epidemic that killed more people than World War I combat.

The 36th Division was not unique in dislike of even relaxed wartime conditions during peacetime. Complaints were filtering back to the United States, pressuring the secretary of war, who began pressuring General Pershing. Particularly popular among the morale-building measures was a newspaper called *Stars and Stripes*. When the paper was recreated in World War II, George Marshall announced that "The morale, in fact the military efficiency of American soldiers . . . will be directly affected by the character of the new *Stars and Stripes*."[6]

The 36th Division started a paper called *The Arrowhead*. In the first issue, General Smith explained why American divisions would have to remain in France even though fighting had ended. The armistice was not a peace treaty. Conferences for such a treaty, convening in January 1919, were expected to be successful, however, and troops were already being processed to be sent home. The 36th would be the seventh sent home of the nine divisions in the American 1st Army. Smith did not know when.

On Wednesday, April 9, 1919, General Pershing paid an inspection visit to the 36th. He was tough in these inspections, but the 36th did well enough to get a letter of commendation. What they did not know was that Pershing conducted these inspections just before divisions were designated to be sent home. The next day, 1st Army headquarters notified the 36th that it would start home on April 27. Army bureaucracy, and the sheer massiveness of the logistical problem, made going home a relatively lengthy process. By early July 1919, however, all the men had been returned to Texas and mustered out. General Smith can be said to have been the last member of the World War I 36th Division. He was mustered out of the national army on July 15, 1919, reverting to his regular rank of colonel. He retired from the army in 1932 at the regular rank of major general. William R. Smith died in 1941.

★ ★ ★ ★ ★

The all-volunteer army was reborn at the end of World War I. The National Defense Act of 1920 was signed into law by President Woodrow Wilson on June 4 of that year. Extensive debate had preceded the passage of this bill. "For the first time after a war, Congress debated at length the peacetime role and reorganization of the Army."[7] The act was also the product of decades of discussion over military reform, particularly after the Spanish-American War. Government and military officials were still trying to formulate the proper place for the "modern" military in American society. What balance of regular army, militia, draftees, and wartime volunteers would serve peacetime and wartime military needs?

The act was debated during the period that came to be known as the "return to normalcy." Five months after its passage, American voters elected the most conservative, least internationally minded president and Congress in twenty years. Wartime idealism was quickly replaced by an anti-military, anti–foreign policy reaction. Virtually no one in this debate supported a standing army sufficiently large to fight all possible foreign wars without expansion. The size of the Cold War military would have been appallingly unthinkable in 1920. The army would have to expand when war came.

Related to the debate over army structure was the debate over universal military training, frequently referred to as UMT. Should all able-bodied males receive some military training? The most common UMT variant was to call for all males eighteen or nineteen to receive six months of military training, and then owe a ten-year obligation in the reserves. Some supported this as good for character, some as an obligation of citizenship. Growing interest in assembly-line mass production, particularly after Henry Ford's success, added appeal to UMT as being good for business. Disciplined men would make better workers, it was assumed. Just a week after the World War I armistice, the National Association for Universal Military Training summed up the pro-UMT view by stating that "Universal Military Training in time of peace and equal national service in time of war . . . is the lesson of the world war, now victoriously ended."[8] The "bottom line" position opposing universal military training was basic anti-militarism, especially when UMT was linked to War Department proposals for an army of 500,000 men, nearly twenty times as large as the army of 1898.

The National Guard was ignored in these early War Department proposals. A large standing army would lessen reliance on the National Guard. Guard leaders, and the National Guard Association, opposed proposals for a large army. They supported a small regular army, to be augmented by universal military training and the guard—more in keeping with American tradition, and less expensive.

The large standing army idea was finished off by military reformer Col. John McAully Palmer and, in particular, by John Pershing. Both favored a smaller regular army, but with complete units able to meet emergencies immediately until reinforcements could be mobilized. Particularly damaging to UMT was opposition in the South and Midwest. Though pro-military, southerners preferred a strong National Guard under state control. White southerners would not tolerate UMT's implication of arming and training blacks. Farmer and labor organizations also opposed UMT. Traditional anti-military feeling worked against UMT. As a Florida newspaper put it,

> Congress knows that the people are for peace, not war, and the protests of the regulars and the ammunition makers will not be sufficient to cause the placing of an unnecessary burden on the country at this time.[9]

The final version of the National Defense Act of 1920 eliminated even voluntary University Military Training. Standby conscription authority for the president was also eliminated. Strength of the regular army was increased to a maximum of 280,000 enlisted men. The army did not come close to this figure until 1940, due primarily to economically motivated congressional cuts in authorized strength. The largest between-wars strength increase was in 1936, when legislation allowed the army enlisted strength to expand from 126,000 to 154,000.

The 1920 act presented the ideal, but in the years between 1920 and 1939, the army slipped into an expansible design, with more skeleton divisions less ready for combat. By 1939, when Germany invaded Poland, the United States had the seventeenth-largest army in the world.[10] This was an army in which, as was later written of one junior officer, the soldier's "social duties . . . were as demanding as his military ones."[11] It was an army which had only recently replaced divisional horse veterinarians with divisional psychiatrists. A veteran of the 36th Division between the wars later remembered that "Division organization called for a horse doctor a long time after they lost the horses. They finally dropped him and picked up a psychiatrist! From the effort being concluded, we needed him."[12]

Under provisions of the 1920 act, the National Guard and the organized reserves would back up the regular army. The president could not call out the guard unless Congress had authorized such a call in excess of regular army strength. Regular army assistance would be provided the National Guard and reserves, increasing their effectiveness in times of crisis. The National Guard was given greater authority in deciding its own structure. The Militia Bureau was placed directly under an assistant secretary of war. The legislation restated that the guard, when federalized, was a component of the army of the United States, along with the organized reserve and regular army.

The 1920 act did not clear up the question of integrity of guard units while in federal service. The act also did not solve the question of whether the guard could serve outside of the continental United States. Were such service impossible, it would be very hard for the United States to undertake any foreign interventions. This act also did not solve the question of how the guard would be taken into federal service. Though units were kept together in World War I, members of the guard were technically drafted into federal service as individuals. Units could have been broken up. Guard regiments were renamed, or renumbered, consecutively, with no state designations. The original commander of the 36th Division was one of 501 guard officers removed as unfit for command. Resignation was permitted for another 638, while 341 others were reclassified. "Guardsmen charged that the Army perpetrated many of these removals simply to provide promotions for regular officers who took the place of Guard officers."[13]

The 1933 National Defense Act restored the guard to its position as the first federal reserve after the regular army. The guard was given a dual status—state during normal peacetime, federal during times of emergency. Units would be federalized as complete units, though the question of overseas service remained unsettled.

★ ★ ★ ★ ★

Virtually the entire Texas National Guard was federalized in 1917. In 1918, before the armistice was signed, Texas was instructed to organize three more brigades (two cavalry, one infantry) for active service on January 2, 1919. When the armistice was signed, however, the order was rescinded. The customary American antimilitary reaction took place quickly with these units. The Texas adjutant general later reported that "interest declined, men moved away from their home stations, the holding of regular drills or assemblies was largely discontinued, and many of these organizations, though showing a large paper strength, had an actual effective strength very near zero."[14]

Texas guard units hastily reorganized for the annual federal inspection. North Texas units had trouble finding places to hold meetings. The oil boom had distracted community interest and taken over most of the available buildings. Congress made things more complicated by adjourning before enacting the Army Appropriations Bill for 1920. The Militia Bureau notified various adjutant generals' departments, including in Texas, of the need to exercise strict economy. The appropriations bill was finally enacted in June of that year.

Under provisions of the National Defense Act of 1920, in order to obtain federal certification and federal money, each National Guard unit was required to take part in a fifteen-day encampment or maneuver each year. This field training was designed to provide guardsmen with intensive training under conditions similar to actual service. Texas units began to receive this training in 1920. Training was conducted by National Guard officers, regular army instructors, and regular army demonstration groups.

A few officers and enlisted men were able to attend regular army service schools. Participation in these schools was voluntary. The guardsmen offered attendance were chosen, according to the Texas adjutant general, "on the basis of their educational and military qualifications and their ability to serve as instructors in the organizations to which they belong."[15] The federal government paid all expenses while the men were attending these schools, including paying the guard attendees at the same rate as regular army soldiers.

Though training seemed to go well, problems did arise. At the 1922 infantry camp, the United States Property and Disbursement Office did not have shoes in enough different sizes for all the men. Many soldiers ended up with sore feet from poorly fitting shoes. At another camp, the air service had to fly in the cash to pay the troops.

The third major part of guard training, the second compulsory part, was regular armory drill, usually once a week. These drills included activities such as instruction in military courtesy and discipline, tactical instruction, close-order drill, target practice, and use of special arms and equipment. A training schedule was published each year, with each unit commander responsible for carrying out the schedule.

The Texas adjutant general explained the purpose of armory drill in his 1924 report to the governor:

> In the armory drill training program the objective sought was to develop discipline and a reasonable proficiency in the elementary subjects of military training . . . The thorough grounding of men in these fundamental subjects during the armory training period [allows the Texas Guard] to devote all of the field training period to the advance phases of training . . .[16]

Some anecdotal evidence, regretfully undated but probably from the earlier part of the between-wars era, exists as to some of the things accompanying, or rumored to accompany, this weekly drill and other guard functions. A veteran of this period described one occasional practice.

According to regulations, if fewer than two-thirds of the men were present, the officers didn't get paid... The two-third attendance requirement resulted in some men being credited for drills they didn't attend. Men were carried on the rolls and counted as present after they have moved away. A few commanders even carried "dead men" on the rolls and pocketed the pay.[17]

The adjutant general, in his report of the year before, had explained the purpose of the annual regular army inspection of each guard unit.

> Each year the training, efficiency and general fitness for service of the National Guard is determined by a careful inspection being made of each National Guard unit at its home station. This inspection, which is known as the armory inspection... is conducted by officers of the Regular Army under orders from the War Department, who report their findings direct [sic] to the War Department, such report later being referred back to the State for necessary corrective action in the case of deficiencies noted.[18]

The division veteran quoted above wrote that at least one unit, Company C of the 143rd Infantry, rented two hotel rooms when the regular army inspector came to town. The second room was kept well stocked with alcoholic beverages, easy to find despite Prohibition. On one occasion, however, the inspector announced, "Your whisky isn't going to do you any good. I'm going to inspect this outfit."[19] The veteran does admit that despite jokes about doing so, he knows of no one who actually brought fishing tackle to the fifteen-day guard summer camp.

Results of the inspections in the 1920s showed a steady improvement in Texas units. Hopefully, at least some of the improvement was real.

Despite at least two well-known uses of the Texas guard in this period for law enforcement, disaster relief was a more common use of the guard. This included helping people cope with hurricanes, tornadoes, and even a grasshopper invasion. One later historian has written about such efforts,

> While duty in communities suffering as a result of some natural disaster was demanding, it also had its rewards. Unlike participation in some incidents, guardsmen responding to disasters were welcomed by members of the community and gen-

erally found people cooperative. There was a spirit of teamwork and cooperation between military and civilian organizations that must have made the duty more pleasant, and undoubtedly the guardsmen appreciated the words of praise and the 'thank you' they received from the local officials and citizens.[20]

The Texas National Guard was still undergoing reorganization when, in September 1919, it assisted in hurricane relief in the Corpus Christi area. On September 15, 1919, the day after the hurricane hit, Corpus Christi mayor Gordon Boone sent a call for help to Gov. William D. Hobby. The telegram, asking for two companies of guardsmen, included the line "Conditions here deplorable, and immediate help needed."[21]

The governor declared martial law in the affected areas. Adjutant General W. D. Cope was sent by the governor to take personal charge of the relief efforts. Over 500 guardsmen were eventually mobilized in this effort.

The immediate problem was to provide food and medical aid to storm victims. Due to floods and the condition of the roads, especially outside of the city itself, most supplies and soldiers had to be transported by boat. The American Red Cross worked with the military, and after the immediate emergency passed took over relief stations. The guard concentrated on keeping the streets secure, rebuilding bridges and roads, recovering and identifying bodies, burning dead animals, and killing rattlesnakes that had wandered into settled areas.

Guardsmen patrolled the Corpus Christi downtown business area. Citizens had been asked to stay off the streets after dark. The *Dallas Morning News* reported on these efforts:

> The city is under martial law and through the dark, wreckage-strewn streets sentries challenge sharply every passer—and will shoot if the call be not heeded . . . There has been very little looting. There will be very little.[22]

The guardsmen sent to the area were demobilized in late October 1919. Both they and the Red Cross were praised for their work during the disaster. The Texas National Guard would not be mobilized for another natural disaster until May 1930, when a tornado hit the town of Frost.

The guard would be mobilized for a major law enforcement

problem in May 1920. This use was an outgrowth of a controversial use of the Texas Rangers.

Though initially created as military units to protect border areas, the Texas Rangers had taken over statewide law enforcement functions in Texas in the 1870s, with the end of Reconstruction and the abolishment of the State Police. This role become more significant during World War I. In February of 1918, the Texas state legislature created a 1,000-man "Loyalty Ranger Force," to generally enforce state laws, and specifically enforce House Bill 15, the "Hobby Loyalty Act." This bill was designed to ensure patriotism among Texans. However, "In their enthusiasm to enforce the law, the rangers, who had always been a bit headstrong, used tactics that seemed a big extreme to some citizens."[23]

Nothing was done about this behavior during the war. However, early in 1919, a legislative investigation concluded that the rangers had been guilty of illegal acts, including murder. The rangers were limited to a maximum of seventy-six men, and new procedures were established to file complaints. The Texas Rangers were temporarily discredited as an effective law enforcement force.

About a year later, in March of 1920, a dock strike, part of a national walkout, spread to Galveston. Violence erupted when the main shipping line tried to bring in replacement nonunion workers.[24] Shots were fired at vehicles carrying strikebreakers out of Galveston. Other incidents of verbal and physical abuse of replacements were recorded. Galveston businessmen believed the lack of arrests came from police sympathy with the strikers. Rumors spread that the Galveston city government was too heavily influenced by the political power of the unions to take effective action to protect replacement workers.[25] The city government had recently been elected with strong support from organized labor.

Matters were made more dangerous by the possibility of racial conflict. Many strikers were black, strikebreakers white. The fear of race riots was used as an argument in favor of state action.

Two months into the strike, on May 13, 1920, Galveston city officials asked Adjutant General Cope to send Texas Rangers to assist the local Galveston police. The rangers, however, were not able to control the situation.

On June 2 a committee of businessmen, from Galveston and from elsewhere in Texas, met with Governor Hobby. Hobby was told in detail about violence and lack of protection for the replacement workers. Hobby, a businessman, had become governor in 1917 when his pro-labor predecessor was impeached and removed from office. Hobby could be expected to understand the economic problems the strike was causing and to sympathize with the businessmen.

At the end of the meeting, Hobby sent a telegram to Galveston mayor H. O. Sappington, giving him three days to ensure effective police protection of movement of cargo or the state would step in. Guard mobilization began. The Galveston city government was unable to give assurances that freight could be safely shipped, so the governor declared limited martial law on June 7, affecting the Galveston waterfront and bay. Within three days, freight was moving.

City government leaders opposed martial law, fearing that it would hurt tourism at Galveston-area beaches. Brig. Gen. Jacob F. Wolters, commanding the Texas National Guard forces in Galveston, addressed this issue when he later commented, "The fact is that probably more people visited Galveston and enjoyed its surf bathing and the attraction of the beach than had during any previous summer up to that time."[26] Local businessmen not only agreed with Wolters, but used martial law as a tourist attraction. A June 20, 1920, newspaper article was an example:

> Upstate folk who want to find out what a mild form of martial law looks like should come to Galveston for a weekend or a vacation visit. This island not only has the finest surf bathing in the world, but also the nicest martial law on earth. People don't have to go to bed at all, if they desire to stay up, and all the beach pleasures are encouraged rather than interfered with by the Lone Star soldier boys . . .
>
> [Their camp] is an attractive reminder of the well known war recently demised.[27]

In addition to tourist promotion, the guard units also undertook law enforcement activities, including some efforts to suppress opposition to martial law itself. On July 14, the governor suspended the chief of police and the police department. The mayor and other civilian officials were suspended from law enforcement functions. By October 8, after only a few serious incidents, martial law was declared over. The strike was broken, "morals" laws revived, and the longshoremen's union weakened. An open-port law was soon passed by the Texas legislature to make it illegal to interfere with any persons loading, unloading, or transporting commerce within the state.

As to the role of the guard,

> In this role as a law enforcement agency the Guard enjoyed the support not only of the governor who ordered them into the situation, but the approval of the Federal courts which upheld the legality of the activities carried out by the military. The ability of the Guard to enforce a wide variety of laws had been demonstrated.[28]

Galveston would not be the last time Texas guard units were used in law enforcement. Elements of the 142nd Infantry would be dispatched to Denison in 1922, just a few months after the division was revived, to enforce the open-port law during a nationwide railroad strike.[29]

★ ★ ★ ★ ★

On November 5, 1920, all states in the 8th Corps area sent their adjutants general and other representatives to a conference at Fort Sam Houston, San Antonio, to discuss plans for reorganizing the National Guard. According to a 1940 history of the Texas National Guard, prepared by the guard itself, "The principal change in the allotment of troops to Texas was the substitution of an infantry division for the cavalry division previously authorized. This was considered a very fortunate change, in view of the great expense necessary to maintain cavalry organizations."[30] This change was formally authorized by a December 3, 1920, letter from the chief of the militia bureau. A request was made to designate the infantry division as the 36th Division, "to preserve the memories and traditions of the National Guard division which represented Texas so gloriously during the World War."[31] On December 14, the state's request was approved.[32]

The Texas National Guard was somewhat jumbled at this time. Sixty-nine units existed on paper, but the majority were inactive. Starting in 1921, the Texas National Guard was totally reorganized. Provisional infantry companies were disbanded. Guard cavalry, however, was reorganized quickly and efficiently. Gen. Jacob

Wolters was given credit for this achievement.[33] During 1921 Wolters even used his own money to keep the cavalry brigade executive officer on duty to inspect and help train cavalry units.

Major efforts at organizing the revived 36th Division continued in 1921. The state was divided into brigade areas. The 71st Brigade was assigned the western section, the 72nd Brigade the east. Each section was further divided into recruitment areas for the seventy-eight separate units (including organic air detachments) required to complete the two brigades. By the end of 1921, twenty-seven infantry companies were federally recognized, with another fifteen in the process of organization. In the first eight months of 1992, federal recognition was extended to fifty-seven new infantry and artillery units. By March 1923, the combat elements of the 36th Division had been organized. The division was a recognized unit of the National Guard, with a divisional headquarters in Houston. Maj. Gen. John A. Hulen was appointed division commander. Hulen was a former Texas adjutant general, veteran of the Spanish-American War and Philippine Insurrection, and the World War I commander of the 72nd Brigade. In an interesting commentary on the connection of a National Guard division to state political life, as well as military needs, Hulen's business and civic leadership was cited, along with his military experience, as qualifications for his position.[34]

★ ★ ★ ★ ★

Economics never ceased to play a role in the military between the world wars. By 1924 the rapid growth of the National Guard throughout the nation had created a shortage of funds for the Militia Bureau. The bureau was forced to curtail the growth of the guard, announcing that until more funds were appropriated, no more guard units would receive federal recognition. By August of 1924, the Texas guard had a total strength of 8,451 officers and men in 147 units and staff departments.[35] A year later, the Texas guard had added only a net twenty-two men. However, a letter from the militia bureau of January 1925 limited the Texas guard to 8,230 men. Nothing was done until 1928, when the 1927 increase of 600 men put the Texas guard over 9,000. The militia bureau ordered a reduction, and by August of 1928, 1,300 men had left the guard. The adjutant general reported to the governor of Texas that:

While this forced reduction in strength of the National Guard has its unfortunate aspects, it is believed that the general state of training of the National Guard is higher now than before due to the fact that only those men who take an active interest in the National Guard and attend drills regularly are now carried on the active rolls.[36]

Guard recruitment ceased to be a problem. Units in larger Texas cities had waiting lists. On the surface, this glut of possible guardsman contradicted the national trend. According to a recent historian of the National Guard, "Recruitment in the Guard rose and fell with the business cycle."[37] In the United States as a whole, it was necessary to recruit 6,500 men each month to maintain strength. Guard officers resigned at 25 percent per year, compared to just under 3 percent for the regulars. During the 1930s, the situation turned around drastically. More people wanted to join the guard than there were funds available.

Why the difference? Texans are often seen as southern militarists, or combative western frontiersmen, depending on the inclination of the analyst.[38] This may have had something to do with it, especially as the Texas guard already saw itself as a direct descendent of Texas Confederate units, the victors of San Jacinto, and the defenders of the Alamo. It could also have been economic, as many rural areas of the Southwest, including Texas (even with the new oil industry), never really shared in the 1920s boom. The Depression arrived early for many. When the "real" Depression arrived, the Texas National Guard benefited. Recruitment was even easier, enlisted turnover decreased, active participation increased. However, a desire for economy led to a decrease in state and federal funds for the guard. The guard became more popular in the late 1930s, with ominous world events and the increasing chance of American involvement in a war.

The Texas legislature never seemed totally happy paying for the Texas National Guard, the same problem that had plagued the guard's ancestors. In 1938 the Texas adjutant general reported the size of the Texas guard as sixth in the nation, behind New York, Pennsylvania, Illinois, Massachusetts, and Ohio. Texas was fourth in federal funds received, fourteenth in state appropriations, and twenty-fifth in per capita cost.[39] Various reports show the state as paying only a small portion of the cost of maintaining the Texas guard. The state was responsible for the costs of using the guard on state service, costs usually met by general guard appropriations. The only occasion between the wars when a special appropriation was required was for the Galveston duty. The largest increases in state funding came in 1931 and 1932, when the guard was on duty in the East Texas oil fields. Ross Sterling, the governor of Texas, "later claimed that the State had actually made a profit of more than a million and a half dollars from taxes on the increased price of oil resulting from production controls enforced by the military."[40]

★ ★ ★ ★ ★

In 1930, units of the 144th Infantry intervened in Sherman to help control a deadly situation. A race riot started with the arrest of a black man for the rape of a white woman,[41] the situation in the Old South with the most potential for violence. Three days after the Saturday, May 3, arrest, a large and hostile crowd gathered near the jail. The officers guarding the black prisoner, George Hughes, had to fire shots to disperse the crowd. That same day, the district judge asked for four Texas Rangers to be sent to guard Hughes at his trial, starting three days later.

Texas Rangers were present when the trial opened on May 9. The alleged victim was taken into court on a stretcher—legal procedures were more flexible back then. A jury was quickly selected, and the first witness called at 11:00 A.M. He was not able to finish before the crowd burst its way into the courtroom. The jury was sent out of the room, the prisoner taken to the district clerk's vault on the second floor. The rangers filled the courtroom with tear gas, causing others in the court to have to flee to the second floor. Firemen eventually rescued these people.

By 2:20 that afternoon, a guard cavalry unit in Dallas was ordered to Sherman. Nearer units did not have rifles and bayonets. A few minutes later, the courthouse caught fire. According to a local newspaper:

> A frenzied crowd grew larger and larger, watching the burning of the Grayson County Courthouse which was a solid mass of flames at 3 o'clock Friday afternoon. With the hoses cut by a crazed mob and firemen being held off by members of the crowd, some of whom attempted violence, buildings on the

north and east sides of the square were being endangered by the ever increasing flames.[42]

The rangers, fading as an effective force in this incident, left town, ostensibly to be able to call the governor. About 6:00 P.M. that afternoon, ten to fifteen guardsman from Denison, from the 144th Infantry and the 36th Signal Company, arrived in Sherman. They tried to reach the courthouse but were forced to retreat to the safety of the jail. "The mob was now in complete control of the city."[43]

The Dallas guard detachment, forty-nine men and eight officers, arrived about a half-hour after the men from Denison. About 100 had been detailed, but only roughly half were sent, due to expense. The detachment surrounded the remains of the courthouse, to try to protect county records in the vault. The crowd briefly hesitated and then began to press the guardsmen back. One officer at the scene told of "an element of criminal bitterness manifested as some mature men began to worm their way to the front to join the boys."[44] The guardsmen were hesitant to shoot, and did not see how the crowd, armed with and throwing rocks, bottles, and bricks, could be controlled without shooting. So the order was given to move back to the jail. The mob pursued.

The shower of missiles pummeling the guardsmen worsened. Merchants along the route to the jail provided bottles for members of the mob to throw. Some in the mob were breaking off the bottoms of the bottles in hopes of causing severe injury.

A private about sixteen years old grew separated from the crowd and was in danger of being killed. Capt. James E. Dunlap went to his rescue. He found that the private had lost his rifle and ammunition, had been knocked down, and was being kicked and beaten with a plank. As Dunlap told the story,

> I rushed to his side, drew my pistol and told the man who was wielding the plank to quit or I would kill him. At this moment a brick or something crashed into the side of my face. Everything went black, and I started going down. I knew that if I ever went completely down, I was lost; so somehow I managed to pull myself together enough as I sank to my knees to fire twice into the air. Other officers came running and saved us.[45]

Someone fired a rifle at the guardsmen when they reached the jail. The guardsmen were ordered to fire two volleys into the air, but the officers fired their pistols at mob leaders. At least two men were hit. The jail was barricaded, with riflemen and machine gunners posted at the windows. Other machine gunners were on the lawn outside. The jailer briefly locked the door, but one of the officers realized this might trap the guardsmen outside and forced the jailer to unlock the door. Several sticks of dynamite were thrown from passing cars, but none caused any casualties or damage. Col. L. E. McGee, leader of the guard detachment, called the governor for more troops from inside the jail.

The mob had returned to the courthouse, where they found and blasted open the vault. George Hughes' dead body was found in the vault; he was probably killed by the fire. His body was dragged through the streets by the mob, to be hanged from a tree. It would eventually be cut down by the guard and given to a black undertaker for burial. A white eventually had to handle the job, as both black undertaking facilities had been burned by the mob.

At 2:15 A.M., a guard detachment arrived, including members of Company H, 144th Infantry. A further detachment from the 144th Infantry, and 124th Cavalry, arrived an hour later. Though most spectators had gone home, several hundred rioters apparently prepared to charge the guardsmen. The sound and sight of bayonets being fixed to guns and order to prepare to fire being given seemed to have been enough to at least temporarily disperse the crowd. The guard then sent patrols to the black section of the city, which the mob was trying to burn down. Black citizens were also escorted to their homes. By the next morning, 330 guardsmen were on duty and things were quiet.

That same morning, May 10, Texas governor Dan Moody acceded to the request of some Sherman citizens and business leaders and declared martial law. Things remained quiet, though with some necessary arrests, until martial law was ended on May 24.

A military commission almost immediately started to investigate what happened. Primarily, the commission wanted to know whether the guard had been ordered not to shoot. Such a report had spread during the riot, emboldening the rioters. Several witnesses at the hearings testified to having heard such a rumor, though apparently no one had direct evidence that such an order had been issued. A newspaper article in the *London Evening Standard*, May 10, 1930, reported being told by the governor that he had issued such instructions. No telegram or written instruction not to shoot was ever found, and evidence showed that the governor had not spoken to the ranger captain initially in charge until after the courthouse had been set on fire.[46] A check of Western Union and Postal Telegraph records turned up nothing.[47] The issue of whether an order could have been issued verbally, or whether the governor suggested that the guard not shoot, would remain conjectural.

Some aspects of the case are bothersome from a 1990s perspective, such as taking the alleged victim into court on a stretcher and the suspect only being given six days to prepare for trial. From a 1930s point of view, perhaps the governor should have anticipated trouble. However, the district judge had only requested four rangers, as the governor pointed out in a letter to a critic.[48] The first guard unit was notified just twenty minutes after the riot began and the judge called the governor. As to the conduct and actions of the guard on the scene, one has to agree with an editorial that declared:

> From out of this horror and disgrace of rioting and incendiarism at Sherman recently there came one outstanding performance of merit for which Texans may well be proud. That was the conduct of her National Guardsmen.[49]

The guard could handle law enforcement. However, it must have come as a relief when after 1932, increasing local law enforcement professionalism, a restored role for the Texas Rangers, and the creation of the Texas Highway Patrol drastically decreased the guard's law enforcement functions. Between 1932 and 1940, units of the guard were mobilized only twice for law enforcement.[50]

The guard, including elements of the 36th Division, was mobilized once more before 1940 for a major disaster, at a school in New London, Texas. School officials had tapped an oil company residue gas line, in an effort to cut costs. Gas had escaped and accumulated under the building's floor and in the walls. On the afternoon of March 18, 1937, the building blew up, killing 294 children. The guardsmen controlled crowds at the disaster site and helped in rescue efforts.

Three and half years later, the Texas National Guard's focus shifted back from state to federal.

CHAPTER 7

Getting Ready for War Again

"Basic training, especially in the infantry, is the toughest training a soldier gets . . . time flies so quickly you hardly know what's cooking . . ."[1]

—PVT. BRUCE KOUSER, 1943

According to Senator William E. Borah, the isolationist-minded Republican senator from Iowa, in a July 19, 1939, statement, "There is not going to be any war in Europe. At least not soon. Germany is not ready for it. All this hysteria is manufactured and artificial."[2] Just about six weeks later, early in the morning of September 1, 1939, the German military implemented Adolf Hitler's April 1, 1939, order and invaded Poland. War had been expected in the higher reaches of the American government but came as a shock to the American people. Two days after the German invasion, Britain and France declared war on Germany.

President Roosevelt declared American neutrality but specifically stated that it was not necessary for Americans to be neutral in thought, word, and deed. Americans could and would make a moral choice. Americans, however, still hoped to avoid the military choice, and to avoid becoming involved in the second major European war in twenty-one years. The six-month period of *Sitzkrieg*, phony war, the virtual quiet in Europe which followed the fall of Poland, increased the American conviction that staying neutral was desirable, and that it was possible.

The man who was to lead the American army in World War II, George Marshall, was sworn in as chief of staff on September 1, 1939. Marshall began his new job watching the awesome speed with which the Germans smashed into Poland. With the rest of this country, Marshall learned a new word—*Blitzkrieg*, "lightening war." The United States Army would have to learn how to deal with blitzkrieg.

The army that Marshall inherited was slightly bigger than it had been a few years before and was twice the size of the American Army in 1914. The regular United States Army still only had about 190,000 men. (The National Guard had about 200,000 men enrolled, in various states of training, supply, and readiness.) The army referred to itself as the seventeenth-largest army in the world.[3] Congress was interested in saving money, especially during the economic crisis of the Great Depression. Congress was also interested in responding to public revulsion against World War I casualties, the strong desire to never again become involved in a European war.

The military was paying the price. In the army, the officers corps was described as having almost a country club atmosphere, with social obligations frequently as important as military obligations. Promotions during this period seemed glacial. George Marshall, for example, was a lieutenant colonel for eleven years. John Pershing's personal intervention was needed to get Marshall his first star as a brigadier general at age fifty-six, while Marshall was still young enough to gain further promotions. Enlisted men lived in a different world, with minimal contact with officers. There was one hopeful sign for the future, however. Army planning had, since 1936, been quite realistic. This planning included detailed plans for expansion to roughly ten times the size of the regular army and the National Guard.

Proposals for a draft to meet the expected need for a large army were being developed. When the political situation made expansion possible, not to mention vital, the army would be ready. On September 1, 1939, this time had arrived. "The maintenance of our peace and neutrality in the midst of the present troubled world demands that our agencies for national defense be immediately placed in a higher state of efficiency,"[4] the secretary of war wrote President Roosevelt.

The National Guard was bigger than the regular army in 1939. However, it suffered from many of the same problems as the army. The National Defense Act of 1920 had authorized the guard's strength at 400,000 men. Half that many were actually enrolled. Enrollment had increased during the economic hard times of the 1930s. Men needed the extra income membership in the guard could bring. In 1933 amendments to the National Defense Act formally recognized the guard as part of the American military structure. However, the guard seemed more useful as a state backup tool for disaster relief and occasional law enforcement. The guard also did not receive the benefit of early army reforms. For example, the conversion of the older "square" four-regiment divisions into more mobile, "triangular," three-regiment divisions was not introduced to the guard until after mobilization into federal service.

The National Guard did not even get the federal funding it would need when changes were proposed. Late 1939 proposals called for an increase in the effective strength of the army and the guard. However, though the federal government would pay some of the guard's expenses, it declared that the responsibility for National Guard recruitment belonged to the individual states.

Domestic politics intervened to limit increases in the size of the National Guard, whatever the source of the needed money, and the regular army. The author of a study of Roosevelt's foreign policy writes that

> Roosevelt resisted pressure for substantial increases in national defense forces and rapid industrial mobilization. Fearful that these actions would agitate suspicions about his peaceful intentions and make neutrality appear as a step toward involvement, Roosevelt temporarily limited expansion of the regular Army and National Guard to less than 50 percent of what defense chiefs asked and suspended economic and fiscal preparations for war.[5]

The guard seemed to be suffering all the disadvantages of the national neglect of the military, with none of the army's planning advantages. The National Defense Acts, most recently in 1933, had adjusted the structure of the military, not attitudes within the military and toward the military. One veteran of the regular army, and the 36th Division, wrote that "There are many dedicated soldiers in the military service. There were few in 1933. It was a dogged fight to overcome the indifference of the public and the lethargy and inertia peculiar to the military system."⁶

The National Guard, more closely connected to local politics, suffered from the same lethargy. One veteran of the 36th Division, Armin Puck, remembers what passed for an "innovation" by the 36th Division's commander in 1937, requiring standard uniforms.

> His famous general order number one was a uniform requirement for all officers in the Texas National Guard . . . up until 1937, in other units, some field officers as well as company grade officers wore enlisted men's clothing out of the supply room. And they didn't wear regulation boots as was the custom in those days. They wore these damned old engineer boots that you could lace or hook up to here. You'd think they were going to chop wood instead of to command a battalion or command a company.⁷

General Puck had several uniforms made when he was appointed a second lieutenant, not only meeting the requirements for numbers of uniforms but having the uniforms "properly tailored and fit."⁸ He also spent $30 for Lucchese boots, the best he could find, to go with the uniforms.*

As late as the summer of 1940, army ground forces commander Leslie J. McNair, returning from an inspection trip, was unsatisfied with guard training. He recommended that guard units be given at least two months' training before receiving draftees. "Such a procedure was necessary in view of the extreme inadequacy of provisions made for the field training of the larger units of the National Guard in time of peace."⁹

Another, more controversial issue intervened before the precise role of the guard could be worked out.

Passage of the Selective Service Act of 1940, the first peacetime federal draft in American history, can be said to have started on May 8, 1940, the day before the Germans invaded the Netherlands, during a dinner at the Harvard Club in New York City. Eight men, gathered for a dinner meeting of the Executive

---

*This compares to 1999 Lucchese dress boot prices, which usually start at about $300.

Committee of the Second Corps Area, Military Training Camps Association (MTCA), had gotten into a discussion of the upcoming twenty-fifth anniversary of the Plattsburg Training Camp Movement. In the wake of the sinking of the *Lusitania*, before the United States' entry into World War I, these eight men had been among those who asked for military training for businessmen and professionals. Grenville Clark, a prominent New York attorney from an old-line family, had been one of the three who went to the military commander of the Eastern Department, Leonard Wood, to ask for help in establishing the camps. Clark was known for his activism, though he preferred to remain behind the scenes as he volunteered support of a wide range of causes. One of his last causes, before his death in the 1960s, would be support of freedom rides for civil rights in Alabama and Mississippi.

Clark agreed with Wood about "the moral organization of the people, an organization which creates in the heart of every citizen a sense of his obligation for service to the nation in time of war . . ."[10] The aristocratic nature of the Plattsburg leaders was attacked, both in 1915 and by later historians. When World War I ended, universal military training proved unpopular in the increasingly pacifist and isolationist climate in the United States. However, "If the Plattsburg program of universal military training proved unacceptable to American institutions and values, they were willing to work within the system and abide by the results."[11]

Clark was already on record, with a journal article, as recognizing the danger of German conquests for United States national security. At the meeting, Clark suggested that rather than concentrate on anniversary celebrations, the group should do something practical. Why not campaign for peacetime conscription? The others at the meeting agreed and planned a larger meeting at the Harvard Club for May 22, 1940. Clark would chair the planning committee for this larger meeting.

Clark also telegraphed President Roosevelt on May 16, 1940, telling the president what the group intended. Clark asked whether Roosevelt felt it was politically and publicly opportune to put forth the idea of compulsory universal military training at that time. Roosevelt's quick response seemed supportive, but in a later comment on the project, he warned about the political problems, that the deciding factor on measures to take was what could be obtained from the Congress. The practical effort to gain public support for peacetime conscription started at the meeting. Among the attendees were retired Brig. Gen. John Palmer, a military theorist from World War I; Henry Stimson, soon to become secretary of war; Robert Patterson, soon to become assistant secretary of war; and William J. Donovan, who, as head of the Office of Strategic Services, would be the chief American spymaster of World War II.

The MTCA conscription committee was advised to check with the War Department about its plans in this regard. The army was assuming a draft, but that, like in World War I, legislation would not pass Congress until the start of war. The War Department officially was still thinking defensively. As two annual reports of the secretary of war put it, "We visualize only the possible necessity of armed defense of our own domain,"[12] and the army was "purely defensive in character and nonprovocative in outlook."[13] After inclusive meetings with army representatives, a face-to-face meeting was held, on May 31, between George Marshall, Grenville Clark, and another member of the conscription committee, Julius Adler. Marshall wanted to concentrate on improving the training and equipment of the regular army and was even hesitant to call for mobilizing the National Guard.

Marshall recognized the importance of a draft but felt it was not his place to advocate a legislative policy the president had not authorized. Clark responded, accurately but quite undiplomatically, that a professional man had a responsibility to give advice to his client. If Marshall believed that compulsory training was necessary, he should advocate it to the president. Marshall refused, and the meeting quickly ended.

Marshall later explained his reluctance to actively support conscription. After complaining about Clark's tone, Marshall said he had been reluctant to get the army involved in politics. He felt, at least in retrospect, that if the army had began to work for conscription,

> I would have defeated myself before I started, and I was very conscious of that feeling. So if I could get civilians of great prominence to take the lead in urging these things then I could take up the cudgels and work it out.[14]

Roosevelt seemed to support the idea of a draft, but did not indicate that he considered it urgent. One reason may have been the overwhelming success of the German *Blitzkrieg* against France. To most observers, German success seemed heavily dependent on the use of machinery, tanks, and airplanes, rather than on the use of manpower. The combined armies of France and Britain heavily outnumbered the attacking Germans. The United States military was small in 1940, and also underequipped. Roosevelt considered the most pressing need to be for more equipment for the American military, and for use in aiding Britain. This required congressional funding, and Roosevelt was not anxious to use up vital political capital. Additionally, Roosevelt was planning on firing the strongly non-interventionist secretary of war, Harry H. Woodring. Roosevelt anticipated some political problems, and did not want to add to them with a call for a draft.

The other factor delaying Roosevelt's active support of conscription was straight political. He overestimated the continuing power of the isolationists—though in 1941 the draft would be extended by just one vote in the House. Roosevelt generally preferred to use presidential influence when he could be sure of success. Politics is the art of the possible, and Roosevelt did not yet consider conscription politically possible. A few days after the meeting with Clark on conscription, George Marshall wrote that "[Roosevelt acts in] each instance almost as rapidly as public opinion would permit."[15] Roosevelt's attention was focused on the upcoming November 1940 presidential election. He was strongly leaning toward running for an unprecedented third term in office.

What would become known as the Burke-Wadsworth Bill was introduced into the United States Senate on June 19, 1940, and into the House of Representatives on June 20, 1940. The next day France surrendered to Germany. The basic bill had been drafted by the National Emergency Committee (NEC), an offshoot of the MTCA, founded to work for conscription and for the bill. Republican representative James W. Wadsworth of New York, a former senator, was the NEC's first contact. Wadsworth had written the 1917 draft bill. He had also received a May 20 telegram from Maj. Gen. John F. O'Ryan, the retired commander of New York's 27th Division, calling for mobilizing the National Guard. One line in the telegram applied to a draft as well as the guard: "Our man power urgently requires mobilization and training as defense measure for eventualities."[16] O'Ryan had attended the May 22 Harvard Club meeting.

Wadsworth immediately agreed to sponsor the bill. Most Democratic senators contacted, however, refused to commit themselves to introduce and support the bill, without some word from the White House. Senator Edward R. Burke, Democrat of Nebraska, finally agreed to sponsor the bill in the Senate. Burke, a Roosevelt opponent, had already lost a primary for reelection. A few days after its introduction in the Congress, the text of the bill was sent over the major wire services. Conscription became a major political issue.

Several events caused Marshall and the army to change their position on the draft. At the end of June 1940, Henry L. Stimson, a Republican, was nominated as secretary of war to replace Harry Woodring. Frank Knox, also a Republican, was nominated as secretary of the navy at about the same time. Stimson's appointment would improve the administration of the War Department.

Marshall became aware of O'Ryan's May 20 telegram. Marshall was still primarily concerned with training and equipping the army, so he opposed federalizing the guard. Marshall knew federalization might eventually become necessary, since a conscripted army could not come into being immediately. The army, therefore, began to plan for a step-by-step mobilization of the National Guard. On June 4, Marshall testified before the House Committee on Military Affairs in support of a bill to permit limited guard mobilization at the behest of the president. At Marshall's urging, Roosevelt added a clause to his defense message of May 31, 1940, stating that:

> There is a specific recommendation I would make in concluding this message, that before adjournment this Congress grant me the authority to call into active service such portion of the National Guard as may be deemed necessary to maintain our position of neutrality and to safeguard the National Defense...[17]

On June 17, Marshall met with his staff directors. The war situation in Europe looked even more dire. Paris had been captured by the Germans and the French were on the verge of surrender. The possibility of Germany gaining control of the powerful French fleet—not anticipating that a few weeks later Winston Churchill would order the British navy to sink the French fleet—was a grave national security threat to the United States. The meeting decided to support mobilizing the guard, and to support the draft.

Wendell Wilkie became the Republican nominee for president in late June 1940. In mid-July, Roosevelt engineered his own political draft and received the Democratic nomination. He was still hesitating about publicly endorsing the Burke-Wadsworth Bill. Gen. John J. Pershing endorsed the bill in an early July letter to the Senate. The National Guard Association endorsed the bill, once amendments secured the place of the guard in the national-defense structure and allowed trained draftees to complete their military obligation in the guard. On July 29, 1940, Roosevelt sent identical letters to the president of the Senate and the Speaker of the House, with texts of the proposed joint resolution allowing him to federalize the guard. The next day, Senate Joint Resolution 286, allowing mobilizing the guard for year but limiting it to use in the Western Hemisphere and American possessions elsewhere (primarily the Philippines) was introduced. On August 2, in response to a question at a press conference, Roosevelt publicly endorsed the idea of a draft. Wendell Wilkie endorsed a draft later in the month.

The National Guard bill and the conscription bill were considered simultaneously. Strict isolationists opposed both bills on the grounds that they were militaristic measures designed to prepare for possible war. Other opponents of both bills thought the legislation would give the president, or the government as a whole, too much additional power. Some isolationists actually supported mobilizing the National Guard. They though this would head off a draft and decrease the chance of American involvement in Europe, by strengthening Western Hemisphere defense. Guard mobilization was more popular than a draft with the general public for similar reasons.

More ardent preparedness advocates supported both bills. They said failure to pass either bill would immediately endanger American national security. Slightly differing House and Senate versions of the bill were reconciled and passed by August 22. On August 27 the resolution authorizing federalization of the National Guard with limitations was passed. Guardsmen could be federalized for only one year. The president was also only allowed to deploy the guard in the Western Hemisphere, and in American possessions around the globe, including the Philippines.

The Selective Service and Training Act of 1940 became law on September 16, 1940. The 1940 Draft Law, the first peacetime national conscription law in United States history, did not start out as much of a hardship for draft-age Americans. Army planners felt that men in their late teens and early twenties made the best soldiers and wanted to start drafting from this group. Political realities, however, made it necessary to limit draft liability to ages twenty-one to twenty-five, and the term of service to one year. Despite total manpower in this age group estimated at 17 million,[18] the 1940 law limited the total call to 900,000.

By the end of 1940, several new terms and ideas had entered common use. "The draft" was the term given to compulsory military service. The Selective Service System was created to administer and give overall direction to the system. Local "draft boards" of prominent citizens carried out the draft. Gen. Lewis Hershey became head of the Selective Service System, a job he held until well into the Vietnam War.

Soon after the law was signed, the Gallup organization surveyed draft-age men and found they were generally supportive of the system. The men of 1940 felt the United States to be worth defending and were willing to fight for their country. Bruce Kouser, a draftee undergoing replacement training at Camp Blanding, Florida (and this writer's uncle), in late 1943 and early 1944 wrote in a letter to his family a few days before "graduation" about his thoughts on seeing an American flag. "It sure was a thrill to see *why* [sic] I was down here..."[19]

On October 16, 1940, a total of 16 million men registered for the draft. Five more registrations took place into 1942, when the draft went to the "modern" system of having men register on their eighteenth birthdays. The initial draft was designed to expand the regular army, and to incorporate the new manpower into the regular army, rather than use the Civil War system of raising entirely new units. The Selective Service System administered the draft on a national level. Individuals were drafted with a complicated lottery system used first in World War I. When a man's "number was up," he would go before his local draft boards, literally, or figuratively in the form of his records. The services had the final say as to whom they accepted, but the draft boards decided whom the services got to consider. Selectees could appeal, but in 75 percent of the appeals, draft board decisions were confirmed.

The draft board members were locally prominent, usually business and professional men in their forties and fifties. The presidentially appointed board members served without pay. Many veterans were appointed to the boards, on the assumption that they would not be ordering young men to do something they had not done. The board members received advice and guidance from Washington, but the local boards decided who went into the army. Anger against alleged system abuses would be directed against members of the community, sparing the federal government. When the United States entered the war, in smaller communities draft board members would personally know men from their community killed in battle.

Virtually no blacks served on southern boards. Blacks, however, could not be called up at a higher rate than their percentage of the population. The army leadership felt blacks to be of limited usefulness and were unwilling to take the lead in changing American social attitudes.[20]

The 1940 law continued the World War I practice of seriously limiting exemptions. The army itself could not always get people drafted whom it felt were needed. It could not always keep people out of the army whom it considered more useful as civilians. At the start of the draft, local boards were very generous in exempting married men. By 1944, however, many boards were filling half their quotas with married men. Physical and age standards for induction were lowered as the war progressed and the army began to run into a manpower shortage. Ironically, medical care improved as physical standards for induction were lowered. By 1945, for example, 20 percent of all soldiers wore glasses, a frequent reason for earlier exemption.

"Greetings," the notice in the mail would read, ordering the young man to report to his local draft board at a certain time and date—often before dawn. After a careful pre-induction medical and physical examination, the army would usually send the accepted men home for two weeks to wind up their affairs. The army quickly learned to swear the men in first. If not, accepted inductees often went and joined the navy.

Chapter 7: Getting Ready for War Again   107

After the two weeks ended, and another early-morning meeting at the draft board, the new inductees were sent to an induction depot to be classified and receive their first introduction to the military life. The classification test the men received was supposed to measure "usable intelligence." Along with interviews, this process was designed to determine the best way to employ the individuals in the military. Contrary to stories, and to such cases as assigning a civilian butcher to the medical corps and making a garbage man a cook, the army, at least at the start of the war, tried to assign military specialties related to what a person had done in civilian life.[21] At least two civilian "ham" radio operators, for example, were assigned by the 36th's 141st Regiment to operate a unit radio station, both for training and for entertainment of the troops.[22]

The army was already advertising that it was trying to teach useful skills. A 36th Division press release, a few months after the division was federalized, advertised that "Today's soldiers have better opportunities to learn trades that will benefit them in civilian life than did those in 1917."[23] This was backed up ten days later, when another press release announced that forty-two members of the division would be attending ten training schools.[24]

Thing would rapidly change. In November 1940, a letter writer seemed bored. By May of 1941, he was tense, writing his wife that "If we are to fight, then everyone ought to know . . ."[25] In November, the press release on the two radio operators seemed proud that the "longest contact that they have made was a navy station on Guam Island, in the Pacific Ocean." In a few weeks, both stations would be too busy to chat.

There was a major "kicker" to the army's interest in assigning its recruits, as much as possible, to a military occupational specialty (MOS) matching their civilian occupation. The central military specialty, combat rifleman, has no civilian counterpart in either profession or hobby. As General Hershey once put it, shooting at people is not an approved civilian activity. At the start, combat riflemen came from inductees listed "for any arm or service." By the last year of the war, virtually all inductees were sent to the infantry.

This included the 36th and other National Guard divisions. The 36th's commander wrote in his diary that "We just received 4,000 recruits from the northern states . . . They are serious-minded . . . and are making good progress. I will not be surprised if some of them are better soldiers, after they complete basic training, than some of our men who were with the Division before induction and have formed careless habits that still dominate them."[26]

For some recruits there was a certain amount of what today we call culture shock. The Baptist chief chaplain of the 36th Division wrote of his assignment to and arrival at the 40 percent Baptist division:

> The Chief of Chaplains decided that this was an ideal spot for a Baptist Lieutenant Colonel, and I was ordered for duty with the 36th Division. (They didn't realize that to Texans the place of a man's birth was more important than his religious designation.)
> I reported for duty on April 19, 1942. On April 21st a division parade was held. While talking with a group of staff officers I asked what the occasion for the parade might be. They told me, "This is San Jacinto Day!" In my ignorance I asked, "What is San Jacinto Day?" They were aghast. After a few minutes silence, Lieutenant Colonel Moseley (the Division Judge Advocate) replied, "What the Battle of Lexington and Concord is to you damnedyankees [sic], San Jacinto is to us Texans." Then I understood, but for a long time I was the Damnedyankee Chaplain."[27]

Division commander Gen. Fred Walker welcomed new men on their arrival to the Texas 36th Division. "You are expected not only to absorb the Texas fighting spirit of this Division, but to improve upon it. You should strive to show our Texans that you are even better fighters than they are . . . We want you to be proud to serve in the Texas Division just as the Texas Division wants to be proud of you."[28]

★ ★ ★ ★ ★

The 36th Division was not among the first guard units ordered inducted on September 16, 1940, the day Burke-Wadsworth become law. The four divisions taken into federal service were the 30th, the 41st, the 44th, and the 45th, the latter partially from Oklahoma. (The 36th Division was federalized on November 25, 1940.) The army had been preparing since July for inducting the guard. Each guard unit was to get ten days' notification before induction, believed sufficient to allow the guardsmen to settle their affairs. The units would then remain at their home stations for up to a week, before heading to army training camps. The army planned to have the guard in camp twenty days before the first draftees arrived, for incorporation into the usually understrength guard divisions.

Marshall had the guard discharge men who met expected congressional exemptions. Divisions also lost experienced men to provide cadres (frameworks) for new divisions. Divisions would consist of professional or guard veterans, and new draftees (including officers) to bring the unit up to strength.[29] "Shortly after induction, we were advised that we would receive about 6,000 men under selective service, to begin arriving about January 13, 1941,"[30] a division veteran later wrote.

However, "the prewar expansible army reached its limits in the autumn of 1941 . . . If a need developed for further expansion in the next year or so, it would have to be accomplished by all-draftee divisions."[31] These would be formed by an experienced cadre of officers and enlisted men taken from a trained division, new or "old" (regular army or National Guard) to form the basis for the new division. Officer Candidate School (OCS) officers and draftees enlisted would be added to this framework. This caused some complaints, including a journal entry by the 36th Division's commander:

> The War Department is still taking away my best officers for instructors at schools, for cadremen for new units and for special assignments . . . This cannot be helped because good officers are urgently needed for expanding the army, but it is tough on the 36th Division and me.[32]

Marshall was also taking away guard officers deemed unfit for service. Aware of the sensitivity of guardsmen's feelings, and the power and influence of the National Guard Association with Congress, Marshall was moving carefully. In June 1941 he established a board of active and retired officers to advise the secretary of war on officer removal. This board was also charged with finding a use, when possible, for officers not qualified for combat duty but still able to provide some other service to the army. "By reclassifying relieved officers Marshall and the Board preserved potential talent while weeding out unfit commanders. It also avoided open political warfare."[33] However, when Marshall felt a guard officer had to go, the officer went.

★ ★ ★ ★ ★

Dwight D. Eisenhower once described the Normandy Invasion, and the campaign that followed and ended World War II in Europe, as a "great crusade." However, research into attitudes of soldiers in World War II showed that most would have laughed at this description. They fought to get the job done and to get home alive. They fought to help their buddies get home alive. When personal survival is in doubt, philosophy is a luxury.[34]

This was the attitude *after* the United States entered World War II. The *New York Times*, in August of 1941, conducted a study of morale and motivation for the War Department. The results were never published, but the document is a good reflection of conditions in the military in times of international tension. The study sheds light on Roosevelt's reluctant public support of the draft, his recognition of the need to "sell" even necessary national security measures to the public.

Lt. Gen. Leslie McNair, in charge of organizing and training army ground forces, in a memo to Army Chief of Staff Gen. George Marshall, described the study's revelations as "astounding. I have known from the training viewpoint that leadership was bad in many cases . . . but had no idea that *discipline* [underlined by hand in the copy this writer examined] was in such a shocking state."[35]

Discipline and morale problems in the military were of vital concern to McNair, Marshall, and others of the 1941 political and military leadership who sensed they were running out of time to get ready for war. In light of the history of later American wars on the "home front," however, some of the reasons the executive summary of the study gives for the morale and discipline problems are particularly interesting:

> . . . from coast to coast conviction as to the gravity of the international crisis is lacking. The men of all branches . . . do not believe the United States is imperilled, even should Hitler defeat the British and bring about the collapse of the British empire . . .
> If the most careful and conscientious sampling of enlisted men's opinion indicates anything that may be relied on, it

Chapter 7: Getting Ready for War Again

points consistently and doggedly to their conviction that "This is England's war"... They are firm in their stand that we should not become involved...

... an overwhelming majority of the representative units and individual soldiers... believe, assert and insist that they are being "dragged into a war which is none of our business," and that the sacrifice demanded of them is not justified by any menace, present or potential, to the United States...[36]

The report goes on to describe the 1941 generation, in words the "baby boom" generation will find particularly amusing applied to their parents, as

a different breed of cat... bereft of national unity, of the crusading spirit then [in 1917] present... The present breed... is questioning everything from God Almighty to themselves... They do not feel like fighting for what they have because they don't know what they have: the simple, sturdier values of American life have not been impressed upon them, either at home or in their schools or colleges—much less in the new Army of the United States.[37]

This was the first mass expansion of the military in peacetime. Some of the problems would never be fully worked out. However, with the draft law, and with federalizing the guard in peacetime, the United States bought itself more than a year to prepare for war, valuable time which it might not otherwise have had.

★ ★ ★ ★ ★

About halfway through the year of federalization, on July 30, 1941, Chief of Staff Marshall wrote all guard division commanders directly, summarizing the state of guard training to that point. The letter is a significant reflection of the attitude of General Marshall and the army high command *before* the famous 1941 Louisiana maneuvers.[38] The letter was also a warning to Guard commanders that Guard divisions had better improve, and quickly:

I am convinced that the next three months will be the most critical period in the development of National Guard divisions. Some of these divisions have already reached a high standard, but efficiency varies from excellent to poor, and the general average appears to be far too low. The training program so far has developed certain definite weaknesses which must be promptly corrected. Accordingly, I am writing directly to all National Guard division commanders, without reference through Army and Corp headquarters, and I would like to you consider this letter as most confidential.

A second paragraph of the letter discusses the wide variety of sources Marshall used to reached his conclusions. The general goes on to say that

> It is apparent in the less advanced divisions that the younger officers have not had enough tactical training or general education to enable them to conduct instruction in an efficient or at least in an interesting manner; that noncommissioned officers suffer from the same deficiencies; that the standards of discipline are too low and reflect the unwillingness of leaders who knew their subordinates in civil life to hold them to a strict compliance with military orders.[39]

Differences were easily spotted in the regular army and National Guard/civilian perspectives. A newspaper article, filed in August 1941, entitled, "36th May Be Unconventional But Sure Can Get Results," discusses these differences. According to the reporter, who clearly had not read Marshall's letter, regular army officers "admit that the [36th Division members] have perhaps the best morale and are the 'fightenest' men in all of Uncle Sam's armed forces." The regular officers, however, "fail to see how the officers can be 'chummy' with the men under them and then expect compliance the next day when they issue orders."

The reporter adds, "army or no army, a Texan is going to be independent." The article seems to assume that Texans were going to continue to look unsoldierly and march "like a bunch of gypsies." The National Guard general then commanding the 36th, Claude V. Birkhead, "seems to be convinced that morale is contingent upon his soldiers being happy. He also seems convinced that comfort, comradeship and family life help to keep the lads smiling."

Correspondent Summers agreed, writing that Birkhead's "ideas seem to be okay, because the 18,000 devil-may-care Texans are sure [they] could go over and kick the pants off Herr Hitler without any help . . . And what's more," Summers concluded, "they probably could do it."[40]

Marshall recognized the importance of morale, but also of skills. He complained in his letter that "General McNair has told me that the basic training of some units has been carried out so ineffectively as to necessitate its repetition."[41]

There is a underlying tone in Marshall's letter that he felt that National Guard officers and men were not taking things seriously enough. Near the end of the letter he commented on reserve officers and how well they were doing the job with regular units. Reserves were officering a high percentage of regular units. Marshall thought reserve officers were handling "friends from home" better, due to lack of embarrassment and a higher level of training and skill. "They are intimately acquainted with the training methods used in our service schools and with available training pamphlets, and this familiarity enables them to carry forward their instruction with continuing interest."

Earlier in the same letter, Marshall wrote, "It is evident that the reclassification of many officers is urgently needed, that a large number are too old for their grades, and that it would be desirable in many instances to transfer officers from their own units." The letter was meant as a warning to guard commanders. "The times are so serious that I must be positive everything possible be done to produce a dependable force. It is imperative that the general standards be raised immediately."[42]

However they had entered the army, soldiers had to be trained. Commanders of the World War II army were determined that its men would know what to do in combat. The yearlong (on average) training cycle began with individual basic training and progressed through multidivisional maneuvers. Newly formed divisions would need more unit-level training, but the process would be similar for all American ground forces. Training included specialized schools around the country, such as a cavalry school in Kansas, a field artillery school at Fort Sill, Oklahoma, and an engineer school at Fort Belvoir, Virginia.

At least one division commander, Maj. Gen. John Sloan of the 88th Division, adopted a system similar to one used in volunteer regiments during the Civil War. The junior officers would learn each phase of the training program and then teach the draftees. When they needed to know tactics, the junior officers learned and then taught tactics. The junior officers gained leadership experience during this period, a kind of learning by doing. They appear to have been better at this than the officers Marshall complained about in his letter to Birkhead and other guard division commanders.

The original plan of the army had been to send trained men from replacement centers to the divisions. It was soon decided that the new divisions themselves could best train their own men, who arrived directly from reception centers. Regular army and National Guard divisions would train their own draftees. Initial days at camp were devoted to teaching the "army way" of doing things. Better-fitting uniforms were issued, along with appropriate "GI" haircuts. Shoe and brass polish became much-used substances. The draftees were shown the proper way to make a bed, and to roll up their extra socks. They shaved every day, whether or not they needed to shave every day. Physical conditioning was a daily part of the training routine.

In the words of Pvt. Bruce Kouser, "Basic training, especially in the infantry, is the toughest training a soldier gets . . . time flies so quickly you hardly know what's cooking . . ."[43] Examining G-3 Operations Records of the 36th Division during training shows that the pace of training picked up during the war.

Close-order drill, marching, quickly became a part of life. Soldiers received twenty hours of formal instruction in drill. This was not just marching in unison but included maneuvering and changing direction with up to battalion-sized groups of men. Close-order marching was considered an effective way of moving large groups of men. The army also considered drill a method of psychologically conditioning the draftees to think of themselves as part of a group. Like the uniforms and the standardized haircuts, drill was a way of subordinating individual personality to the needs of the group.

Drill was a time-honored part of military life. During the Civil War it had a practical application, when military tactics included drill-like maneuvers under fire. In World War II, however, when combat troops were surveyed, they listed drill as one of the least useful skills from basic training. Close-order marching could actually be dangerous under some circumstances, outside of being under fire. Soldiers have to break step when crossing a bridge to avoid the potential for vibrations to crack the bridge. Classroom training gained a similar reputation to drill, even after the introduction of films. The men felt that classroom subjects were either irrelevant or better taught "hands on."

Infantrymen had to qualify with the M1 rifle, the basic infantry weapon, but they also familiarized themselves with the automatic rifle, the light machine gun, and the 60mm mortar. All members of the division, even in noncombat positions, had to gain at least some familiarity with the M1.

No matter how well trained a division, there would still be casualties from various causes. Casualties had to be replaced. Though regiments could directly recruit their own replacements, the Civil War system, for political reasons, called for veteran units being allowed to decrease in strength and new regiments to be formed—and new colonelcies to be given out as political plums.

The World War II army preferred to keep existing units as close to full strength as possible. This required a method of properly training replacements able to fill combat losses in existing divisions.[44] A draftee undergoing the seventeen-week individual and small-unit training course (raised from thirteen weeks) would receive similar instruction to that given the riflemen of the 36th Division, the 88th Division, and other divisions trained as units. There were some differences. Replacements would be trained in all weapons used in the infantry platoon and company. Infantrymen being trained by a new division only learned their specialties, aside from the M1. The assumption was that though replacements were likely to become combat infantry, they had to be able to fill any needed position.

Replacements had very high casualty rates. They would, logically, be needed in regiments and divisions in combat areas, where most casualties were occurring. It was not feasible to give the replacements the year of training given new divisions, training which taught the men to function as part of a team—a major combat skill, equally important to being able to use particular weapons. The men might be thrown into intensive combat situations without the experience gained by the veterans. A recent study has found that

> Replacements, often poorly received until they proved themselves to veterans of their new unit, felt alone during their baptism of fire and typically suffered higher casualty rates in their first engagement than did units composed entirely of untried soldiers.[45]

Evan E. Voss, a veteran of the 36th Division, was one of several replacements sent the 142nd Regiment in January 1944. They arrived in Italy on January 1, 1944, to spend a week in Naples. Two weeks later they participated in the Rapido crossing (of which more later). They received a few days of what Voss, in a 1987 article, described as "river crossing" training and then headed to the front lines.

The sergeant of the 3rd Platoon, Company C, briefed them on the crossing:

> The Sgt. then asked if we had any questions. My God, did I ever had a question, but I was too petrified to talk. I wanted to run as far as I could, but the discipline and training I had been taught told me that was the wrong thing to do.[46]

An hour later, the platoon leader returned from headquarters with a rumor that two entire regiments, the 141st and 143rd, had been wiped out. His company moved out. Voss continued, "I guess I realized that I was about to be killed, it would be soon and there was no reason to worry about it."[47] They were near the river when word came that the attack was cancelled. Voss was lucky. He was discharged in August 1945 as a staff sergeant, re-enlisted for the Korean War, and retired in 1973 as a lieutenant colonel.

The United States Army's talent for improvisation was shown in the way some divisions, such as the 36th Division and the 88th Division, commonly provided further training for replacements in Europe—often as part of continuing training for veterans. This was seen as a way to incorporate the replacements into the unit, developing, as much as possible, the sense of working with a team. This was also seen necessary to correct some loss in readiness almost inevitable with the long period of time it frequently took replacements to arrive at their division. Replacement depots, the "distribution centers" for replacements, were, as a recent historian put it,

> often poorly managed and lacking in training and recreational facilities ... Combat skills deteriorated from idleness. The lack of training, recreation or purpose combined to reduce the replacement's morale and effectiveness just before he was sent to the front.[48]

The first five days in combat were the most dangerous. This was especially true for replacements sent to divisions actively engaged in combat, with the most need for fresh troops. "Soldiers sent to divisions already in combat and suffering heavy losses did not receive additional training and were put directly into the firing line where more often than not they became instant casualties."[49]

Stephen Ambrose is equally harsh:

> If victory required replacements, some of them would have to be expended. One had to be tough. The problem here is that the Replacement System was guilty of the worst sin of all in war, inefficiency. It was paying lives but getting no return. It was just pure waste and the commanders should have done something about it.[50]

The higher commanders were too far from the front, with too little understanding of front-line conditions, to take proper measures. Nothing can prepare you for combat, as most veterans surveyed by the army after the war put things. Ambrose continued that "The only way to learn about combat is to experience it—but surely the Army could have done more to promote passing on information to replacements and to new divisions, from division staffs down to front-line squads."[51]

Voss and his associates seem to have done better than most, with the 36th Division making a good effort to properly incorporate replacements. But the war still had a fifteen months to ago after he arrived in Italy.

★ ★ ★ ★ ★

The day after Summers' amusing article on early phases of the maneuvers in Louisiana (before the famous September 1941 "army" versus "army" exercises), the Associated Press reported, "'Annihilation' of 36th and 45th is Feared," as Maj. Gen. George S. Patton's 2nd Armored Division conducted a simulated armored attack "unparalled in history of U.S. Army."[52] General Birkhead, commanding the 36th, would pay a price for "losing" his division. On September 13, 1941, Brig. Gen. Fred L. Walker, a regular army officer from Ohio, received the following telegram:

> WASHINGTON, D.C., 3 PM, SEPTEMBER 13, 1941. BY DIRECTION OF THE PRESIDENT, BRIGADIER GENERAL FRED L. WALKER IS HEREBY PLACED IN COMMAND OF THE 36TH DIVISION REPLACING GENERAL CLAUDE V. BIRKHEAD.[53]

Three later newspaper articles on Walker mentioned that Birkhead had been retired for age,[54] which would have been consistent with Marshall's policies. However, an unpublished memoir of the period gives the reasons for Birkhead's relief, according to Third Army commander Walter Kreuger, as being that "No man worked harder at his job or was more dedicated. But he was spending too much time doing jobs that belonged to Corporals and Lieutenants and not enough time being a General."[55] This would not be inconsistent with Marshall's implied complaints about what might be called the "constituent service" prewar National Guard approach to being officers.

Birkhead was not pleased with his relief. His first words to Walker, after Walker arrived at the 36th Division's temporary Louisiana headquarters, were, "I've been expecting you, and I want to tell you you are not welcome here."[56] Walker tried to explain that the relief was not personal. After asking where he and his aides could spend the night, he was told that there was no visitors' quarters, "But you may put up your tents in the open space in the rear."[57] Just a few days later, Walker led the division into the September 1941 maneuvers, in which the division performed better than the month before. With the conclusion of these maneuvers, the division returned to Texas and settled down for a long period of training.[58] This training apparently began slowly, easing off from the intensive maneuvers. "Life this far isn't half as hard as it was in Louisiana since we eat all our meals ... and have twice as many men to do about half as much work ..."[59]

As their training progressed, the 36th moved from Texas to train in other areas of the country, including Camp Blanding in Florida (later a replacement training center) the Carolinas, and New England. Maneuvering in South Carolina in late July, just before heading for Massachusetts, the division got "new" helmets, apparently the now very familiar design used from World War II through Vietnam. One officer wrote, "We are to receive the new style helmets in a few days and do they look queer."[60] They protected better, he admitted.

On April 1 and April 2, 1943, the 36th Division embarked for Africa, and "field" training.

In evaluating the World War II training and replacement sys-

Chapter 7: Getting Ready for War Again   119

tem, and offering comments as to how it might have been done better, one has to remember one thing—we won. World War II soldiers met their initial combat tests as better-trained and more proficient soldiers than those who fought the Civil War and World War I. The replacement system, however, made things unnecessarily hard.

The major problem with the World War II system was enabling the individual replacements to become part of a cohesive unit. Whenever possible, receiving divisions gave the men additional training before using them in combat. Gen. John Sloan of the 88th Division, for example, would try to incorporate new men into his division while his units were retraining behind the lines. The 36th used a similar system. The hope was that the new men would become accustomed to their new unit before going into combat.

A unit-replacement system was considered but initially rejected.[61] According to the official army World War II history, the "Green Book" series, "The War Department considered the creation of more combat units to be impracticable within exist limits of time and strength."[62] By the time the war ended, an idea to combine more frequent unit rotation with using replacement teams of two to six men was still in the planning stage.[63] In some ways, the Civil War system of unit replacement was a better idea. A division needed a year to get ready for combat in World War II. A better alternative to individual replacement would have been replacement on a small group level, the conclusion the army itself had reached. The small group would have developed some cohesion and confidence. Incorporated into a unit with veterans, the replacements would still have been able to learn from the veterans. A similar system, on the battalion level, was adopted in Italy to incorporate black battalions into white divisions, an unofficial early example of integration.[64]

Still, the first five days of combat remained particularly hazardous for almost everyone. A man's chances of survival in combat would grow with time, as he gained experience and survival skills—before basic exhaustion began to take its toll.

Most of the World War II training compromises, especially with replacements, grew out of manpower shortages. In 1942 the army could afford to give the 88th Division a year of training. It was over two years between federalization and the time the 36th went overseas. By the end of 1943, when this writer's uncle entered the army, it could only provide seventeen weeks of training for replacements. The Americans were running out of troops, partially due to the use of so many men in support positions. Both possible solutions, cutting down on support services or drafting women for support positions, were politically unfeasible. So the army made do with the resources it had available.

CHAPTER 8

Salerno

"The damn crazy Americans won't stop and fight. They're just walking through us."[1]
—German soldiers at Salerno, 1943

Despite the failure to trap the bulk of German troops, the Allied liberation of Sicily was going well. What next? The Americans wanted an invasion across the English Channel into northern France, and a strike directly at the heart of Germany, as soon as possible. This would provide the pressure-relieving "second front" the Soviet Union wanted and needed. This was also seen as the best way to bring the full might of American materiel power against Nazi Germany.

The British, particularly Winston Churchill, liked the idea of an attack through the "soft underbelly" of Europe, the Balkans, attacking Germany through the back door. The Soviets, and some Americans, suspected another motive—keep the Soviets out of the Balkans.[2] The British remembered the heavy casualties they took in World War I, manpower losses still not fully made up in 1943. The British had more men killed at the Battle of Ypres, for example, than the United States lost in all of World War I. Despite his concerns, however, on September 11, 1943 (two days after the invasion at Salerno), Churchill wrote South African Field Marshall Jan Smuts, "There can be no question whatever of breaking arrangements we have made with the United States for 'Overlord,'"[3] the invasion of northern France.

The Germans had not yet built the "Atlantic Wall" defenses in France, so getting ashore *might* have been a little easier.[4] In 1944 German Field Marshall Erwin Rommel told an aide that "the first twenty-four hours will be decisive . . ."[5] Getting ashore was not enough, however, even if the beachhead survived. In 1943 ". . . the Germans could have concentrated ground and air forces against us quickly and massively, which they were unable to do in 1944 . . ."[6] In 1943 the Allies were not ready for the massive effort required to invade across the English Channel to France, survive in France, and accomplish something once they got there. Even before the attack on Sicily, the cross-channel invasion was put off until 1944. Though it took a while for Winston Churchill to fully give up on the idea, invading the Balkans was dropped as too peripheral.

Out of a desire to keep pressure on the Germans by tying down troops, to help ease pressure on the Soviets (and possibly even keep them in the war), to keep Allied and British troops in action, to knock Italy out of the war and, perhaps, from sheer momentum, an invasion of Italy was the next step. One force would cross the narrow Straits of Medina. Another force would invade Italy somewhere midway up the "boot." This chapter begins the story of that force.

★   ★   ★   ★   ★

"In Italy the Anglo-Americans hack away without imagination and with slight success," a former German ambassador to Italy wrote.[7] From the start of the Italian campaign, the Americans and the British often seemed to lack imagination, showing an unwillingness to gamble strategically for possibly great gains.

Just before the invasion of Sicily, it was assumed among Allied planners that an Italian collapse and withdrawal from the war would cause the Germans to withdraw from Italy,[8] something confirmed by monitoring German communications. Though his common practice was to order "no retreat," Hitler had decided not to defend southern and central Italy after the expected Allied invasion and expected departure of Italy from the Axis. Field Marshall Erwin Rommel, commanding in northern Italy, urged such a strategy on Hitler. Field Marshall Albert Kesselring (a Luftwaffe, or German air force, officer), commanding in the south, thought defense of most of the country was both possible and advisable.[9]

Directly striking at Rome was very tempting to Allied planners. However, Rome was outside of the range of ground-based aircraft operating from Sicily. The Naples area was as far north an invasion site as seemed feasible. Naples was a large port. Once the expected German sabotage, and any battle damage, was repaired, Naples would be a valuable port for bringing in supplies and troops.

The Gulf of Naples itself was rejected as a landing site as being too well fortified and too easy for the Germans to defend. The Gulf of Gaeta, just to the north, was better. Particularly appealing was the lack of dominating high ground around the beaches. Allied armor would be able to quickly land and deploy to surrounding flat plains. However, Gaeta was also heavily fortified. The beaches were too soft. A sandbar off the beach would create major problems, as most landing craft would not be able to get to the beach. Gaeta was also beyond the effective range of Sicily-based aircraft. Though still being considered as late as August 13, Gaeta was rejected as the invasion site. Gen. Fred L. Walker, 36th Division commander, was among those happy about this decision, writing that "I thanked my lucky stars we did not land in that area."[10]

The Gulf of Salerno was selected, for positive advantages as well as by default. The town of Salerno was a small port, which would be helpful until Naples could be captured and put back into working order. A second smaller port existed in the area. No shoals blocked the sea approaches. Underwater gradients were good and would allow ships to come in close to shore. No one guessed how vital this would prove during the battle. Salerno coastal defenses were just fieldworks, not permanent defenses. Montecorvino, an excellent airfield, was close to the beach. Finally, fighter craft from Sicily could provide cover to Salerno—though at the limit of fighter range. General Walker, at least, thought this was the main reason: "because our air capabilities could provide a more certain and effective support for a landing in SALERNO BAY [sic] that site was selected."[11]

There was several notable problems. The Salerno beachhead was far enough away from the planned Allied crossings at the Straits of Medina to make mutual support impossible until one force could advance. The Sele River cut the plain at Salerno, and any beachhead, into two sections. The riverbanks were sufficiently steep to make bridging necessary, as well as hampering movement between the two sections. Mountains enclosed the Sele plain. Finally, only two narrow gorges opened from the mountains through to the direction of Naples.

The Allied commanders decided that units of the 5th Army, under the command of Lt. Gen. Mark W. Clark, would land on September 9, 1943. The British 10 Corps, with the 46 and 56 divisions, would land north of the Sele. The British were to seize the towns of Battipaglia and Salerno and the Montecorvino airfield. United States Rangers and British Commandos, operating independently, were to seize the Sorrento Peninsula northwest of Salerno. The northern force would then attack toward Vietri Pass, and Naples. Plans were to capture Naples on D+5.

The American VI Corps, commanded by Maj. Gen. Ernest J. Dawley, would invade the southern part of the beachhead. The 36th Division would lead, with the 45th Division (National Guard like the 36th, but primarily from Oklahoma) in floating reserve. The Americans would seize the high ground and establish a beachhead, protect much of the right flank of the British operation, and keep enemy forces out of the Sele plain.

There was later some question about whether the 36th, a well-trained but combat-inexperienced division, was the most logical choice to lead so complex and difficult an operation as a contested amphibious landing.[12] The 36th had been selected for use in Sicily but was removed when the 45th Division arrived in Africa at the right time in combat loaded ships.[13] General Walker, though anxious to get into combat, did not like 7th Army commander George Patton. Walker was actually happy with the decision to delay the 36th's entrance into action until the Salerno landing.[14]

The 36th had been training to regular army standards since its November 1940 federalization. It had been in North Africa several months. Walker, still on friendly terms with Clark, had urged that the division be given such a chance.[15] Walker and Major General Dawley had worked together well in North Africa. Finally, there is reasonable speculation that Clark wanted to show that well-trained troops, without combat experience, could perform well in a difficult combat situation.[16] "The 36th Division is a good outfit,"[17] Clark wrote. It seemed simple enough. But with several weeks to go, there was still time for political complications.

Things became more complicated on July 25, 1943. Benito Mussolini, ruler of Italy for twenty-two years, was overthrown. Marshall Pietro Badoglio, Mussolini's successor, immediately declared that Italy would remain in the war as a member of the Axis. Kesselring believed him. Rommel did not. Rommel was right. Virtually simultaneously, Badoglio opened secret negotiations, on Italy's surrender, with Eisenhower and his staff. The Italians wanted Allied protection if they left the war, fearing German wrath once the surrender was announced.[18] Allied troops would protect the Italians.

If the Italians surrendered, why invade? Why not just let Italy leave the war and take her army with her? There were several reasons this was not considered feasible. Surrender negotiations with the Italians were somewhat messy. The possibility always existed that something could go wrong. The Allies must also have considered the possibility, however remote, that the Italians could switch sides a second time.[19] Perhaps most importantly, landing in Italy was still seen as necessary to tie down German units and keep them away from the Soviet front or northern France. Ultra (the name for Allied efforts decoding German military transmissions) intercepts indicated German intention to withdraw from most of Italy.[20] Not only did planning continue for the invasion, but the Americans and British now considered Italian surrender necessary for the invasion to succeed.

The Allies were not the only ones unsure about the Italians. Adolf Hitler did not trust his ally, though he did not know about the Italian negotiations with the Allies. Hitler decided, on September 7, 1943, to send an ultimatum to the Italian government. The ultimatum demanded greater freedom of action for the Germans in Italy. It required the Italians to resist expected landings up the "boot," as well as the British force already slowly moving up the coast (since September 3) from the southernmost port of Calabria. Failure to respond favorably would allow the Germans to take any steps needed to protect their troops. Hitler ordered that a draft of this ultimatum be prepared for his signature by September 9, 1943.

The German plan in case of invasion or Italian surrender was basically a fighting withdrawal to the north, saving as many men from the newly activated 10th Army as possible. Rommel and

Kesselring would disarm Italian troops in their sectors. Generaloberst Heinrich von Vietinghoff genannt Scheer, commander of the 10th Army, was not to start his withdrawal early, to avoid giving the Italians an excuse to defect. Withdrawal was so set a plan that von Vietinghoff told the commander of the Italian 7th Army in the south that the Germans would not support a proposed Italian counteroffensive against the British in Calabria.[21]

German plans were not very detailed, just to get back to the Rome area with as little damage as possible. They would then withdraw to a "permanent" defensive line in the Apennines. There was no plan of action in case of an invasion *and* an Italian surrender.

The Germans intended Rommel to take overall command of the portion of Italy the Germans would hold, with Kesselring to be transferred to another front. However, on September 12, Kesselring was kept in command until his forces approached Rommel's area.

The Bernhardt line of German defenses south of Naples, also called the Winter line, was originally a temporary position to enable Rommel to finish work on more permanent defenses north of Rome. Kesselring's success at delaying the Allies, and at convincing Hitler he could continue to delay them, led to Hitler's decision to hold as much of Italy as possible, in the expressive words of a television documentary, to "force the Allies to bleed for every inch of Italian soil."[22] A November 6 order placed Kesselring in charge of all the Italian defenses. Rommel was sent to France.

The Italian surrender was announced on September 8. German contingency plans were put into effect. The surrender announcement was broadcast to the men of the 36th Division, a day away from the landing at Salerno. Many cheered, thinking the invasion would now be a cakewalk. Some regretted missing out on action. Others pointed out that the far tougher Germans would now be defending the beaches, and the men would get all the action they could possibly wish for.[23]

★ ★ ★ ★ ★

The American/British invasion fleet arrived off the Salerno area the night of September 8, 1943. At 11:00 P.M. the ships sounded general quarters. Soon after midnight on September 9, the assault and landing craft were lowered, and the men boarded. When the craft were ready to go, the boats headed toward the appropriate British or American rendezvous area, three to five miles from shore. From there they would head to an unknown reception on the shores of Italy. Roughly three hours elapsed before all the troops reached the rendezvous areas. Landing craft are not comfortable, so it was fortunate the sea was calm with little wind.

At 2:00 A.M., German shore batteries opened fire on the northern section of the invasion fleet, the British 10 Corps and supporting ships. The warships began returning fire. Precisely at 3:10, on schedule, an American ranger force under the command of William Darby landed near the tiny port of Amalfi, the first step in cutting a possible route for German reinforcements from the direction of Naples. Five minutes later, a direct prelanding barrage opened at the British invasion beach. At 3:30, again virtually on schedule, the first assault waves of British 10 Corps troops landed.

Elements of the 141st and 142nd infantry regiments, 36th Division, landed at the same time to the south of the British beaches. There was one key difference—no preliminary bombardment. Mark Clark had thought the bombardment of the British beaches to the north would provide a diversion, allowing the 36th Division to slip in virtually unopposed.[24] Clark also had some hesitation about firing on "neutral" territory.

General Walker had reached the same conclusions:

> I studied the latest air photos of the beaches and surrounding high ground. I could not find any fixed or organized defenses near our beaches, nor elsewhere within our sector. There was a three gun-railroad artillery battery opposite our beaches but our intelligence reported it to be Italian property, obsolete and unmanned. It was within rifle range of our first waves ashore and if gun crews should exist they could be quickly driven off by our infantrymen. As it turned out these guns were obsolete and were not manned. I could find what might be a few old emplacements back from the beach. I knew that the 16th Panzer Division was in the area but I did not think they would be strong in any one sector. Italian soldiers previously in that area would not be there. The British landing to the north would have a naval bombardment for some time before we were to hit the shore. This, I felt, would definitely indicate the main landing phase to the Germans and the location of our landing beaches would remain uncertain to them until we actually hit

the shore . . . I did not think a preliminary bombardment would be helpful tactically . . . I therefore requested that no preliminary naval bombardment be fired . . . If I would have it to do over, I would have requested the preliminary bombardment; not so much for its tactical value, though it would have helped some; but because it would have been an encouragement to the first waves of troops to hit the shore.[25]

Walker and Clark were quickly proven wrong. The 142nd Infantry first encountered trouble on the left of the 36th Division sector. The regimental record of events reports that "Suddenly flares went up from the beach, machine guns started opening up all along the beach, artillery started falling, mines were encountered and the wave group found themselves met by a very stubborn enemy defense system which was dug in waiting for them."[26]

One division veteran describes landing a jeep in the invasion:

I came ashore at Paestum, well I didn't exactly come ashore dry. The operator of the landing craft that I was in was a little nervous and hit a sand bar some distance from the beach. He decided he could not get any closer to the beach so he let the ramp down and ordered us to disembark post haste since the Germans were shelling the Hell out of the area. The Jeep that I was in was the first to come off, the driver and I were completely submerged in a matter of seconds. The Jeep had been waterproofed in Africa with snorkel tubes, so we stood up to get our heads about water and continue to drive the vehicle on the bottom toward shore.[27]

The 1st Battalion, 141st Infantry, landed on the extreme right of the division (and entire landing) front. The first two waves hit the beach about 500 yards south of the designed Blue Beach, working their way inland about a mile to a railroad near the Solofrone River. The third wave, hitting the correct Blue Beach, ran into heavy enemy resistance. Later reports indicated that only 300 men of a German panzer grenadier regiment were defending this area. However, they were well dug in and backed up with machine guns, mortars, and the infamous 88mm antiaircraft gun being used as antitank artillery.[28] German fire even managed to delay the landing of supporting American artillery.

The 3rd Battalion of the 141st Regiment landed on Yellow Beach, just to the north. It ran into resistance from the beginning but was able to move inland about 400 yards. The elements of the 1st Battalion, 141st, pinned down on Blue Beach were cut off by a German counterattack, with five tanks, which hit the right flank of Yellow Beach about 7:30. The attack was fought off, with rifles, grenades, machine guns, and one 40mm antiaircraft gun used for ground fire. One company was overrun. Those of the company who took shelter in ditches were usually unharmed, those in open fields shot or run over.

At this point, Blue Beach was abandoned and the rest of the regiment brought in at Yellow, though the 1st Battalion remained isolated. Col. Richard Werner, commanding the 141st, tried to call for naval gun support. The ships were out of range in his sector, though a rocket boat was soon able to offer support, and decrease German fire, off one of the 142nd's beaches. Naval guns did not open fire until 9:00 A.M.

Communication problems the first day, initially a total lack of communications between the beaches and the ships, denied the higher command levels any precise information. Walker, at about 7:00 A.M., reported to Dawley that heavy enemy gunfire was preventing vehicles from landing and the beaches being cleared—probably referring to the 141st at Blue and Yellow beaches. He later noted that on landing he could see that Red and Green beaches were clear. "I looked for the troops that had preceded us. Most of the personnel within view were those landing at the same time, part of the 143rd Infantry. The beach was clear of personnel and equipment."[29]

Despite being in the middle of a major battle, the men onshore had little idea of what was happening outside of their immediate area. This seeming isolation was a common occurrence in World War II. Even General Walker was initially isolated from the action. Landing just after 8:00 A.M., Walker made his way to a house inland. "While at the section house I looked over the area in all directions. I could see no activity by our own troops anywhere . . ."[30]

Initial uneasiness on the part of the on-the-scene commander—even though the impression quickly changed—combined with extremely difficult communications to give higher-level commanders the impression that the Salerno invasion was having major problems.

Some Germans had initial success with at least one tactic.

They would fire rifles in the general direction of 36th Division elements, but without any real hope of hitting anyone in particular. When fire was returned, revealing the American positions, the Germans would call in more accurate mortar and artillery. When the Texans figured out this tactic and began to copy the trick, the Germans withdrew. By that evening, though the 1st Battalion of the 141st Regiment was still cut off, the regiment as a whole was in Italy to stay.

The initial waves of the 142nd started taking fire before they even reached the beach. Communications problems were exacerbated by destroyed radios. Fire soon became so intense that some boats, under inexperienced pilots, were chased back offshore. The regiment's records of events reports that "These were quickly turned back and landed the men, but this action caused a break in the intricate time schedule set up and from then on the waves were not landed properly as to time or place."[31]

One sergeant was shot in the chest and shoulder before his boat landed. The ramp stuck when the boat reached the beach. In spite of his wounds, the sergeant kicked and pounded until the ramp dropped and he was able to lead his section to the beach.[32] He was hit again, and killed, after reaching the beach. The sergeant would earn a posthumous Distinguished Service Cross for his actions.[33]

Sgt. Manuel S. Gonzales, of Company F, earned the same medal about the same time. He discovered the position of a German 88mm cannon that was creating havoc among the landing craft. The German crew saw him creeping forward and opened fire. Gonzales' backpack was set on fire. Gonzales was wounded by grenade fragments during an exchange with the crew, but managed to kill the crew and destroy their ammunition.[34]

Lt. Robert H. Carey, from Maine, soon after reaching shore, was fired on by three Germans with machine pistols. Carey shot back, killing one German, before his weapon jammed. Carey then grasped his carbine as a club, advanced at the two Germans—who were still shooting—and bashed a second. The third was tackled and taken prisoner.[35]

The commander of the 2nd Battalion of the 142nd, Lt. Col. Samuel S. Graham from Huntsville, Texas, told his command that everyone was supposed to be on top of their objective, Mount Soprano, by nightfall that day. The mountain, an extremely visible objective, was 3,200 feet high. "This simple order had a magnetic effect on the troops . . . Disorganized by enemy fire and darkness, men operating alone or in twos and threes pushed forward despite the German fire . . . Their tactics of individual infiltration applied more pressure than the enemy could cope with."[36]

The 2nd Battalion's executive officer, Maj. James T. Padgitt, found himself, in the darkness, in a small gully with a soldier from Company E. Padgitt later wrote that the soldier, who spoke some German, translated what the nearby Germans were yelling. "The damn crazy Americans won't stop and fight. They're just walking through us."[37]

The poetically minded member of the regimental staff who kept the record of events wrote, "Our men steadfastly moved ahead in the face of the intense fire and got off the beach as soon as possible."[38] At 8:45, General Clark informed his boss, Gen. Harold Alexander, that the entire 36th Division was ashore.

The overall picture was still confusing. The Germans proved to be every bit as tough and resourceful as their reputation indicated, making things very hard for the landing troops. German counterattacks, especially those using tanks—with the delay in getting American artillery onshore—created problems. The Germans in the immediate area of the beach, though, were heavily outnumbered. Their attacks were repulsed.

The first day ended with the 36th Division, and the British and American invading forces in the 10th Corps, successfully ashore. There would be further crises, but the key first day of the invasion had passed successfully. The task, now, was to expand the beaches and begin the advance northward.

Gen. Dwight D. Eisenhower was convinced, however, that a German counterattack was likely. Ultra intelligence indicated that reinforcements had already been ordered to the Salerno area.[39] Eisenhower was concerned that Allied sealift capacity would not enable a full division to immediately reinforce the beachhead. Gen. Bernard Montgomery and his British troops would likely take several days to get from the toe of Italy to Salerno and link up with Clark.

A third landing in Italy, the amusingly titled Operation Slapstick, was carried out on September 9, the same day as Salerno. Thirty-six hundred British paratroops were sent by ship to successfully capture the port of Taranto, on the eastern coast of the "heel" of the Italian boot. Little diversionary value was served, however. The German division commander in the area realized his troops were too limited and too dispersed to offer effective resistance. The Germans withdrew northward.

September 10, 1944, D + 1: Action continued at Salerno on the second day, September 10, on the British front. The German 16th Panzer Division, after spending September 9 opposing the entire invasion, withdrew from the VI Corps front and concentrated against the British 10th Corps to the north.

The northern portion of the heights surrounding Salerno blocked the quickest move to Naples and were considered more important. Two German divisions just north of Salerno were being rehabilitated. Two others, to the south, were heading north. Montgomery's 8th Army was doing little to delay the withdrawal. The only real delay was from the Germans stopping to delay Montgomery.

Despite continued attacks on the British front, and the divisions approaching from the south, the 36th Division was not busy fighting and was able to push the beachhead forward. The right flank was strengthened.

As General Walker wrote on September 10, "The enemy has not materially interfered with our landing. My plans are going well. The beaches are free from enemy artillery and machine gun fire. I am pleased."[40]

September 11, 1943, D + 2: The Germans reinforced the 16th Panzer Division's concentration against the British 10 Corps. They were exerting heavy pressure against left flank of the British section, the left of the entire Allied beachhead. The German 10th Army captured 1,500 British prisoners on September 11 alone.[41] Luftwaffe air attacks, concentrated against Allied shipping, were meeting some success. Targeting shipping was an effort to eliminate, or at least decrease, damaging Allied naval gunfire. Vice Adm. H. Kent Hewitt, the American admiral in charge of naval support for the invasion, found it necessary on September 11 to take his flagship, the *Ancon*, out to sea for the night.

Available German air force strength over the beachhead was greater than numbers alone would indicate. Though the Luftwaffe was heavily outnumbered, its bases were close to the beachhead.

German planes could fly several sorties a day, something not the case with Allied planes coming all the way from Sicily. Fighter coverage from Sicily had actually decreased on September 11, though an Allied carrier force was able to stay in the area. The Montecorvino airfield, captured on the 11th by the British, would prove useful once German infantry and artillery within range could be chased out.

On September 11, two regiments of the 45th Division were sent in to try to close the gap between the British and the American parts of the beachhead. The 142nd Infantry was assigned to support this effort. The target was called the "Sele-Calore corridor," the relatively narrow floodplain of the Sele and Calore rivers. The area starts roughly twelve miles inland, near the village of Serre, at the edge of rugged hills. The rivers flow gently downhill until they meet, five miles from the shore.

The problem of the gap seemed to have been solved, simply and logically, in planning. British 10 Corps would seize the heights around Battipaglia and then around Eboli. The VI Corps would take Hill 424 near Altavilla. They would then meet at Ponte Sele, quite effectively cutting off the corridor.

Unfortunately, things did not work out as planned. The Germans held Battipaglia against the British. The Americans first focused on their right flank, establishing a defensive barrier against German forces moving up from the south. With the right flank seemingly secure, VI Corps would concentrate on closing the gap to its left.

The center and left of the VI Corps attack made little progress. The right of the September 11 advance was an attack by part of the 142nd Infantry on Hill 424 and the nearby town of Altavilla. As one participant later put it, "Altavilla commanded the heights overlooking the Gulf of Salerno. It was like all our objectives in Italy—it was a height of some sort with the enemy looking down on us."[42]

This attack went well. As described in the comprehensive army history series the "Green Books,"

> In contrast with the opposition met by the two regiments of the 45th Division, a battalion of the 142nd Infantry [the 1st] took Altavilla and the nearby hills with no trouble at all. Troops entered the village during the morning and occupied dispersed positions on the heights without resistance. That af-

ternoon, when patrols reconnoitered eastward as far as the Calore River, they found no German forces. American domination of the Sele-Calore corridor from the south now seemed established.[43]

By 3:00 P.M. on Saturday, September 11, D+2, Altavilla and the nearby high ground had been occupied without a fight. One squad of C Company occupied Hill 344, off the right flank of the battalion. The rest of C Company was on the right of the main battalion front, on or near Hill 424. Company A was in the center, Company B to the left. However, the nature of the terrain was creating problems. Many ridges and gullies, and areas of brush and trees, made Hill 424 a hard place to defend. Detachments were not in sight of each other, giving the units (and the men) the feeling of fighting alone. Limited visibility decreased effective fields of fire and made central command difficult. The 1st Battalion, 142nd Infantry, held the most advanced position of the VI Corps front.

September 12, D+3: The Germans realized the 1st Battalion would have trouble defending its area. That night, the Germans started efforts to retake the Altavilla area. They had an estimated three battalions of infantry, and tanks, in the higher ground north of Altavilla.[44] German and American patrols skirmished most of the night as the Germans infiltrated gaps in the dispersed 1st Battalion's positions. Soon after daylight on September 12, the Germans opened fire.

The Germans seemed to be attacking from all directions, but with a particular focus on Company A, in the center. A serious predawn attack was defeated with hand grenades. Heavy artillery bombardment after daylight interrupted already strained communications within the American battalion. Just after noon, Company A was finally ordered to disengage and withdraw to a new location on the right flank of Company C—also under attack. Company A was not able to accomplish this maneuver. Company B also found it impossible to move to assist Company C.

One private in Company B observed an enemy machine gun crew moving on the extreme left of the company position. The private, Clayton T. Tallman, took off alone to intercept the Germans. Tallman was spotted moving along a stone wall, and the Germans opened fire. He then climbed on the wall, carefully took aim, and shot each German through the head. A few moments later, Tallman destroyed a second German machine gun crew.⁴⁵

About 1:00 P.M., the commander of the 1st Battalion, Lt. Col. Gaines J. Barron, moved forward to the area Company A was supposed to occupy but had not been able to reach. His executive officer would supervise the withdrawal of the battalion, when a few hours later the Germans had broken through and retaken Altavilla. By 9:00 P.M. the American survivors were back at the regular division line, and the Germans were fully in control of Altavilla and Hill 424. Walker, who had already ordered a battalion to reinforce Altavilla, before he learned of its loss to the Germans, was planning to go back, something also ordered by General Dawley. "I wanted to retake the hills at ALTAVILLA because they gave the Germans excellent observation over our whole area,"⁴⁶ Walker wrote in his journal.

September 12, Mark Clark arrived and took over formal command of the beachhead. Admiral Hewitt had been technically in command during the invasion and initial buildup. Clark was concerned by the German strength in the Persano and Battipaglia areas, precisely at the weakest point in the Allied line. He was also growing concerned over whether VI Corps commander Dawley, Walker's immediate boss, was up to his job. At least one later commentator feels Clark was wrong in leaving Dawley in command, if he had lost confidence in Dawley.⁴⁷

Clark rejected the idea of issuing an order reminding his troops that in case the Germans broke through to the beach, the men should destroy supplies and equipment. Clark thought such an order would hurt morale.⁴⁸ He was probably correct.

September 13, D+4: Walker was ordered to try and retake Altavilla. Only two battalions, the 3rd of the 143rd and the 3rd of the 142nd, could be spared to make the attack. The resulting effort failed to capture Hill 424, dominating the town, and was only able to hold the town for one day.

Pfc. Charles E. Kelly earned the first Congressional Medal of Honor awarded to ground forces in Europe. Kelly, with three other men, had been dispatched to destroy a German machine gun nest. They ran into a seventy-man German force and had to fight off this force *and* the machine gun. After using his Browning Automatic Rifle to kill the gun crew, Kelly opened fire on one group of Germans from fifty yards. He emptied a magazine, reloading while the other men fired. Kelly then resumed firing. The four men inflicted heavy casualties before retiring, unharmed.[49]

Kelly was not done. A little while later, he and some other men were defending a building in Altavilla. Kelly used several different weapons to protect the house. Kelly finally went onto the balcony of the building. He took 60mm mortar shells, removed the pins, tapped their bases to arm them, and then threw them down on the Germans. Landing nose first, they exploded, killing some enemy and discouraging further close-range attacks on the building.[50] Kelly's method of exploding mortar shells was so unusual that it had to be authenticated before he could be awarded the Medal of Honor.[51]

★ ★ ★ ★ ★

"There was failure at Altavilla, but in the Sele-Calore corridor the situation came close to disaster,"[52] is how one historian described events north of Altavilla. The 2nd Battalion of the 143rd had moved into position the night of September 12, trying to plug the gap between the 36th and the 45th. The 45th, north of the Sele River, had not moved to the right, and there was still a gap of roughly three kilometers. General von Vietinghoff, commanding the German 15th Army, was planning a massive attack on the beachhead but had originally not intended to move until September 14. He did not expect to have enough men on hand until then. He even scheduled a planning conference with one of his corps commanders for the evening of September 13.

That morning, the Germans found the gap. Von Vietinghoff thought the Allies had voluntarily split themselves into two sections, part of a plan to evacuate the beachhead. He assumed the heavy German pressure was too much for the Allies, that the additional shipping in the area was preliminary to an evacuation. The possibility that such a gap had been allowed to appear through American carelessness never occurred to von Vietinghoff. An attack was ordered in midafternoon.

At about this time, the 2nd Battalion, 143rd Infantry of the

36th, assumed a defensive position two and a half miles northeast of Persano. It set up antitank guns and a few quick minefields. The American battalion sent out patrols on both flanks, to try and find the nearest American units. However, there was no contact. A three-mile gap separated the battalion from the portions of the 36th busy fighting at Altavilla. The 157th Infantry, on the left, was two and a half miles away to the rear, protecting the crossing of the Sele. Walker wrote in his diary that when he was told by Dawley to send a battalion,

> I was a little surprised at this because I understood that a regiment of the 45th Division was advancing between the rivers and was protecting our left flank. Dawley stated that units of the 45th Division would be just across the Sele River opposite the left flank of the position the battalion is to occupy.[53]

Walker took Dawley's word and later was surprised at learning how isolated the battalion had been.

About 3:30 that afternoon, elements of the XIV Panzer Corps struck the 1st Battalion, 157th Infantry, at a tobacco factory captured earlier by the Americans. More than twenty German tanks hit the defensive positions around the factory. The Americans fought back, but part of the 1st Battalion was quickly forced back to 3rd Battalion (157th Infantry) positions. This unit was also under attack. The Germans pushed forward to the Persano crossing, driving the rest of the 1st Battalion from the factory.

The Germans crossed the Sele River into the Sele-Calore corridor, striking the left flank of the 2nd Battalion, 143rd Regiment. Other German units, entering the corridor near Ponte Sele, hit the battalion's right flank. The battalion was not well deployed, occupying poor ground. Patrols were not sent out, missing a chance to detect the German buildup. An artillery spotter was present with the 2nd, but he was killed early in the action. Though the battalion was located by a reconnaissance troop,[54] the troop was unable to get sufficiently precise targeting information to division artillery.

By 6:00 P.M., enemy artillery was in place around Persano. The 2nd Battalion, 143rd Regiment, was at least temporarily out of existence as an effective unit. Combat losses were 47 killed, 36 wounded, 425 missing. Only 9 officers and 325 men made it back, though some of the missing returned later.[55]

Colonel Martin's force at Altavilla was pinned down about the time of the near-disaster at the Sele-Calore corridor. Two field-artillery battalions stood between the Germans and the sea. Both opened fire at point-blank range across the Calore. A few tank destroyers from the 636th Battalion, coming ashore that afternoon, were sent immediately to the Calore. Other artillery helped when possible, but the German attack was equally strong, though not as disastrously successful, around Altavilla. Clark's headquarters, just a few hundred yards behind the lines, had some of its clerks and other staff provide a hasty firing line. Others were preparing to evacuate.

Dawley ordered Walker to pull out of Altavilla and to form a defensive line on La Cosa Creek, between the Calore River and Mt. Soprano. This front was eight miles long and would have to be planned and occupied at night. Walker commented, "It was not my intention to completely stop a crossing at the river, but it was my intention to meet a crossing at the river and to dispose of any enemy tanks that should succeed in crossing, as soon as they should appear on the open ground."[56]

The front was divided into three sections, each under the command of a brigadier general. Walker felt this was the best way to coordinate different units which had not worked together before. Gen. William H. Wilbur, attached from 5th Army Headquarters, commanded the left; John O'Daniel, also on loan from the 5th Army, was in charge of the center. Assistant division commander Otto Lange was given the right section, though Col. John D. Forsythe, CO of the 142nd, ran this section until Lange could finish duties on the beach. All available troops, of whatever formal assignment, were put in or near the line. The 36th Division was not to retreat from this position.

Clark had ordered elements of the 82nd Airborne to jump in to reinforce the beach, even before the German attack at Persano. Admiral Hewitt, at the same time, was preparing for possible withdrawal from the beach. However, he objected to the possibility, citing the technical difficulties in loading landing craft on the beach and then getting them back into the water. Another admiral also protested, as did Richard McCreery, British corps commander. Walker and 45th Division commander Maj. Gen. Troy Middleton also saw no need for withdrawal. Dawley formally objected.

Though some historians would later accuse Clark of having panicked at Salerno,[57] Clark was also seeking ways to reinforce the beachhead. Despite pressure from the combined chiefs of staff and from Winston Churchill himself, Montgomery's 8th Army was neither advancing very quickly nor doing a particularly good job of delaying the Germans. Other units were on the way by sea, including the 3rd regiment of the 45th Division. Clark also arranged for the 82nd Airborne to jump in to reinforce the beachhead, even granting Ridgeway's request to ensure no antiaircraft fire at the time the 504th Regiment was scheduled to arrive. Planes had been shot down, and men killed, by "friendly" fire at Sicily. Clark and Ridgeway were willing to risk German attacks to ensure it did not happen again. By midmorning of the 14th, the regiment was in line with the 36th.

The Germans continued attacks at various points, on September 14, which were fought off. An 8:00 A.M. attack inadvertently let itself be taken in the flank by elements of the 179th Infantry. A few hours later, naval gunfire helped defeat an attack against the 157th. At least two other attacks were launched against the 45th Division front, but to little effect.

The same pattern of small- to medium-size attacks was repeated on the 36th Division front. A larger attack near the Calore River was defeated by the 36th Division, and naval gunfire. Walker did not think the naval gunfire had been necessary. Clark, however, a few days later wrote the navy: "Please convey to Admiral Cunningham my deep appreciation for the splendid and wholehearted cooperation and support given by the Allied navies during our operations in this area. Their naval gunfire had been most effective..."[58]

Late that day, Brig. Gen. Otto Lange was told to leave his beach duties and take command of the south sector of the La Cosa Creek line. Walker told Lange to take command at once, to make sure that his sector would be ready for a possible attack the next morning. Early the next morning, Walker found Lange just getting up. "I discussed with General Lange my instructions of the night before and he admitted he had not complied with them because he was worn out and needed a rest. Everybody needed a rest."[59] Walker then told Lange he would recommend to Dawley that Lange be transferred out of the division. Lange was replaced by William

Wilbur. Lange was reduced to his permanent grade of lieutenant colonel and sent home.

September 15, D + 6: September 15 was in many ways a duplicate of the day before. Most of the day saw repeated thrusts by German armor, with heavy artillery support. Division forces were constantly moved back and forth to meet these attacks, to provide the strongest defenses where needed. At one time during the day, Maj. Gen. Lucian K. Truscott, Jr., commander of the 3rd Division, was visiting the front. His division was scheduled for deployment. While visiting General O'Daniel's sector, they saw three men headed back to the rear. When O'Daniel asked where they were going, one said they were going for rations. O'Daniel said that two had to go back to the lines, with only one going for the rations. Truscott called this "a dramatic underscoring of how desperate had been the need for men."[60]

The next morning, the division received a report of approximately forty German tanks hidden behind a hill about a mile southeast of Altavilla. Artillery fire dispersed the tanks, but they remained a threat to any attack aimed at retaking Altavilla.

The Allies were beginning to think the extreme danger might be ending. It appeared that the Germans might be slowly pulling back, though Ultra did not confirm this until two days later.[61] This had always been the German plan, to do as much damage as possible to the invasion—including defeating it, if opportunity arose—but to count only on defending Italy further north.

Despite two attacks against the northern section, the Germans were detected as having pulled back from the Sele and Calore river junctions. A battalion of the 179th Infantry entered this area, and easily advanced several miles. Reconnaissance pilots reported seeing no major German troop concentrations in offensive posture. Colonel Forsythe, commanding the southern section of the 36th Division front, reported to Walker an absence of activity the morning of September 16. Walker suggested to Dawley that the VI Corps assume the offensive. Walker suggested an attack on Altavilla that evening, using two battalions of the 504th Parachute Infantry with a company of tank destroyers and a company of tanks. Dawley agreed, and the 504th was ordered to seize the hills in the

Altavilla area, particularly Hill 454. All division artillery and naval gunfire was to support the advance.

The 1st and 2nd battalions of the 504th reached the hills by 8:30 A.M. of September 17, D+8, to be later reinforced by the 504th's 3rd Battalion. Fighting was very heavy, with a strong German counterattack. The Germans brought up reinforcements from below Altavilla. Within two hours, Col. Reuben H. Tucker, commanding the 504th, was suffering heavy losses and in need of assistance. VI Corps decided the area could not be held and ordered Tucker to withdraw that evening. Naval gunfire attacked Altavilla that afternoon. Tucker later informed the division that he had secured and could hold Hills 315 and 424. He was given permission to remain.

That same morning, General O'Daniel briefed his battalion commanders that, though the September 16 German counterattack had not taken place, a serious attack was still possible. The Germans had been reported as reinforcing the La Cosa front, and they were attacking heavily at Altavilla. The reinforcement report appears to have been wrong. The counterattack against the 504th appears to have been to cover withdrawal.

By the end of September 19, virtually all Allied forces were advancing. The Germans were gone. The 36th Division was withdrawn from combat on September 20, to spend the next several weeks guarding the beachhead. The division engineers were sent to help repair Naples after its capture on October 1, well behind Clark's initial schedule.

One historian has written:

> The 36th Division Infantry fought well at every attack point during the critical days from the landings on September 9 until the Germans' withdrawal . . . The view of the American high command that untried troops could meet and defeat the Wehrmacht had been vindicated. The strains of battle, however, had revealed a number of weaknesses in the staff and command organization of the division.[62]

He might have added in the VI Corps and the 5th Army, but the evaluation seems reasonable. Even the 2nd Battalion of 143rd

Regiment, mangled by the Germans at Persano, was not totally responsible for its mess. Another historian feels that it did not deploy properly.[63] The battalion called for artillery help but did not realize that their artillery observer had been killed, or did not understand the consequences. Division artillery was not able to give sufficient support, for fear of hitting American troops.

However, the battalion had been placed in a virtually untenable position. Walker had been given incorrect information about the situation the battalion would face and did not have the information checked, an error he admitted when asked after the war.

Resisted seaborne landings are one of the hardest military maneuvers to successfully pull off. The Germans had been undecided about their strategy. But never did they plan to let the Allies land without some cost. The 36th Division, and the rest of the 5th Army, had managed to gain a foothold against German troops of various levels of readiness, but generally good, well-commanded, and tougher opponents than the Italians. All commanders made errors, including, as he admitted, Walker. Not all commanders kept their jobs. Departing were Lange; LTC McShane, Division G-3 (replaced by Fred L. Walker, Jr.); and Colonel Forsythe, commander of the 142nd (transferred by Clark to 5th Army headquarters).

VI Corps commander Ernest Dawley was also fired. On September 20, John P. Lucas replaced Dawley.

Walker seemed more distressed by Clark's plans, on the fourth day of the battle, to withdraw the forces. He never thought this was necessary.

Montgomery's forces were moving too slowly to be of much help. Possibly the Germans departed when they learned of his arrival, but this had been their intention from the start. Montgomery himself wrote in his memoirs, "I have never thought we had much influence on the Salerno problem."[64] Locally gathered intelligence should have given Clark some indication of what he was facing, even without Ultra intercepts. While reading sections of Clark's war diary at the Citadel, this writer *found* only one reference to intelligence reports. The diary appears to have been kept by staff, and security would have prevented mentioning Ultra. However, Clark did not appreciate Ultra until after the Anzio landing, four months later.[65]

The September 13 German attack was stopped by unsupported American artillery and was a close call. Kesselring's chief of staff credited the still split German command arrangement, with Rommel in charge in the north, with preventing two more possibly decisive divisions from being sent to exploit the potential breakthrough of September 13.[66] Again, this split command should have been known to Clark—though having and acting on intelligence are two different things. Harold Alexander later thought the Germans lacked the strength to do any real damage if they reached the beach, aside from embarrass the Allies.[67] Adm. A. B. Cunningham, naval commander in the Mediterranean, thought that any tanks reaching the beach would have been highly vulnerable to naval gunfire.[68] As to the 36th,

> The Division has been ordered into 5th Army Reserve, and the first phase of "Avalanche" Operations was now closed for the RCT. The program now was to police the area of vehicles and the dead and to reorganize during an expected seven day period with intensive training to be held on the very ground where weak points had been observed.[69]

CHAPTER 9

San Pietro

"Each mountain had to be taken, each valley cleared, and then there were mountains ahead and still another main defense line to be broken."[1]

—Official 5th Army History, 1945

The slow slog for San Pietro was more reflective of the Allied campaign in Italy—not to mention echoing the First World War—than were most of the 36th Division's other battles. One participant described the San Pietro campaign as

Rain, rain, rain. Military operations are always being conducted in the rain. The roads are so deep that moving troops and supplies forward is a terrific job. Enemy resistance is not nearly as great as that of Mother Nature, who certainly seems to be fighting on the side of the Germans . . . This is a heartbreaking business. An advance of a few miles a day, fighting a terrible terrain and a determined enemy, building bridges, removing mines, climbing mountains. The men get punch drunk.[2]

When the 36th Division was withdrawn from action, September 21, the Allied forces were just a few miles inland from the Salerno beachhead. It was not until October 1 that the Allies captured Naples, substantially later than the September 14, D+5, capture called for in Clark's initial invasion plan. The 111th Engineers helped put the port in working order, enabling Naples to serve as a major supply center for Clark and Montgomery's armies. The 36th Division as a whole spent this time recuperating and re-equipping behind the lines at Salerno beach.

Von Vietinghoff, and his temporary replacement, Lt. Gen. Hans-Valentine Hube, withdrew the 10th Army northward, eventually heading to prepared positions on what would come to be called the Winter line.

The Allied campaign, in the face of heavy German resistance, took two months to move the roughly eighty-five miles from the Salerno port, in the beachhead center, to the Mount Cassino–based Winter line. Particularly difficult was the mid-October attack across the Volturno River, about twelve miles north of Naples, which took several days of hard fighting. About this time, the idea of another landing behind German lines first became attractive.

The Allied advance slowed down in mid-November on reaching the Winter line. The main defensive line hinged on the Garigliano River, the Rapido River, and Mount Cassino, ending with the Apennines. What can best be described as a "temporary" salient left the main defensive line at the Garigliano to include Mignano Gap, the town of San Pietro, and the nearby area. A German historian later wrote that it was "really vital to hold the line across Italy's waist."[3] In the words of an official United States military history:

> However temporary in original plan, the Winter line provided a formidable barrier to General Clark's army. It was a succession of defenses in depth, and no single key position presented an opportunity for a brilliant stroke that could break the entire system. Each mountain had to be taken, each valley cleared, and then there were mountains ahead and still another main defense line to be broken.[4]

The 36th Division re-entered the American battle line between November 15 and 17, 1943, relieving the 3rd Infantry Division. The 36th was given a five-mile stretch of line in a mountainous area around Mignano. The division would spend the next two weeks building up supplies, patrolling, and getting familiar with the sector. The weather was cold, rainy, and muddy when it wasn't snowing and frozen—a typical winter in the Italian mountains. A division veteran later described this trip:

> On 15 November 1943, the 36th Division pulled up tents

and left the apple orchard outside of Naples. We loaded into trucks for the ride near the front lines, got as close as the drivers could take us, and began climbing one of the tallest mountains I had ever seen. The trail was narrow, slippery, very rocky and it was raining at the time . . . We finally reached a clearing near the top and took a break. It was raining hard now. To make matters worse, there were bodies scattered all over the clearing.

When daylight was almost upon us we took cover in the brush and vines. We stayed hidden all day and when it was dark enough we took off again. In the early morning hours we moved into the area where the 3rd Division had vacated earlier and relieve them. It rained for the next 4 or 5 days and the foxholes were full of water.[5]

The weather made supplying front-line units extremely difficult. Supplies could go only so far by truck, or even by Italian mule. The final stage of supply had to be carried out on foot, under enemy observation and often under fire. Another division veteran described the supply expedition:

Chow call sounded and while we were eating our C.O. announced our next mission. We were to carry food, ammo and water to the 504 Parachute Battalion who were fighting the Jerries on top of Mt. Sammucro. This seemed easy enough. Take them what they needed then back to our area for more sack time. We would carry no weapons.

We moved through the village of Ceppagna about 5:30 P.M. I remember chickens, pigs and grape vines as we moved through peoples' back yards. There was still enough daylight to get a good look at the slope we had to climb. It didn't look easy, about 1205 ft. high and steep.

A guide was there to lead us as we started the climb. In about an hour we reached a plateau. Here we would get our load of rations . . . The moon was bright and the evening fall chill began to set in. In the moonlight I noticed some dead paratroopers on the ground. I realized this must be a collection point for everything going up or down. I always wondered why dog tags had a notch in them. Each dead soldier had a dog tag propping his mouth open and the notch wedged between the teeth. A gruesome sight in the moonlight.

As we started our difficult climb, a cloud settled over the mountain and the moonlight disappeared. The footing was slippery. Three steps forward, one backward. We reached the top about 1:30 A.M. unloaded and rested a while.

We started our trip down. You had to be careful in the dark. We reached bottom at break of day. Another mile and we'd be back to our bivouac area and rest. We'd been gone from sunset to sunrise.[6]

Just after lunch, the men were ordered to return to Mount Sammucro to relieve the 504th.[7]

The rest of November was just a planning stage, a "quiet" period for the 5th Army as a whole, as well as the 36th. The 1st Special Service Force, commanded by Col. Robert T. Frederick, arrived in Italy at this time and was attached to the 36th Division. This combined United States and Canadian force, with the innocuous-sounding name, was actually a highly trained combination of rangers, paratroopers, mountain troops, arctic troops, and a few other specialties. The force—with an unusual rank structure where no one was lower than corporal and the three battalions were commanded by full colonels—had originally been recruited by Frederick for a strange pet project of Winston Churchill's, using airdropped "snow bulldozers" (to come as close as possible to describing the machines) to sabotage Norwegian power stations. When this project was cancelled, the force was diverted to more practical enterprises. After service in the Aleutians, it arrived in Italy.[8]

Frederick recruited soldiers used to the outdoor life, such as lumberjacks and National Park Service forest rangers. He also searched army stockade records. All members of the force had to be able to pass extremely tough training, probably the toughest training in the army, in a variety of skills:

> Braves were not simply unusually fit, aggressive and skilled, they were also of above-average intelligence. Even more than paratroopers, they were expected to think for themselves in combat . . . Not only were they parachutists and explosive experts but they were highly trained mountain troops.[9]

The 1st Special Service Force, sometimes called the "Devil's Brigade," was later the subject of an interesting, though not fully accurate, William Holden/Cliff Robertson film.[10]

Compared with some of the other time the 36th Division

spent in Italy, most of November was routine and quiet. But, as one veteran later put it,

> The rest of the month of November was rainy, muddy, cold and quiet, though the Division lost more than 550 officers and men from the heavy enemy artillery fire and from patrol actions. Non-battle casualties brought on by the severe weather and poor living conditions were three times this amount.
> But this was the lull before the storm.[11]

★ ★ ★ ★ ★

The Liri Valley, the apparent "Road to Rome," was the focus of Allied attention. Operational goals were centered on three objectives—breaking through the Winter line, breaking into the valley, and capturing Rome before Christmas. The first Axis capital to fall would be a valuable propaganda coup. Additionally, the Allies would soon start to lose strength in Italy as men and equipment (including landing craft) were transferred back to the United Kingdom to prepare for the cross-channel invasion of France.

Gen. Harold Alexander's plan, as of early November, called for a three-phase attack. The 8th Army would attack in eastern Italy, its last offensive under Montgomery, before he was also sent to Britain to prepare for Overlord. The 8th would take Pescara, and then swing west on Highway 5 toward Rome. Though they would still be 150 miles from Rome, Alexander thought a major attack in this area would draw German troops from north and west of Rome.

The second phase would have the 5th Army attacking through Cassino into the Liri Valley. In the third phase, the 5th Army would launch an amphibious assault as close to Rome as possible.

Montgomery attacked in the east on November 20. By December 28, in the words of the Green Book,

> With his units seriously depleted and his troops extremely tired, with mountains deep in snow and roads impassable, General Montgomery brought his attack to a halt. He had driven the Germans from strong positions and had inflicted heavy casualties, but he failed to make a strategic breakthrough . . . the back door to Rome was still closed.[12]

Clark changed his plans several times in October and November, responding to changing conditions. The final plan, issued on

November 16, called for a three-phase attack. A thrust on the left of the Mignano Gap would open the attacks. The British 10 Corps and the American II Corps (including the 36th) would cooperate in the first phase. Their goal was to secure the Camino-Difensa-Maggiore mountain mass, on the left of the Mignano Gap. VI Corps would try to divert enemy attention. After 10 Corps seized Monte Camino, II Corps the other two mountains, 10 Corps would take over the entire area. II Corps would be free for the second phase, the capture of Mount Sammucro. Diversion would be provided by 10 Corps along the lower Garigliano, VI Corps to the immediate north and northwest of Cassino. Finally, VI Corps would seize the high ground behind Cassino, II Corps would attack along Highway 6 toward the Liri Valley, and 10 Corps would protect the left by crossing the Garigliano.

The 5th Army would have air support, particularly bombing of key areas on the Winter line. However, resistance was expected to be heavy, no longer the rear-guard actions common since Salerno. Intelligence had already spotted large-scale German concentrations and defensive construction behind the Garigliano and Rapido river lines. The river line was estimated by Allied intelligence to be one of the strongest natural defense positions south of Rome. The Germans were making the positions even stronger by blasting defensive positions out of solid rock.

Diversionary attacks by 10 Corps started at dusk, December 1. The 46th Division aimed at objectives near the village of Calabritto, on the lower slopes of Monte Camino. The attack stalled the next morning within 200 yards of the village. The 56th Division attacked that evening, in an effort to seize at least the southern half of Monte Camino, an effort both diversionary and in support of the 36th. The attack made excellent progress after dark that day, and before sunrise on the 3rd. British troops reached the top of Monte Camino that morning but were forced back. An attack the next morning regained the crest but again was forced off. Monte Camino was finally taken on the evening of December 6. The 46th Division then was able to take Calabritto. By December 10, the Camino hill mass was clear of German troops.

The II Corps also undertook diversionary activities. General Geoffrey Keyes, the corps commander, scheduled patrolling, artillery shelling, and bombing. The 3rd Ranger Battalion was moved to the right flank of the corps sector and told to let itself be observed making conspicuous preparations for an attack. Clark, when warned of difficulties ahead, had optimistically told Alexander not to worry. "I'll get through the Winter line all right and push the Germans out."[13] This optimistic attitude was reflected in Walker's diary entry for November 18, 1943. "Clark and Keyes visited my CP today. They have ants in their pants and want to get going now that they have a 'fresh' Division in line."[14]

Walker wrote of a visit the same day by the commander of the British 56th Division:

> Major General Templer, who commands the 56th British Division on my left flank, called with Clark today, and we discussed plans for possible operations on our fronts and ways of cooperation between our divisions. Templer was not enthusiastic about doing anything on his part, but quite enthusiastic about following up on any successes I could attain. I shall cooperate with him in any way possible, but I don't think he is going to do anything except talk.[15]

Intelligence indicated that the Germans had a battalion on the Difensa-Maggiore mass, another on Monte Lungo, and a third in the area of San Pietro. One or two were in reserve in the Mignano, and one on each flank. The German defenses had artillery support and were arranged in depth.[16]

Walker's detailed plan called for the 1st Special Service Force to advance to the top of Monte la Difensa during the night of December 2–3. The 142nd Infantry would follow part way but turn north to take Monte Maggiore. The 141st Infantry would provide covering fire between Monte Maggiore and Monte Lungo and be ready to take the latter. The 143rd was to be ready to move right into the attack on San Pietro, if the first phases went quickly.[17]

Heavy artillery fire opened the attack at 4:30 P.M., December 2. In one hour, 22,000 rounds of artillery hit Monte la Difensa. "Not since El Alamein had such an array of artillery signalled an attack."[18] An estimated 75,000 shells were fired by corps artillery during the first forty-eight hours, in support of the 36th, including the first sustained combat use of the new eight-inch howitzer. The German troops, in foxholes, were apparently relatively unharmed by the shelling but were cut off from outside support and supply.

Artillery fire ended with darkness, and the 1st Special Service Force moved out. Two of its six battalions had been assigned the assault duties. The men climbed all night up the steep heights of Hill 960 of Monte la Difensa, without manmade trails. They reached the top by daylight. During the day on December 3, the lead battalion advanced toward Monte la Remetanea, Hill 907, beyond Hill 960. Units of the 142nd Infantry, however, were heavily shelled on the 3rd, following the special service force partway up Hill 960 before turning to Monte Maggiore. The growing German resistance to the special service force, and its casualties on Monte la Difensa, prompted Frederick to ask that his force be relieved. This was impossible before the force was able to expel all German troops on December 8.[19]

As the special service force was attacking Monte la Difensa, the 142nd went after and captured Monte Maggiore. Conditions were just as bad for them as for the special service force. Even after Maggiore was taken and held, units on supply duty required a twelve-hour round trip.[20] Conditions were also bad for the Germans, with supply in even worse state. Fighting was pretty much over by December 8, when the 142nd Infantry relieved the 1st Special Service Force on Monte la Difensa. Two days later, the 56th Division, which had performed better than Walker expected in its attack, relieved both American units. The VI Corps had not been able to make good progress on the right. The II Corps was not able to take Monte Lungo or San Pietro.

The 3rd Ranger Battalion, also attached to the 36th, had begun the attack on San Pietro during the night of November 29, under orders from General Keyes, to determine German strength in the town. If the battalion met heavy resistance, it was to withdraw. If they met light resistance, they were to take the town. Their initial approach got them pinned down by heavy fire throughout the next day. The battalion withdrew, having learned little about the defenses. Walker criticized the general idea, a reconnaissance in force, and its execution by the battalion commander. Walker, as a division staff officer later recalled, "assumed that the Battalion Commander knew enough not to get his outfit out into the open and get it shot up."[21] There is no indication in Walker's diary that he questioned the order or that he made clear to the battalion commander that what appears to have been a direct, frontal attack was a bad idea.

Ranger patrols got closer to San Pietro on December 2 without problems. Keyes informed Walker that the 36th Division should be able to get through San Pietro without pause and then advance to San Vittore. A ranger patrol on the night of December 4 did not provoke enemy reaction. However, a 143rd Infantry patrol reported the town full of enemy soldiers. The strength of the enemy was unclear when operations continued.

The 1st Battalion of the 143rd Infantry reached the top of Monte Sammucro at dawn, December 8. A few Germans were blasted out of emplacements, and the battalion controlled the mountain. The battalion held the hill, spending several days fighting off German counterattacks, inflicting heavy losses on the Germans. The first enemy attack, though, on December 8, almost recaptured the hill. The battalion

> fiercely clung to the granite crags of 1205 despite fanatical counterattacks by an enemy determined to hold the Winter line. In one day alone nine counterattacks were launched and beaten back, and the swirling snow banked around the bodies of once proud "supermen" of Hitler.[22]

Just after the 1st Battalion attack, the 3rd Ranger Battalion turned right at the foot of Mount Sammucro and climbed Hill 950. Spotted at less than 1,000 yards by two German machine gun outposts, the rangers were able to take the hill at daybreak. Later that morning, three heavy volleys of German artillery forced the rangers off the hill, to dig in on a small hill nearby. On December 9, aided by mortars lugged up the hill, the rangers were able to retake and hold Hill 950.

The 2nd and 3rd battalions of the 143rd began their direct attack on San Pietro at 6:30 A.M. on December 8. "Moving from the southeast," one historian has put it, "both units immediately came under murderous fire from the brilliantly conceived defenses in front of the town."[23] Two days of attacks achieved little against the very strong enemy defenses. At one point during the attacks, one officer and platoon leader, 2nd Lt. Richard A. Stewart, from Minneapolis[24] (fewer and fewer T-Patchers were from Texas), tried to jump barbed wire and throw a grenade through a pillbox slit. Stewart was unsuccessful and was killed in the attempt.[25] Efforts against the town temporarily ceased on December 11. San Pietro was the

36th's first encounter with built-up areas fortified by the Germans. Elements of the division advanced only a few hundred yards before the attack was halted by intense automatic fire, barbed wire, and mines.

A fourth attack that morning was directed against Monte Lungo, another key peak in the area. The 1st Italian Motorized Group, about 5,500 men, was assigned to work with Walker in the attack on the San Pietro area. (With the special service force, the rangers, the 504th Paratroop Battalion, and the Italians, Walker had virtually his own corps.) The Italian king, and Premier Badoglio, had offered to help fight the Germans. The Allied command felt it was important for Italian units to help liberate Rome and participate in the battles leading to the fall of Rome.[26] Keyes told Walker he wanted the Italian unit to succeed in its first combat action. Monte Lungo seemed lightly defended, and a good target.[27]

The Italian force relieved a battalion of the 141st Infantry near Monte Lungo the night of December 7. The next day, the Italians would attack. The commanders ordered no ground reconnaissance, nor did they send out combat patrols. The one patrol sent out at all, a security patrol to the flank, failed to return. The morning of the 8th, after a thirty-minute barrage which they thought knocked out all resistance, in a heavy mist which acted as a smoke screen, the troops moved out two battalions abreast. Things went well, briefly, for this dangerously nondispersed formation. Then the lead battalions began to take mortar and machine gun fire, which stopped the advance. The Italians became ineffective but managed to stay in position.

Despite being asked for ammunition needs by General Keyes, the Italian commander had underestimated his men's needs. Other division and corps batteries were either out of range or could not fire due to the risk of hitting the Italian troops. By noon, Walker let the Italian commander withdraw what was left of his unit. The commander of the 141st helped restore order and set up defensive positions.

The attacks were very difficult, against well-prepared German positions taking full advantage of natural terrain. The rangers had been thrown off their hill, only retaking it with additional weaponry. The 504th Parachute Infantry helped the 143rd hold Monte Sammucro. Walker also had to tolerate an unwelcome diversion:

During these battles, I was visited by all the brass hats in the theater. Crown Prince Umberto, Assistant Secretary of War McCloy, Clark, Keyes . . . Alexander, Templer, etc. etc. etc. I had a difficult time attending to tactical demands with all these tourists coming and going. Photographers and newsmen were numerous, also. All of them must be received courteously and made to feel welcome. However, they are in the way.[28]

The second effort to take San Pietro and Monte Lungo started on December 1, 1943. A key element in the attack was a December 9 visit to Walker by Gen. Donald W. Brann, 5th Army G-3 (operations officer). Brann told Walker that Clark wanted to make more use of armor, in particular Ernest Harmon's 1st Armored Division. Clark had asked for and received the division and was somewhat embarrassed at not finding any use for the unit. Was there any way at San Pietro that these tanks could be used? Walker did not think so offhand. However, since General Keyes had also expressed an interest in using armor, Walker would try. He asked the tank commander to prepare a company to attack San Pietro with the 143rd Infantry.

Walker was still stuck with the Italian 1st Motorized Group. He was told by Clark that in view of their demoralization to give them an objective they "cannot fail to accomplish."[29] Capturing the eastern part of Monte Lungo, after the 142nd had taken the rest of the mountain, filled this requirement.

Walker was not thrilled with the final attack plan. "Just finished giving orders for an attack to be made on the 15th. The plan definitely does not suit me, but I can see nothing better to do under the circumstances."[30]

The second attack would be large-scale and coordinated, attacking all three of the major objectives—San Pietro, Monte Lungo, and San Vittore. On the night of December 14, the elements of the 504th Paratroops and the 143rd Infantry on Monte Sammucro would attack three smaller peaks about a mile west of Hill 1250. This would outflank San Pietro and threaten the escape route from Monte Lungo. They would also be at a better jumping-off point for the attack on San Vittore.

The main attack would start at noon on December 15. The tank company would approach San Pietro from the east, supported by a battalion of the 141st from the south. In the evening, the entire 142nd Regiment would attack Monte Lungo from the west. After daylight on the 16th, the Italians would be sent against the eastern part of Monte Lungo, hopefully just to "mop up."

Very early on December 15, the 1st Battalion, 143rd Infantry, advanced toward two of its hill objectives. The 504th Paratroop advanced toward the third hill. Unfortunately, the lack of concealing vegetation on the upper portions of Monte Sammucro, and the bright moonlight, gave German artillery and machine gunners excellent visibility. Both units took heavy fire and casualties. By daylight, neither had been able to take their objectives.

The attack on San Pietro began on schedule on December 15. F and G companies of the 2nd Battalion, 141st Infantry, led the assault on the fortified village. Both companies made little initial progress against very heavy German fire. Casualties were heavy. The actions of the commander of Company G, Charles M. Beacham, reflect the fighting. In the words of the 36th Division awards and decorations officer, who wrote up the successful citation for Beacham's Distinguished Service Cross:

> Wounded in the face by shell fragments, he continued to lead the attack. The radio operator was a casualty, and Captain Beacham took up the instrument until blood from his wounds seeped into the radio and impaired its operations. Still refusing aid, he moved over the fireswept terrain, reorganizing his company and preparing to resume the attack. Another shell sprayed his entire body with fragments. Weakened from the loss of blood, he turned the company over to his executive. He refused to be carried from the field, but aided another wounded officer and two wounded men through the heavy fire to the battalion aid station.[31]

What was left of the companies was finally able to build up a firing line behind a stone wall closer to the village, but not close enough for the American infantry fire to make any real impact. They could not call in artillery fire. They were unable to identify precise German locations, and American tank crew members had taken refuge in the village.

Part of the sixteen-tank company attacked down the Ceppagna road toward San Pietro. The rest of the tanks would come down a well-defined and accessible mountain trail from

Monte Sammucro. The tankers made things even more complicated, with half of the "trail" platoon to provide covering fire, the other half to move below the town to cut off escape. A company of infantry accompanied the tanks, as did what was called a Valentine tank, a bridging device. This was expected to be needed, as the Germans would have mined, if not destroyed, any bridges.

The tanks moved out at noon, two tanks, the Valentine, and then the rest of the fourteen tanks. The road itself was tough, with a straight drop to the left and a stone wall to the right making it necessary to stay on the road. The move up the trail proved not to be viable, as the lead tank had to cut a path to San Pietro across terraces. For three hours the tank worked its way to the village, destroying several machine gun nests and a German command post. At the end of the afternoon, in sight of San Pietro, the lead tank got word to return.

The second tank was damaged after crossing the fifteen-foot culvert, surprisingly still standing. The crew stayed inside, shooting at targets of opportunity. The Valentine crew pulled out of line at the culvert, to let the remaining fourteen tanks pass. Three more tanks passed the Valentine, the culvert, and the disabled tank. The first started to turn up another trail, and then noticed it was blocked by a destroyed German tank. The commander radioed back to the company commander, who instructed the remaining tanks to stay on the road. Each of those three, after crossing a thirty-five-foot bridge, were destroyed by German antitank fire.

The next two tanks struck mines trying to bypass the Valentine. The next tank tried to push the tank in front off the road and struck a mine. The tank after that, unsuccessful in trying to push the two preceding, was unable to climb a terrace. The remaining tanks were ordered to follow the path the lead tank had taken. The first flipped over on its side. The next threw a track. The third slipped off the embankment on the left. Another collided with a disabled tank. Two of the last three, trying to find different paths to San Pietro, threw tracks. When the company commander withdrew his tanks, four of the sixteen Sherman tanks returned, with nine more crews and some of the Valentine crew. The bridging tank itself was blocked.

Company E of the 143rd Infantry was also unable to reach the town. The 1st Battalion, 143rd Infantry, was too heavily engaged on

Mount Sammucro to attack San Pietro from the rear. Similar problems emerged on the right flank of the division, on the attack coming down Monte Sammucro. The attack down the side of the mountain, by the 143rd's 2nd and 3rd battalions, was also heavily cut up by German fire.

The 2nd Battalion of the 141st Infantry, stuck at the wall south of San Pietro, was pressed to attack that night. Some men made it into the town, but the survivors were quickly forced back to the stone wall. A dawn attack on December 16 was not successful. The same was true of the attack of the 2nd Battalion of the 143rd down Monte Sammucro. This element of the 141st was withdrawn to Monte Rotondo that afternoon. The next day the 1st Battalion of the 141st Infantry replaced the 143rd element on the mountain.

During the night of December 15, the 2nd Battalion of the 142nd and the 3rd Battalion of the 143rd moved westward across Highway 6 and around the southern end of Monte Lungo. They climbed Monte Lungo's western slope, chased out elements of the defending 29th Panzer Grenadier Division, and reached the top of the mountain by dawn on December 16. By midmorning, virtually all the mountain was in American hands. The Italians captured what remained. That evening, the Germans evacuated San Pietro.

They still held San Vittore and sections of Mount Sammucro. Mount Sammucro was captured in a Christmas Eve attack by the 1st Special Service Force, the 504th Paratroop, and parts of the 141st. One day later, Monte Sammucro, and all the high ground overlooking San Vittore, was clear. The 36th was relieved by the end of the year, for a rest. The Germans held San Vittore until January 6, 1944.

One veteran of the San Vittore fighting wrote later of having to help recover bodies, on Christmas Day, for graves registration. He was not in a mood to eat turkey dinner that night. "A couple of weeks later we moved to San Angelo D'Alife which is in the area of the Rapido River."[32]

CHAPTER 10

The Rapido River

"When I saw my regimental commander standing with tears in his eyes as we moved up to start the crossing, I knew something was wrong..."[1]
—Capt. Zerk O. Robertson, 143rd Infantry, Company L, 36th Division, 1946

"... the action to which the Thirty-sixth Division was committed was a necessary one and ... General Clark exercised sound judgement in planning it and in ordering it..."[2]
—Secretary of War Robert C. Patterson, 1945

"Nothing is too tough for Texans. And that is why they would prefer to forget the battle at the Rapido. It makes them uneasy. Like the Alamo, the Rapido will always be connected to Texas and its traditions ... [But] Texas—and the other Americans who fought on the Rapido—need have no shame."[3]
—Historian Martin Blumenson, 1970

The Rapido crossing was the offspring of the Anzio landings. The crossing was intended as a diversion for the Anzio strike directly at Rome. Anzio was designed as a way around the stalemate of the Italian campaign at the Gustav line, about eighty miles south of Rome. The Rapido crossing can be described, therefore, as a straight forward push designed as diversion from a flanking maneuver made necessary because the previous straight forward pushing was not working. As one historian put it, "For the men who had to break through it, the Gustav line was a Calvary, a grisly martyrdom brought upon them not by the failings of their own but by the mistakes of others."[4]

"There is no doubt that the stagnation of the whole campaign on the Italian front is becoming scandalous,"[5] is the way Winston Churchill described the Italian situation in late 1943. This stalemate had not been expected, even though Ultra intercepts of German communications gave the Allied high command indications that the German withdrawal had ended, that they intended to hold at the Winter line as long as they could.[6] The slow Allied progress made it unfeasible to stage an end run around the German lines, with a landing at Anzio, to "finish off" German resistance in the Liri Valley. On December 18, at Mark Clark's suggestion,[7] the proposed landing was cancelled. Seven days later, at Winston Churchill's urging, Clark was informed that the Anzio landings should be carried out sometime in the last week in January.[8]

Churchill ensured that sufficient landing craft would be held in Italy until February 3, 1944. The Anzio landings would be carried out sometime in the last week in January.[9] This would allow 5th Army commander Mark Clark just over a week to bring in troops and supplies before losing his landing craft. There was serious doubt as to whether the 5th Army could break through the Gustav line in time to link up with the beachhead. With Albert Kesselring commanding German forces, any surprise achieved by Anzio was not likely to last long.

Clark liked the idea of the Anzio landing itself but had strong doubts about the constraints under which he had to operate. Other Americans had even stronger doubts about the feasibility of the operation. VI Corps commander John Lucas wrote of the plan, apparently after his corps had been assigned to the landings, "This whole affair has a strong odor of Gallipoli [an unsuccessful amphibious landing Churchill ran in World War I] and apparently the same amateur IS on the coach's bench."[10] The planning went on, however.

Clark's war diary describes him as supporting the plan sufficiently to take units from the Anvil invasion of southern France, but, according to Clark's diary, kept by his aides, in effect, a pistol was being held at his head because he was told, totally to his surprise, that if he was to engage in SHINGLE it must be done with inadequate landing craft, that the craft would be available for only two days after the landing, and that no resupply or reinforcement troops would be available. In effect, therefore, he was asked to land two divisions at a point where a juncture with the balance of Fifth Army was impossible for a long period, thereby leaving the two divisions in question out on a very long limb.[11]

Clark began to look for diversions, ways to attract German troops away from the beachhead. Clark soon thought he found one. He wrote General Alexander that

> I intend to attack in greatest possible strength in Liri Valley several days in advance of SHINGLE [Anzio] with the object of drawing maximum number of enemy reserves to that front and fixing them there. In that way and in that way only can the SHINGLE force exercise a decisive influence on the operation to capture Rome.[12]

The same 5th Army history in which Clark's letter is quoted would cite the attack into the Liri Valley, the Rapido River crossing, as being in "accordance with the 15th Army Group directive to maintain maximum pressure against the Gustav line."[13] Ultra intercepts of German communications, however, indicated no real effort by the Germans to put more troops in the general area of the Anzio invasion, aide from a battalion of forty tanks assigned near the southern portion of the planned beach.[14] Yet Clark still felt the need for his diversion from a diversion.

The Liri Valley advance, in itself, would have several phases. The first step was to get to the line of the Garigliano and Rapido rivers. By January 13, the river line had been reached. The river lines, however, were integral parts of the main lines of German defenses. The tributary of the Garigliano River, the Gari, called the "Rapido" river in reports and histories, was a particularly strong defensive area. "The ideal site for any type of river crossings has never been found, but the Rapido River was used by the Italian Army schools to represent the perfect defense position,"[15] as a division officer later put things. There was some dispute about the precise strength of the German position. However, a first-rate German division, the 15th

Panzer Grenadiers, was defending the position. According to testimony at immediate postwar congressional hearings,

> The Germans had prepared a defensive position on the west bank of the river strongly fortified and thoroughly organized. This would have been a most difficult position to attack successfully even if there had been no unfordable river as an obstacle along its entire front.[16]

The right flank of Clark's effort, an attack by the French Expeditionary Corps, made progress for several days but was stopped at the river line. The British 10 Corps was supposed to make two crossings, one near the mouth of the Garigliano, one by 46 Division, to cover the immediate left flank of the attack directly across the Rapido.

No one really liked attacking directly across the Rapido, not even Clark.[17] Planning, however, proceeded for a one-division attack. Diversion from Anzio was not the only reason for this attack. If the attacking division could break through into the Liri Valley, the 1st Armored Division was available for immediate followup. After the war, II Corps commander Geoffrey Keyes expressed the view that this was a significant motivation for Clark's favoring the direct Rapido crossing. Clark had "more or less scrounged the division out of North Africa and thought he had to put it to some use."[18] Clark had also wanted this armored division used at San Pietro, for much the same reasons.

For much of December, Lucian Truscott, Jr.'s, 3rd Division was assigned to make the attack. Truscott had expressed doubt about crossing unless the heights above the area were in friendly hands. He is quoted in one history of the Anzio landing as having described the Rapido crossing plan as "a terrible idea that would cost most of the division if it succeeded."[19] By early January, Truscott's 3rd Division had been assigned to the Anzio landings.

Americans were not the only ones to criticize the plan for a frontal assault across the Rapido. After the war, the overall German commander in Italy, Albert Kesselring, was highly critical of the plan.[20] Gen. F. von Senger u. Etterlin, commanding the 14th Panzer Corps at the Rapido, discussed the Rapido, in interviews partially held at lunch with General Keyes. According to the summary of the discussion, von Senger felt that

The American attack across the Rapido ran in to the strongest positions in the German line. The 15th Panzer Grenadier Division, commanded by General Rodt, was *von Senger's best division.* [Underlined in copy consulted]. Machine gun positions were placed along and back from the river in a zig-zag pattern with interlocking fields of fire. Thus, even when the American troops were able to land . . . they were bound to be caught in a cross fire by the positions on either side and directly in front of them. When they got across the river they went ashore at the base of a triangle at whose three points was sited a machine gun position.[21]

In December, Keyes pointed out to Clark problems with the Rapido attack. A bridgehead at San Angelo, just across the Rapido, would be under close observation (and presumably fire) of the enemy at Cassino. Armor would be committed over poor roads and would soon be out of range of friendly artillery. Keyes appears not to have felt actually crossing the river to be a particular problem. However, he did offer an alternative solution. He suggested that both the II Corps and the British 10 Corps attack across the Garigliano River, and mountains, and move into the Liri Valley from the south. Richard McCreery, commanding 10 Corps, objected, saying that his troops were neither trained nor well equipped for mountain fighting. Fifth Army Chief of Staff Alfred Gruenther accepted McCreery's objections, and Clark later agreed with Gruenther. Keyes' plan was dropped. However, Keyes was later told by McCreery that the plan might have worked.[22]

Gen. Ernest Harmon, commander of the 1st Armored Division, saw no problem in breaking out of a Rapido bridgehead and rapidly advancing through the Liri Valley. He was eager to get his division into action.[23] Once in the valley, Harmon would be on the most direct route to the Anzio beachhead. Clark's desire to use armor appears to be the reason why an alternative version of Keyes' plan, attack with just American troops through the mountains, was not considered, nor was Walker's idea for an attack through the mountains north of Cassino.[24]

★ ★ ★ ★ ★

Does the commander's attitude influence the results of an operation? Just after the war, General Walker was interviewed by a staff officer. The interview summary states that "The division received orders for the attack approximately 2½ days in advance. Therefore there was not sufficient time to prepare for a major attack against a determined [defense]."[25] However, Walker had known about the possibility of such an attack for several weeks. As early as January 6, 1944, he noted in his diary, "Although we are in a rest area, I directed the Division Engineer, Oran Stoval, to prepare a topographical study of the town of Cassino, the Rapido River, and the area on this side of the river, just in case we may need that information."[26]

Walker's professional judgement was that the crossing, as ordered, would not work. He wrote, before the 36th Division received its orders, that:

> I have been giving a lot of thought to a possible plan for crossing the Rapido River just in case we have to do it. My experience at the Marne River against a German crossing in WWI taught me to appreciate the difficulties for which preparation must be made. There are rumors that some Division will have to cross it, but nothing official.[27]

Though not ordered to make the attack until January 16, Walker had expressed his views on the difficulties and the low chance of success earlier. Walker made several entries in his diary on potential problems, and on expressing these problems to Clark and Keyes. Keyes also did not like the plan but was doing his best to carry out instructions. A 36th Division staff officer who had a chance to work closely with Walker wrote that:

> Walker never filed a formal written complaint to Fifth Army about the orders he received about the Rapido. But certainly he objected verbally to Mark Clark and Geoffrey Keyes. He had strong personal feelings about his orders. He told me he spoke as strongly as a subordinate general can speak to his superiors.[28]

Walker thought he was protesting as much as he could within the military structure. However, according to the Green Book, "The extent of General Walker's opposition to a crossing of the Rapido was never apparent to his superiors."[29] "About Gen. Walker's objections to the crossing. . . . Gen. Keyes [at the meeting with von Senger] says that when the plan was first suggested, Walker did protest. However, on the eve of the assault he told Keyes that he was confident his division could do the job."[30] Walker later described one meeting with Keyes, stating, "I explained my plan of attack, pointed out the difficulties in general terms, and tried to be optimistic in spite of my real feelings."[31] Describing what was apparently the same meeting, a 36th Division regimental commander states that Walker, though presenting the difficulties, did not say that crossing was not feasible.[32]

Walker was not sufficiently conveying his feelings to his superiors, though it can be questioned whether Clark could have been talked out of the plan even with a Walker resignation. The combined Anzio assault, diversions, and Rapido crossing were based on an assumption that the Germans could be driven out of the Winter line. However, the Germans had been told by Hitler to hold at all costs. By this time Clark should have known the high degree of skill and determination Kesselring would bring to this task. Clark's headquarters remained optimistic, though.

A more serious error on Walker's part may have been *unintentionally* conveying his doubts to his men.[33] A journalist later wrote about another general that "Knowing that doubts spread through an army like cholera, he kept his qualms admirably concealed."[34] Walker would not have wanted to do so, and in fact noted this concern in his journal. "We are undertaking the impossible, but I shall keep it to myself; however, my staff and regimental and battalion commanders are no fools."[35]

Nonetheless, Walker could have helped convey an air of defeat. Clark later suggested that Walker's attitude might have hurt the attack.

Walker wrote in his journal the day of the attack:

> Tonight the 36th Division will attempt to cross the Rapido River opposite San Angelo. We have done everything we can, but I do not now see how we can succeed. On top of everything else, Army has rushed us into this mission too fast. We should have had ten days instead of three to get ready. But even with adequate time, the crossing is still dominated by heights on both sides of the valley where German artillery observers are ready to bring down heavy concentrations on our men. The river is the principal obstacle of the German main line of resistance.
>
> I do not know of a single case in military history where an

attempt to cross an unfordable river that is incorporated into the enemy's main line of resistance has succeeded; so according to history, we may not succeed. The mission should never have been assigned to any troops and, especially, when both flanks will be exposed when we get across . . .

Clark talked to me on the phone this evening and sends his best wishes; said he was worried about our success. I offered no encouragement. I think he is worried over the fact that he made an unwise tactical decision when he ordered troops to cross the Rapido River under these adverse conditions. However, if we get some breaks, we may succeed. But they will have to be in the nature of miracles.[36]

Walker's concern was shared by others. An observer, a division engineer, noted that "The infantrymen I talked with didn't like night fighting and lacked confidence in their ability to knock out the enemy in a night engagement."[37] According to the executive officer of the 143rd Infantry, "it was common knowledge in the battalions and at regimental headquarters that the units would fail in the crossing operations because the defenses on the far side of the river were too strong for infantrymen to attack and live."[38]

Commanders should never project pessimism. Did Walker make the difference at the Rapido?[39]

★ ★ ★ ★ ★

French efforts on the right had stopped at the river line. The British 46 Division at Sant Ambrosio, on the left, not only attacked late but failed to cross the river. Keyes thought the failure was due to British unwillingness to launch attacks in force until they saw an effort was working. Clark wrote that the 46's failure was

> quite a blow. I was fearful that [the British division commander] had a mental reservation as to the possibilities of success of his operation . . . I flew to II Corps, feeling that it was necessary to discuss with General Keyes the results of this failure, for although the 46th effort would not entirely have protected his [36th's divisions] left flank, its failure would leave it entirely uncovered during his crossing of the Rapido River . . . I sent General Gruenther by plane to see McCreery, who feels that the . . . attack of the 36th Division has little chance of success on account of the heavy defensive position of the enemy

west of the Rapido. I maintain that it is essential that I make the attack fully expecting heavy losses in order to hold all the troops on my front and draw more to it . . . ."[40]

Clark then made a significant error, looking only at the 46 Division's failure, neglecting to examine the left flank of the British attack. According to a historian usually supportive of Clark, "What General Clark did not know was that his attack had succeeded."[41] The British 56 Division had established a bridgehead across the Garigliano River and was holding against German attacks. The local German commander saw problems, and got Kesselring to send him the 29th and 90th panzer divisions. German troops *were* being diverted from possible dispatch to Anzio.[42]

Clark had access to Ultra dispatches. However, there is no indication he took advantage of this information, or more conventional forms of intelligence, to consider the opportunities the British success and German reaction presented. It has been said that "The best information is useless if you don't believe it."[43] Clark did not really begin to appreciate, to believe, Ultra until after the Anzio landings.[44] He was still determined to get into the Liri Valley across the Rapido River, to open the valley to his armor as well as divert troops from Anzio.

★ ★ ★ ★ ★

"No general should leave his flanks exposed."[45] However, with the failure of the British 46 Division, the 36th would attack with its flanks unprotected. The division would prepare as well as it could, but even the terrain on the division's side of the river seemed to be working for the Germans. There was little cover for several miles from the Rapido, leaving any troops quite easily observed from German positions on Monte Cassino. What was not seen, due to night or protective smoke screens—which themselves were too strong, hampering supporting American fire—was heard by the Germans, provoking artillery fire.[46] Night and morning fog, characteristic of the immediate area, proved a mixed blessing. The fog helped conceal attacking forces in the night and morning but also made navigation more difficult. Judging from the damage it was causing, even firing blind or by sound, German artillery seemed pretargeted.

The 141st narrative of operations notes that "Hostile mortar and artillery fire was received from the time the troops left the assembly area until they reached the rivers [sic] edge. It was effective as the fragments punctured many of the boats and bridge equipment before they reached the waters [sic] edge."[47] Men and equipment had to start from the nearest concealed jumping-off points, two to five miles away from the river, behind Mount Trocchio. They would have to make do with a lack of roads over low ground. The ground was still muddy, despite ten days without major rains,[48] during the Italian winter rainy season.

Mines, which the Germans were expert at planting, could only be cleared at night, each engineer probing the ground very carefully with a metal probe. Teams would work for fifteen minutes before switching, leaving the probes in the ground to mark where they had stopped. Their replacements would be very careful not to step out of cleared areas. Cleared paths were marked by white tape. The 141st Regiment's narrative of operations for this period describes a problem:

> Infantry elements moving to the river through lanes that it had been reported as swept, encountered a considerable number of anti-personnel mines. These anti-personnel mines not only wounded a number of men and tended to disorganize the advance but they also destroyed the rubber boats.[49]

The 142nd Regiment would stand in reserve, to help the 1st Armored Division exploit an expected breakthrough, while the 141st and 143rd made the actual crossing. The two assault regiments were in place the night of the 19th–20th, with the attack scheduled for 8:00 P.M. on the 20th.

After dark on January 19, Lt. Col. Aaron W. Wyatt, commander of the 141st Regiment, ordered the 1st and 3rd battalions to move to the jump-off point, between Mount Trocchio and the lower hill at La Pieta. Companies A, B, and C of the 1st Battalion were to cross in rubber boats at 8:00 P.M. the evening of the 20th. The battalion's objective was the high ground about 1,000 yards inland, due west of the crossing site. The battalion would be crossing near an "S" bend in the river, an advantageous defensive position which enabled the Germans to fire on the Americans' flank. Boats would be waiting near the jump-off point, but engineers would have to guide the attacking units. The boats would have to be carried to the river. Unfortunately, also, the boat dumps were within range of enemy artillery. Twenty-nine were apparently destroyed by one artillery volley, as the dump could be seen from German positions on Monte Cassino.[50]

The companies adjusted their attack plan to make do with fewer boats, but trouble continued. The engineer guides were late getting to the boat dumps. The less than fully cleared "safe" paths proved a problem. Company A's guides got lost, stranding the company in a minefield. Company B was led to battalion headquarters, rather than to its planned crossing point. The initial half-hour American artillery barrage was ordered to continue when, at H-hour, the regimental commander received word of the delays. As the 141st narrative put it,

> At 2100 a report was received that all three of the 1st Battalion rifle companies had crossed. This report was later found to be entirely incorrect, as only small numbers of A and B companies had crossed and practically none of C company. A few boat loads of men did succeed in crossing and they proceeded to clean out opposition on the opposite bank thus making it possible for the others to cross at a later hour.[51]

Company B tried to turn in the dark, to head to its intended crossing point. The Americans continued to struggle to the river with the heavy crossing boats. A sergeant present at the crossing later stated that "The biggest mistake of the operation was carrying the boats and bridge equipment."[52] An officer described the effects of the dark and the fog: "You could hardly see your hand in front of your face . . . struggling down this sunken road . . ."[53] The men were not struggling very quietly. As Kesselring later put it, "The noise was so great that as the troops approached the river bank, our artillery was immediately on the qui vive."[54]

Company B lost so many men in the approach to the river that it no longer had enough men to carry its boats. The boats had to be dragged. Some of the boats had holes in them, invisible in the dark. These boats quickly sank, carrying about twenty of the heavily loaded American soldiers with them. The 141st narrative states:

> The attempts to put A, B and C companies across the river

were negative both by hostile machine gun fire directed from the river at points close to the opposite bank and from points in the hostile main positions from which long range fire could enfilade certain sections of the river. Hostile artillery fire, mortar and Nebelwerfer (rocket) fire was continuous in a slow cadence throughout the night. German machine pistol fire was continuous. Small arms fire from the opposite bank of the river increased the number of boat casualties and the boat situation was becoming serious."[55]

Company A was eventually led out of the minefield. Some of the men were able to cross with a few rifle platoons of the 3rd Battalion, on the one footbridge successfully installed, only at 4:00 A.M. the morning of June 21. Most of its survivors ended up as casualties on the other side of the river, with the few men who crossed trying to hang on through the day against heavy German fire.

One of the sources consulted for this book[56] indicates that there is some question as to whether any elements of the 3rd Battalion were able to cross the first night. "Narrative of Operations of 141st Infantry in the Crossing of the RAPIDO river on January 20 to 23, 1944," attached to the divisional narrative of operations for Rapido, may contradict other official records in that on the fourth page it states that at roughly 5:30 A.M. "the regimental commander gave orders that all elements on this side of the river that had already crossed should be withdrawn to their original assembly area." One historian seems to feel that some of the 3rd Battalion did cross. Walker, however, writes that "The 3rd Battalion of the 141st did not cross because the only footbridge was destroyed by continuous German artillery and mortar fire even before all of the 1st Battalion was across."[57] This question reflects the battle's great confusion. Clearly, though, the 141st's attack had not gone well. The few men across the river would have trouble hanging on until the next night.

Walker's diary describes the similar difficulties faced by the 143rd Regiment—in its efforts to cross the Rapido:

> In the left (south) sector, the 143rd Infantry planned to cross at two separate points. The leading elements procured their boats, improvised footbridges, and were led to the river by guides from the 19th Engineers.

The 1st Battalion, 143rd, arrived at the river on time. En route it received rather light artillery fire, but there was some confusion due to the clumsy loads and mud and because of the darkness and fog. It began to launch the rubber boats immediately upon arrival at the river and started Company C across. The artillery and mortar fire which had been falling on the crossing area was increased, and by the time Company C was across, all the boats were destroyed or lost. Major David M. Frazior, the Battalion Commander, sent for more boats and most of the Battalion was across by 6:00 A.M. It was unable to make any important progress. It had to defend itself from the beginning and was forced into a pocket with its back to the river. After daylight, it was fired on continually at short range by German tanks and self-propelled guns in hull down positions, and by small arms and mortars. It was completely overwhelmed by this fire and could not be supported by other troops on our side. Being in the open with little or no protective cover, it was about to be wiped out when the battalion commander ordered what was left of it back to our side of the river. It recrossed on the one remaining, but damaged, footbridge.[58]

One of the members of the 143rd Infantry later recalled the second night of attempts to cross the Rapido:

The next night we WENT. It was bitterly cold, and the closer we got to the River, the colder it got. We couldn't move fast as visibility was about zero, so this made it worse to try to keep warm.

We went down a little horse and wagon road, and on the right side was an embankment about six feet high. We had already picked up our rubber boats, so we scraped against the side as we headed toward the river.

[Across the river, the writer and the second company scout] took off our equipment and start working on the hole. Thank God for the mist and fog from the River, and our smoke screen ... We were about three feet deep when the Germans spotted us, then all hell broke loose. "Screaming meemies," mortars, artillery fire, and machinegun fire about six to eight inches above ground hit us. Our equipment laying outside was blown to hell, the dirt we were piling up was blown back into the hole. We still didn't know how bad off we were because when they stopped firing for a few minutes, we would stand up and try to see what was doing on. All we could see were GIs being lined up and taken prisoner . . .

I made it back to our side, and to the road we came down on the night before. The piles of bodies were gone. I got back to our bivouac area out of artillery range. I laid down completely exhausted, and felt like I had turned into an old man overnight. I know I was never the same person again . . .[59]

However, perhaps because the first night's crossing had driven the Germans back, nearly all the men were across the river by daylight on the 22nd. The 3rd battalion of the 143rd was across the river by 6:30 P.M. but had taken heavy casualties. The 1st battalion of the 143rd was fully across several hours later.

The 141st had started planning the second day's attack at 7:15 the morning of the 21st. It was also initially scheduled to attack at 2:00 P.M., 1400 hours. Boats failed to arrive on time. "The boat and crossing equipment situation was the key to activity throughout the afternoon,"[60] causing several postponements. General Keyes visited the regimental command post at about 3:30. He was

> surprised and displeased to find that preparations for the attack were at a standstill because of the boats . . . The Commanding General, 36th Division, was kept informed of the fact that the boats had not yet arrived and he gave instructions that every effort be made to find them and that the attack be launched just as soon as it was possible to do so.[61]

By 5:30 the next morning, the 141st was across. This was a mixed blessing, as both regiments were under steady German artillery and infantry attack. The Germans began to drive them back to and across the river, inflicting heavy casualties. But the damage was not all one way.

S.Sgt. Thomas E. McCall commanded the machine gun section in Company F of the 143rd Infantry Regiment. The night of January 21 his section took heavy fire crossing the footbridge American engineers had managed to build. They were not able to set up under proper cover. Setting up out in the open led to the guns soon being knocked out by German fire. McCall found all his men either dead or wounded. He grabbed one of the machine guns, which still worked. Supporting it against his hip, he fired on the run to within thirty yards of a German machine gun position. This position was silenced, as was a second position twenty yards to his left. The commander of the rifle platoon following McCall last saw him advancing on a third German position. McCall was captured and was still a prisoner in April 1945 when his father accepted the Congressional Medal of Honor.[62]

German attacks, however, were driving the 36th back toward the Rapido. News of the Anzio invasion, the second phase of the pincer attack, made little impact on the men of the 36th. Just after 11:00 A.M. on the 22nd, the last footbridge back to the American side was destroyed. By 10:00 that night, no American weapons were heard from the German side of the Rapido.

General Keyes ordered that the 142nd Regiment, kept in corps reserve, be used for a third attack the next morning, January 23, at 2:30 A.M. Keyes again ignored Walker's objections to the attack. A few hours later, however, in cancelling the third attack after talking with Clark, Keyes added some clear criticism of Walker's attitude. Keyes has been quoted by Walker as saying either, "You are not going to do it anyway" (about which Walker complained in his diary),[63] or, "You might as well call it off as it can't succeed anyway as long as you feel that way about it."[64] Walker was right in complaining about the unfairness of the first remark, but Keyes finally seems to have sensed the depth of Walker's objections.

Clark's final summary of the Rapido attack was an entry in his war diary:

> In deciding upon that attack some time ago, I knew it would be costly but was impelled to go ahead with the attack in order that I could draw to this front all possible German reserves in order to clear the way for SHINGLE. This was accomplished in a magnificent manner. Some blood had to be spilled on either the land or the SHINGLE front, and I greatly preferred that it be on the Rapido, where we were secure, rather than at Anzio with the sea at our back.[65]

No mention was made of the desire to use armor or get into the Liri Valley.

The day after the battle ended, Clark and Keyes went to see Walker. As Walker described it,

> Clark and Keyes came to my headquarters today for lunch

and conference, and were not in a bad mood. I invited them into my command truck, along with Wilbur and Colonel Walter Hess. The operations of the preceding two days were discussed primarily along the lines of General Clark's initial question to the effect, "Tell me, what happened up there."

The conversation was not one in which there was generally any attempt to blame anyone for the serious check which the Germans have given our operations. In fact, at one point, Keyes made a statement generally to the effect that, from the information available before the operations, it had seemed to him to be a worthwhile operation. Clark at this point interjected the remark, "It was as much my fault as yours."[66]

Walker thought they were accepting blame for ordering a bad operation. Clark was apparently, however, referring to fault for not supervising the command of the operation. However, even when Clark become defensive after the war about attacks on his handling of the Rapido, he never blamed the men of the division for the attack.[67]

Many ranking 36th Division officers were removed by Clark a few days later. On January 24, Walker had witnessed an argument between Brigadier General Wilbur, assistant division commander, and Gen. Harold Alexander's deputy, Lt. Gen. Jacob Devers, with Keyes present. On January 26, Colonel Martin, of the 143rd, who would win a DSC for Rapido, was removed from command, replaced by Lt. Col. Paul Adams of the special service force. On January 29, Clark removed Wilbur; Colonel Werner of the 141st; Colonel Kerr, the chief of staff; Fred Walker, Jr., G-3; and General Walker's aide, his younger son, Charles Walker. This purge was justified by Clark on the grounds that low division morale "was due to a lack of good officers in key positions."[68] Walker thought differently. "I know that Clark would not act in such a manner, if he intends to continue me in command of the 36th. Hence, there is only one inference I can draw, and that is that I am marked for relief from command of the Division as soon as Clark can find an easy way to do it."[69] On June 11, 1944, Clark offered Walker command of the Infantry School at Fort Benning, Georgia. Walker thought he had no choice but to accept. He had been told by Keyes, on March 3, that "you should not be relieved now, not until the division was rested and after a successful operation."[70]

The Anzio invasion got ashore safely, landing the same day as the Rapido crossing ended. General Lucas, commanding the VI Corps at Anzio, took time to consolidate the beachhead and failed to move quickly to capture nearby strategic hills. By this time, unfortunately, Kesselring had sealed off the beachhead. It would stay sealed off for four months, until the Allies finally broke out of the beachhead and through the Winter line.

★ ★ ★ ★ ★

The Texan veterans of the division did not look at command psychology, nor the proper use of Ultra intelligence. They could not know what effect the crossing might have had on the German commanders. Their memories of the severe damage their division took at Rapido, for little effect, would stay long, with some veterans walking out of a meeting in the 1980s in which Clark was introduced.[71] In 1946 the 36th Division Association, at its founding meeting in Brownsville, Texas, requested a congressional investigation of the Rapido crossing.[72] They created somewhat of a stir.

Though he expressed some concern with potential diplomatic problems, Clark primarily wanted to clear his reputation. As he wrote Dwight Eisenhower early in 1946:

> From the messages I received from Al and radiograms from Associated Press, United Press, and from news appearing here, I personally am being attacked for inefficiency, inexperience and useless sacrifice of American youth. These personal attacks naturally, to some degree, tend to damage my position among the Austrians and Allies here, particularly the Soviets, who can not understand such procedures ... These charges are absolutely without foundation. The attack of the Rapido and adjacent coordinated attacks were essential to save many times the number of lives which, if we were heavily opposed at Anzio, would have resulted ...
>
> As I see it, if vigorous denials are not made by the War Department of these false accusations, Congress may be encouraged to investigate with the result that this may be the beginning of a series of similar investigations which may imperil Allied relations, particularly British, who already are charged in articles now appearing with failure to carry their load in Rapido attacks ...
>
> I am in no position to evaluate the seriousness of these attacks. I have been led to believe as a result of my fighting in

Italy that my record was a damn good one and needed no defense by me.[73]

The army command seemed more concerned with squelching any potential congressional investigation. Clark never, at least publicly, blamed the men of division.[74] In his memoirs he describes the division as

> a fine outfit and they performed a high service at the Rapido. Like the guards and tackles of a football team, the 36th was called upon to attack and attack hard in conjunction with other divisions of many nationalities on its right and left, while somebody else made a spectacular end run to Anzio that would have failed if the guards and tackles had not been up to their jobs.[75]

Clark suggested that Claude Birkhead, Walker's predecessor as 36th Division commander, blamed Clark for his relief and saw an opportunity to "get" Clark.[76] The army, however, called Walker the "main actor in the plot" and suggested that the War Department "should put the squeeze on General Walker in a big way."[77] The eventual congressional hearing would agree with Secretary of War Robert C. Paterson that "the action to which the Thirty-sixth Division was committed was a necessary one and that General Clark exercised sound judgement in planning it and in ordering it."[78]

★ ★ ★ ★ ★

Rapido was a diversion from Anzio, but it was also an offensive effort of its own, designed to cut its way through the German lines into the Liri Valley, and then to Rome. It has to be considered as both. Was it an appropriate plan?

The Germans also saw the strategic value of the Liri Valley and the Rapido. "The Germans had prepared for the crossing at precisely the place that the attempt was made."[79] Clark's diary for the Rapido attack, and his later memoirs, reflect his continuing view that the attack had to be tried.[80] But there were clear problems. "Our Division didn't fail in its mission, it was an impossible mission to begin with,"[81] was how a 36th Division officer later remembered the Rapido. Rapido provided no opportunity for any real use of combined arms. The men were unable to construct a tank bridge across the Rapido. Confusion and lack of communications with higher headquarters, as well as damaging the infantry's performance, made it impossible to determine the Americans' precise location. American artillery could not fire for fear of hitting American troops.

There was evidence available at the time, via Ultra, that at least the diversion from Anzio was being achieved elsewhere. Walker's plan for an attack from the north into the valley, and Keyes' plan to attack from the south (effectively extending the British crossing) had better chances of working then the Rapido thrust. Rapido turned out so badly that the Germans did not even realize they had defeated a major effort.

Walker never changed his opinion that the crossing was not justified. However, he never seemed to feel that he might have created problems for implementation. More significantly, Walker also never seemed to accept that Clark and Keyes may not have appreciated his objections nor the depth of his feeling.

> Despite my conviction on the eventual outcome of the attack, despite my inability to impress my superiors with the difficulties involved, I did not show my feelings to my subordinates. I took every possible step to make the operation succeed. As an officer, I had taken an oath to obey orders, and I proceeded to do so to the best of my ability.
>
> At one point I considered the possibility of asking to be relieved from my command in protest. But I could not abandon the 36th. I felt that I could conduct the operation better than a new division commander who would not know the capabilities of my subordinate units and commanders, and who would be inclined to push the troops into extreme and useless attempts.[82]

As Clark later wrote:

> I also have the impression that Walker claims he protested the attack in writing, or at least officially. If so, I never received the protest. Have talked with Keyes who likewise does not recall official protest from Walker.[83]

Walker never seems to have said, "This cannot be done." Walker can be faulted for not objecting strongly enough, and for possibly letting his subordinates sense his concern. But were they decisive errors? In many ways it is unfair to judge past decisions with the 20-20 hindsight of a historian. On the other hand, the bottom-line justification for any military action, move, or plan is—did it work? Walker's actions may deserve some criticism, but Walker was right. Direct assault across the Rapido River, with not even a full division, did not work. This writer has to join an earlier historian of the Rapido in wondering whether a fully enthusiastic Fred L. Walker would have made any difference, and in concluding—"No."[84]

Clark had several alternatives. Perhaps the Rapido attack could have been planned as a pure holding action, with the main attacks elsewhere. This would have been in keeping with classic American strategy and good military command principles. And a one-division (actually two-thirds of a division) frontal assault was not what eventually broke through the Winter line. There were alternatives that should have been clear at the time the Rapido attack was ordered and carried out. Clark did not have to know what we know now, just what he did, should, and could have known then. He needed the flexibility and adaptability to spot the successful diversion part of the British 10 Corps attack achieved. He needed to recognize that both Keyes and Walker provided better alternatives. Clark needed not to be so wedded to his original idea that he failed to recognize a change in the proverbial "facts on the ground."

Clark was not able to spot, recognize, and adapt to a changing situation. The result of his failure is the lesson of the Rapido.

CHAPTER 11

Velletri

"It is indeed ironic that after all the sins that had been visited upon them it was the Texans who gave Clark the key to Rome."[1]

—Historian ERIC MORRIS

"Many outfits deserve the credit for the whole operation, but those of us who were present will always remember the men of the 36th climbing silently in the night behind the enemy, armed with little but their American competence and a personal faith in their quiet, retiring general who had never let them down. If Generals Alexander and Clark received the key to the city of Rome, it was General Walker who turned the key and handed it to them."[2]

—ERIC SEVEREID

Too late for some of the men of his division, Walker's idea for crossing the Rapido north of Cassino town, and coming down through the mountains, was finally being carried out. The British bridgehead over the Garigliano was fighting off heavy counterattacks, so the next effort in the Cassino area was made to the north. Walker wrote in his journal that "Clark has now decided to attack with the 34th Infantry Division over the high ground north of Cassino, which is what I suggested he do a long time ago."[3] Walker was instructed to feint another crossing of the Rapido in the same area where the previous crossing had failed. He was to be ready to use the 142nd Infantry Regiment, undamaged at the Rapido, to cross north of Sant'Angelo. Combat Command (brigade-size group) B of the 1st Armored Division would exploit this opening into the Liri Valley.

Combat was as fierce as it had been around San Pietro, two months before. Keyes soon moved the 36th to the right of the 34th, and both continued to push forward. The 133rd Infantry Regiment, from the 34th Division, actually gained a toehold in the town of Cassino—at the bottom of the same mountain which the famous abbey sat atop. But the Germans were as tough as always, the weather as bad as usual. Both divisions were taking heavy casualties, rapidly losing their effectiveness as combat units. During the second week in February, Alexander sent a deputy to visit the front. The deputy reported that morale was so bad that the men were almost mutinous—probably overstating the possibilities, but not the loss of effectiveness. Another view was given a few months later by a staff officer, an old friend of General Walker from World War I:

> Now the 36th was slugging it out, week after week, month after month with Kesselring... With regiments reduced to the size of battalions, companies with only one officer remaining, most of the highly training NCO's casualties... the 36th was hanging on. Some day the story of those exhausted, tired, determined men, fighting cold, rain, snow, hunger, sickness and death, as well as the enemy, in the mountains north of Cassino, will be made known.[4]

The 36th and 34th divisions would have to be relieved from active operations soon. February 12 was set as the date for them to cease combat operations, to be replaced by the New Zealand Corps. When the fighting ended, units of the II Corps had reached within a mile of Highway 6 in the Liri Valley.

After a February 14 truce to allow the Germans to remove bodies of their men[5]—a rare chance for members of the 36th to meet the enemy face to face without guns—elements of the 36th Division held Monte Castellone for the rest of February until their relief. Portions of the 34th stuck around longer, holding a small piece of Cassino village.

On February 15, the Abbey of Mount Cassino, the highest point in the area, was destroyed by Allied bombers. The Allies had been sure it was being used as a German observation post. This was wrong. After the bombing, however, the ruins made a much better defensive position when they were occupied by the Germans.

After two and a half months' rest, refitting and training near Naples, and an eruption of Mt. Vesuvius,[6] the 36th Division was redeployed to the Anzio beachhead in May, a several-days-long process ending on May 22. This was only a few days after the situation in Italy began to change. The Anzio beachhead was still stuck, though hard fighting had repulsed several heavy German counterattacks.

The Germans were already developing a fallback position, and had been doing so for several months. The Caeser line was roughly fifteen miles south of Rome. The Alban Hills, the VI Corps' *suggested* (the orders were ambiguous) immediate goal after the initial invasion, were a part of this defense system. Kesselring had never planned to hold southern Italy permanently, just to delay the Allies as long as militarily feasible. This would keep the Allies as far as possible from the industrial heartland of Italy, in the north, and from southern Germany. Even the Winter/Gustav line had fallback positions. This line had a forward salient, and a rear salient—the Hitler line. The lines were certainly meeting their objective of delaying the Allies. One can argue that the Anzio stalemate tied down more German divisions than the successful formation of a single Allied line would have tied down.[7] The Italian effort was also tying down Allied troops that might have been used elsewhere.

The attack on the "southern front," Operation Diadem, was scheduled to start on May 11, 1944. A new factor had been introduced in the two months before the attack, the first "draftee" divisions. The American 88th and 85th divisions were going to enter combat in Diadem. These divisions had been built from scratch, unlike regular army divisions and National Guard divisions such as the 36th. The regulars and the guardsmen showed they could function well in combat. Now was the draftees' turn.[8]

Allied attacks on the Gustav line began on May 11, as scheduled. The 5th Army attacked on a narrow twelve-mile front, the British 8th Army on a larger area to the right. Going was slow at the beginning as the Germans, their defenses, and the terrain remained as tough as usual. By May 24, the Allies had broken through the Gustav and Hitler lines, well on their way to a linkup with the

Anzio beachhead. Mount Cassino, even more troublesome and deadly than the Rapido, was finally captured by Polish troops. In their sectors, the 85th and 88th divisions performed well.[9]

The Rapido was finally crossed. The 4th British and 8th Indian divisions began the attempt on May 11. The British 78th Division crossed on May 14, to secure the bridgehead. Three days and three full divisions were necessary to secure a beachhead, another three divisions to exploit this breakthrough. A postwar historian describes the problem:

> ... it was in the Liri Valley that the operation was most threatened. Here ... 13th British Corps spent three hellish days establishing a bridgehead across the Rapido and Gari rivers. The Germans fought with fury, aided by a thick fog that caused utter chaos as the leading assault units of the 8th Indian Division became lost and confused. Even the white tape normally so effective in guiding men forward at night became useless; units were forced to creep forward with each soldier grasping the belt of the man ahead of him to avoid getting lost.[10]

Gen. Fred Walker's diary, diplomatically, records no reaction to the trouble encountered by the far larger British/Indian/Canadian force in crossing and staying across the Rapido.

The Anzio breakout was scheduled for May 23, by which time the attacks to the south were well underway. The Anzio beachhead itself was fourteen miles long, varying from five to seven miles inland. The ground was generally flat, with a slight slope as one moved inland. The land became more rolling inland toward the Alban Hills, ten to twelve miles away. At the closest points, the beachhead was just over thirty miles from Rome, just under fifty miles from the left flank of the Allied position at the Gustav line. The original Allied plan, as conceived at Alexander's headquarters, called for the VI Corps, after breaking out of the beachhead, to continue northwest toward Valtamonte and Highway 6. VI Corps would be in a position to block the retreating right wing of the German 10th Army. The breakout was successful when, on May 25 (the same day patrols from the VI Corps and II Corps made contact along the coast), Clark altered Alexander's plan—without notifying Alexander. The bulk of the VI Corps would turn left, heading toward the Alban Hills and Rome.

Clark wanted Rome so badly that at one point he told Alexander he would have his men open fire if the 8th Army tried to get in first.[11] The whole political issue between the Americans and the British, the feeling that the British were trying to take the credit while the Americans did the work,[12] came to the fore. Clark later called such controversies "conceived in good faith as a result of honest differences of opinion about the best way to do the job."[13] At the time, his reaction was:

> We not only wanted the honor of capturing Rome, but we felt that we more than deserved it . . . My own feeling was that nothing was going to stop us on our push toward the Italian capital. Not only did we intend to become the first army in fifteen centuries to seize Rome from the south, but we intended to see that the people back home knew it was the Fifth Army that did the job and knew the price that had been paid for it.[14]

Clark wanted Rome, he wanted the Americans to take Rome, and he wanted Rome captured before the pending Normandy Invasion.[15] Alexander did not even mention the disagreement to Clark in dispatches.[16] Alexander was frequently inclined toward a somewhat passive style of leadership, especially with an independent-minded subordinate such as Clark.

The breakout worked. Both Allied fronts linked. However, by May 28 they had run up against the Caeser line. The first attacks against this line were repulsed with heavy casualties.

The 36th had relieved the 1st Armored Division south of the village of Velletri. VI Corps commander Lucian Truscott, Jr. (who replaced John Lucas back in February) planned to use the 1st Armored for an attack elsewhere on the front. The 34th Division would attack west of Lanuvio, with the 36th to support and follow up. Walker foresaw yet another operation against entrenched Germans. This time, however, Walker had an alternative.

Monte Artemisio, a major position in the Gustav line, was a ridge, roughly four miles long, about a mile to the north of Velletri, dominating the town and several key roads, including one to Rome. Patrols from the 36th Division had been exploring the area between the mountain and the village. They found what seemed like a gap in the German lines. Had the Germans finally made an error?

The gap might not be there for long. The Germans discovered

the gap the same night as Walker. Their problem was the boundary between the I Parachute Corps and the LXXVI Panzer Corps. The 362nd Division, responsible for the Velletri sector, had suffered severe loses in the Allied breakout from Anzio. The Hermann Goering Panzer Division, to the left of the gap, was partly out of line. It had been drawn in the direction of Valmontone by the initial Allied attacks after the breakout from Anzio. The Goering division commander sensed a problem and sent a patrol to find the left flank of the 362nd. The patrol walked for two miles before finding the men from the other division, at a fork in a road just northeast of Velletri. Gen. Wilhelm Schmalz, the panzer commander, sent an engineer platoon and a patrol to occupy points in the gap. However, American patrols saw no sign of either of these groups.

The panzer corps commander learned of the gap and ordered troops sent to close the gap. These reinforcements did not arrive in time. The commander of the German 362nd Infantry Division also learned of and made unsuccessful efforts to close the gap.[17] Kesselring learned of the gap on May 29 and immediately ordered the 14th Army commander to close it. The commander passed along the order but thought the measures Schmalz had ordered were solving the problem. He even successfully argued the point with Kesselring the next day. That same day, the Germans decided to leave the Goering division in place near Monte Artemisio. Apparently, despite their concern, no one thought to send a daylight patrol into the area to see what the situation actually was and if it had been corrected. Luck, missing at Rapido, was with the 36th at Velletri.

On Saturday, May 27 (apparently very late in the day), Walker wrote in his diary,

> I made a reconnaissance of the front occupied by the 143d Infantry. Patrols have not recently met any strong resistance on this front.
> I asked [division chief engineer Oran] Stovall to determine whether a temporary road can be made over Mount Artemisio for artillery and tanks. If that area is lightly held by Germans, we may be able to break through there.[18]

The next morning, Stovall reported that it was possible to build such a road.

Walker took another look on May 28, flying in a spotter plane over the division front and over Monte Artemisio. "I saw almost no field works. Nor did I see any entrenchments or gun positions. It looks to me like this is the place to break through."[19]

The evening of May 29, Walker attended a VI Corps division commanders' conference with Truscott. According to Truscott's orders, the next day the 36th Division was to move to the rear of the 34th Division, then to its left, and then replace that division in line. Walker saw no chance for success replacing another bogged-down division. Truscott did not seem open to discussion that evening, so Walker decided to talk with him the next day.

After a 9:00 A.M. 36th Division officers' briefing on May 30, on Truscott's plans, Truscott came by the 36th Division command post. Walker spoke with Truscott about the alternative plan. According to Walker, Truscott said Walker might have a good idea and he would call back in an hour:

> I held up the movement of the troops, awaiting an answer. When he phoned at about 11:00 A.M., he said he had talked to Clark, who approved the change, and that I should go ahead with my plans. His last words were, "AND YOU HAD BETTER GET THROUGH." [Capitalized in original].
>
> This is a veiled threat meaning; if you don't succeed, you'll be on your way back to the States; the responsibility is all yours; although I approve of your plan and consider it better than mine, I assume none of the responsibility.[20]

Truscott's description of his approval makes no mention of the implied threat, just of questioning Stovall as to the engineering feasibility.[21]

The division photographer, in a letter quoted in a history of the fall of Rome, tells of witnessing a meeting early in the afternoon of May 30 between Walker and Truscott. He thinks Clark was also present. The gist of the discussion seemed to correspond with Walker's memory, that he would not be ordered not to try his maneuver, but would not be covered by an order from corps or army. Walker would take responsibility, whatever happened:

> By advocating a major change in plans General Walker had laid his professional reputation on the line and his future, he felt at the time, was bound up with the success or failure of the pending operation against Velletri. If he could put T-Patchers up

and over M. Artemisio and get in the enemy's rear, behind Velletri, then the entire German resistance below Rome had to crack.[22]

Almost six months earlier, an interesting and prescient article had been published in *Die Suedfront*, a newspaper published for German soldiers in southern Italy. The Americans were not alone in using the tactics the article described—the Germans had introduced such tactics as far back as the sturmtruppen of World War I—but were particularly skillful:

> The Americans use quasi Indian tactics: They search for boundary lines between battalions or regiments, they look for gaps between our strongpoints, they look for the steepest mountain passages . . . They infiltrate through these passages with a patrol, a platoon at first, mostly at dusk. At night they reinforce the infiltrated units, and in the morning they are often in the rear of a German unit, which is being attacked from behind, or also from the flanks simultaneously.[23]

The 141st had probed into an area just beyond the Velletri-Valmontone Highway at the foot of Monte Artemisio. The 36th Combat Engineer regiment would hold most of the 36th Division front while the 142nd and then the 143rd passed through the 141st, and then up the mountain at night. The 142nd received its orders at 4:00 P.M. that afternoon. The actual infiltration started just before 11:00 P.M.

The 2nd Battalion of the 142nd led, followed by the regimental headquarters company, then the 1st, and then the 3rd Battalion. After passing through the area held by the 141st, the route followed a sunken trail through cultivated flatlands to a deep gully to the principal objectives.

Not only were the men told not to talk, they were told that any Germans they met would have to killed by bayonet, knife, or other quiet means.[24] At least one German soldier was "covered" by this order. A war correspondent, Kenneth Dixon, accompanied the 142nd. A young soldier, crawling past, pointed to a German and whispered to Dixon, "That guy must know you, reporter. He's smiling." Dixon had heard nothing. But when he looked, a German soldier was leaning against a tree "wearing two grins in the moonlight—a white one where his teeth were bared and a dark red one three inches below. I kept crawling."[25] The soldier involved in the incident confirmed that several German sentries were quietly killed, with knives or wires.[26]

Another participant in the move wrote:

> It was a night of confused anticipation; we expected to meet the enemy in every shadow and we did not know what to expect. My squad was near the front; though we did not know exactly where we were supposed to be, we felt that the objective would be reached if we continued to move up the hill.
>
> The crest of the mountain was reached just before dawn and the movement turned to the left. (I still don't know whether we followed the expected route of march, perhaps no one knows)
>
> Early in the morning the advance patrols ran into enemy outposts and other troops moved up to clean out the enemy and we moved on to the road that was our objective.[27]

According to the 142nd regimental records,

> the time of greatest tension happened around 0300 when enemy air action hovered overhead. During the night there had been single planes obvious on reconnaissance but at this time the enemy was over in great numbers. Antiaircraft shot up its sparkling display of defensive fire. The planes hovered threateningly near. Then the inevitable flares were dropped to light up the area as broad daylight. They were not so far off. Or so it seemed. Of course everyone in our foot column hugged the ground and waited. Bombs were bursting and the chatter of strafing action grew louder. But it was soon apparent that we were not the enemy's target that time. In fact, it was very favorable to us as it focused attention away from our sector to over in front of Velletri where it was taking place . . .[28]

By 6:35 the morning of May 31, the 142nd had reached its objectives on the northeastern end of the ridge. One German officer was taking a bath when captured. The 143rd was following the 142nd, with the 141st roughly paralleling the course, moving slowly toward the town of Velletri. By the end of the 31st, all of Monte Artemisio was in American hands. By the morning of June 1, artillery observers were on the ridge, viewing what forty years later would be called a "target-rich environment." The official military history Green Book states that "The only problem was to obtain enough batters to do the firing and observers to direct them."²⁹ Division and corps were asked for every available observer to come help. Soon, as the 143rd's operations report for June put it, "forward observers were sitting around . . . like crows on a telephone line, having a field day."³⁰ General Truscott described the breakthrough as "the turning point in our drive to the northeast."³¹

A sergeant with a tank destroyer battalion described an incident later that morning:

> Next morning we were in a clearing a hundred yards from the road when a German command car stopped at the road junction. The officer got out smartly, pulled the road map out of his case and looked at it for a few minutes. Then he slowly raised his head and stared at us for about half a minute, then studied the map for fifteen seconds, then zoomed back out in the direction of Rome. We were laughing so hard we couldn't fire on him.³²

American casualties were extremely light. However, one of the first two members of the 142nd to reach the summit of Monte Artemisio wrote of exchanging shots with some Germans, and of the cost of even the most successful infiltration. This section of his narrative begins with a German grenade landing sometime in the morning.

> A dud! In the same instance I got a glimpse of steel helmet ducking back behind the shell of the old building to my right. I told Art to watch the plateau to his left and I would watch the right as they were coming up on each side of us. Every time I would see the helmet around the corner, like a turtle sticking his head out, I would drive him back with my M1. The bullets kept cracking in our ears like bull whips—we all know in the infantry that they don't sing like they do in the "westerns." When they break the sound barrier as they pass your ear they crack, nearly deafening you. I told Art to keep lower down for one from somewhere had zeroed in. He said he didn't see how it could come from in front unless he was in a tree or tower of some kind, and since everyone behind us was supposed to be GIs. These were to be Art's final words. As the bullets began to crack in my ears again I told Art to get down. No answer—no movement—still same position. As I drove the helmet back around the corner again and again I told Art to get down but

again no response. I thought "my God, a bullet has come through the soft dirt and hit him in the chest." As I reached to get hold of him his helmet went bouncing over the cliff and I knew he was hit. In panic I lifted him down into the hole. I yelled "Medic, Medic," as everything in Art's head poured out in my hands as the back of his head seemed to explode and all of the brain surgeons in the world could not have helped. Art wore prescription sun glasses and the bullet went through his right eye without knocking off his glasses. He didn't move or moan, never knew what hit him.[33]

By June 1, Colonel Lynch reported to General Walker that the 142nd was in satisfactory condition, despite German artillery and counterattacks—including some nighttime infiltration.

The 36th Division did not just infiltrate over a mountain. Starting on the morning of May 31, division engineers built a road. Feasibility of building such a road, as a route for tanks and tank destroyers, was a major criterion for such approval of Walker's plan as Truscott gave—it was vital to Walker suggesting the plan in the first place. Major Stovall was only making an educated guess when he said he thought the road could be built. The only way to be sure was to try.

The lead bulldozer operator, John Bob Parks, Company B, 111th Engineer Battalion, later wrote an excellent description of the operation. After receiving his assignment,

> I started filling ditches, pushing rocks and trees out of the way; at any time I got it smooth enough to climb over, I moved forward to the next obstacle and other equipment finished the job. At first the job offered few difficulties; but as we started climbing, the job demanded my full attention and I lost track of time and everything else except that damn white tracing tape that was always in front of me.
>   ... I can remember some artillery fire, and at times sniper fire ... small arms fire and scattered artillery fire from above and to our right continued to cause trouble all day ...
> The mountain slope got so steep that we had to level out some more; the rocks and dirt gave way under my tracks and would lost traction. The trees were also getting thicker and larger; the demolition crews had to blow many of them out of the way; others were cut down by men with saws and axes. We reached some cliffs that looked impossible to climb, but by

cable and snatch-block arrangements we were able to pull the small dozers to the top and they cut their way to the lower elevations. This was slow work and dangerous because if the dozer ever tipped over it would have rolled a mile . . .

We tried to make the road as long as possible before turning back because each hairpin turn had to be made wide enough for an awkward tank to turn without losing a track. Working parallel to the slope and moving up as we would and switching back in the opposite direction, we moved up sometime only a few feet above the last cut, but up, up, up all the time.

[After repairs to Parks' machine while the other bulldozers worked] We worked like hell till some time after dark when the Colonel told Lieutenant Powers to stop us because he was afraid we would do more damage to the men and equipment in the dark. The enemy had gotten real nasty and had closed in the darkness.

. . . After a can of cold "C" rations, I lay down by my dozer blade and went to sleep. The last thing I remember was thinking that if I were only four inches longer and weighed as much as 150 pounds I could reach the controls of the dozer from a sitting position on the seat and I would not get so tired.

[After a sharp firefight with some Germans, Parks continued on but ran out of fuel. After being refueled] Captain Crisman reminded me again that I was holding up the whole damned Fifth Army. One of my fingers had been busted earlier and was hurting something awful but I was afraid to tell anybody for fear that I would be replaced; this was one job that I was going to finish.

We finally got to the top; though I had been fully aware that we had done some climbing, I was not prepared for the view from the top. I could hardly see the place we had started from the morning before. Anzio and the sea was visible in the far distance.

I had little time to view the landscape however, because we still had the road to build to the bottom on the other side. We got some long range heavy artillery later in the day. I was used to 88's and other incoming fire and to our tanks and tank destroyers that had been firing over us. I did not know at first that we were their target. Not that it would have mattered a whole lot, because . . . we would have to get to the bottom before dark and . . . the Krauts were lousy shots. One of our R4 dozers had been wrecked earlier, but the rest of us had kept going. Naturally the descent was much easier; we were pushing downhill and we were able to hold the equipment on the slope by that cable on the back and easing off on the winch when I wanted to move ahead.

We reached the bottom and a second class road well before dark and I watched some armored units pass us and disappear in to the valley. We had finished the *road that could not be built.* [Underlined in the original] . . .

Directly moving on Velletri was Walker's next step. At 4:30 P.M., May 31, Walker ordered the 141st Regiment to take Velletri. The regiment assembled and jumped off about four miles east of the town. Vineyards and orchards around the town made observation, and the attack, somewhat more difficult. Velletri itself was on a hill dominating the nearby territory. However, by the time the 141st advanced, at dawn on June l, the two regiments on Monte Artemisio had sealed off the major escape routes to the north. Second Battalion, 141st Regiment, blocked the main German road to the northwest. Part of the 1st Battalion cut the road to the east. The third battalion would attack from the northeast.

The 36th had an additional advantage in the attack on Velletri, aside from having the town virtually surrounded. Division artillery had excellent spotter positions on the peaks of Monte Artemisio. The spotters were able to see far beyond the German lines and accurately control shellfire, bringing it down where needed. They were able to quickly stop shelling of a particular area, as the advance speeded up, to avoid hitting American troops.

The attack on Velletri stepped off at dawn on June 1. The 1st Battalion of the 141st Regiment, in control of the road between Velletri and Valmontone, attacked from the east, aided by tanks and tank destroyers. The battalion covered about three miles that day. The 36th Engineer Combat Regiment attacked from the south. They were not able to make much progress against a strong defense, and ended up in a holding action. The 2nd Battalion of the 141st, with all but one company of the 3rd Battalion in support, came in from the northwest. This was intended as the main attack.

The town was well defended, with the Germans engaging in

frequent counterattacks. The 2nd battalion attack, for example, was held up several hours by a German counterattack. The Germans were repulsed, and the battalion entered Velletri just before 5:00 P.M. Just over a half-hour later, Company F, leading the battalion advance, reached the center of town. The 141st was taking casualties, but German losses were heavier. This included the American capture of one group of five German musicians.[35] Many Germans were killed or wounded trying to escape from the town on June 1.

General Walker was with the 1st Battalion of the 141st all afternoon.[36] Newspaper correspondent Wick Fowler described the general that day:

> General Walker . . . was apparently oblivious of the special dangers accorded a two-star general with insignia plain to any hidden sniper. He moved among the men, his pistol holstered, and directed them as though he were playing chess back at division headquarters. He knew where every unit was and he knew just how he would play the game to take the long-stubborn fortifications.[37]

Walker was with the lead tanks for part of the advance. Lt. Col. Hal Reese, division inspector general and a friend of Walker's since World War I, was with Walker:

> When there was a pause in the German fire, Hal Reese and I walked along the road behind the self-propelled cannon company vehicles of the regiment while the foot troops moved ahead through the cane and vineyards on both sides. As we approached the town of Velletri, the Cannon Company vehicles stopped at a bend in the road. Hal walked out beyond the bend, in front of the leading vehicle.
> In a few minutes, Captain MacCombie came to me and told me Hal was dead. A German antitank shell and a mortar shell had simultaneously struck near the vehicle and a fragment from the shells tore away Hal's left side. He died instantly.[38]

By June 2, Velletri was clear of the enemy. The Caeser line was effectively broken. As Clark commented in his war diary, "Walker has done well."[39]

★   ★   ★   ★   ★

Despite his best efforts, John Bob Parks was not the first American into Rome:

> I got nearly to the walls of the city and decided it would be good to drive the first bulldozer into town, but I got stopped on the idea. The Colonel stopped me and asked where I thought I was going, and when I told him he told me to take cover behind some buildings or I would be the first dead dozer operator in Rome.[40]

By June 3 the German military high command decided to evacuate Rome. Kesselring was ordered to hold south of the city only long enough to evacuate his forces in and around the city.

That same day, reconnaissance patrols of the 88th Division spotted the Roman skyline. Clark, who seems finally to have recognized the advantages of Ultra and reading German communications, sensed Rome was about to fall. He was not the only one. His war diary records that Maj. Gen. Alfred Gruenther, Clark's chief of staff, observed, "The CP has gone to hell. No one is doing any work around here this afternoon. All semblance of discipline has broken down."[41] Aside from sensing the 5th Army was in for the "kill," at least as far as capturing Rome, Clark was concerned the Germans might destroy the Tiber bridges. He ordered II and VI Corps commanders to form mobile task forces to get quickly into the city and seize the bridges.

Records indicate that a patrol from the 88th Cavalry Reconnaissance Troop, 88th Division, entered Rome the morning of June 4, but quickly withdrew.[42] At 6:15 that evening, the 1st Regiment of the 1st Special Service Force, and elements of the 3rd Battalion, 13th Armored Regiment (the whole under the command of now–brigadier general Robert Frederick) entered Rome. Despite some sharp but brief firefights with retreating German units, these units stayed in Rome. All bridges in their sector were secured by 11:00 P.M. Rome was "officially" liberated on June 5, 1944. The next day, American and British troops invaded Normandy.

The 36th Division had several more weeks of fighting in Italy, chasing the Germans north. On the morning of June 26, 1944, the 141st Regiment was relieved by the 442nd Regimental Combat Team, which it would meet again in France, and the entire division was withdrawn to prepare for the invasion of the Riviera.

Chapter 11: Velletri

On June 11, Clark wrote Walker offering him command of the Infantry School at Fort Benning, Georgia. Walker felt he had no choice but to accept. Walker always felt he was relieved as an embarrassment to Clark, after Salerno (Walker thought Clark almost panicked at the Salerno invasion) and the Rapido.[43] II Corps commander Geoffrey Keyes had told Walker, according to Walker's diary, that he was going to be fired after a victory.[44] Clark, however, wrote in his memoirs that Walker was not fired, that Clark thought he would appreciate the change of scene to a more responsible position. Walker officially said farewell to the division on July 7, 1944. A few days later, he flew back to the United States.

★ ★ ★ ★ ★

Fifteen years after the fact, a professional military analyst wrote about Velletri:

> Among the prerequisites for successful infiltration are: thorough planning based on mission analysis, terrain study, and enemy dispositions; an offensive spirit characterized by boldness; training and physical hardness; and periods of poor visibility or other conditions [night!] where the enemy's surveillance is neutralized.[45]

Walker and the 36th met these requirements at Velletri. Clark's plans were "set in stone" at the Rapido. "Facts on the ground" seemed to have little impact on the facts in his mind. Things, fortunately, were different at Velletri. Walker had the presence of mind, the flexibility, and the basic military ability to recognize an opportunity that presented itself. He correctly interpreted the data and its significance—that the gap in German lines existed—and presented an opportunity to break through the German lines. He was able to act quickly to convince his superiors and gain permission to carry out his ideas. (Perhaps Clark was starting to learn the lesson of flexibility.)

What about the men themselves? They seem to have done a lot better at Velletri than at Rapido. If nothing else, at Velletri they moved a lot more quietly. At Rapido they made enough noise to alert the Germans and bring down very damaging artillery fire. At Velletri the opposite prevailed. Germans who heard the initial infiltration, if any, became prisoners—or died. Everything seemed to go wrong at the Rapido. Nothing seemed to go wrong at Velletri.

In the four months between the two battles, did the men of the 36th suddenly decide they liked night fighting? Did men who are supposed to have been reluctant to kill from a distance with a gun suddenly become quite willing to kill with a knife? Or was it more likely the unconscious self-fulfilling-prophecy effects of pessimism and optimism? Do people, without meaning to, try less hard when they feel their efforts will be wasted? Do positive expectations provide just that extra little big that might make the difference between success and failure?

The men of the 36th could not have saved the Rapido crossing. It was a bad plan, lacking in flexibility to respond to changing conditions. However, they could have ruined the Velletri infiltration. But this was a good plan, and they knew it. The men of the 36th charged with the Velletri nighttime infiltration did what they were supposed to do, and the plan worked.

"The successful penetration by the 36th Division on 31 May . . . offered opportunities unforeseen during the past four days, a period which had been marked by grinding, costly, and frustrating fighting."[46] Something that could not always be said about the fighting in Italy (certainly not at the Rapido or Anzio) can be said now—the opportunities were taken.

Training unit at Fort Blanding, Florida, circa spring 1944. Private Bruce Kouser is upper right. —Collection of the author

Private Bruce Kouser, 36th Division, 142nd Regiment,
—Collection of the author

"On the way" will soon be the call, 81mm emplacement for Company D, 141st Infantry Regiment, near Haguenau on March 9, 1945. At the sights is Cpl. John E. Richards of Pompton Lakes, New Jersey, and with the round in hand is Pfc. Ernest Dolloff of Hamilton, Texas. —Signal Corps photo SC 364270

Training unit at Fort Blanding, Florida, circa spring 1944.
—Collection of the author

Bogged-down American tank near Cassino, Italy.
—U.S. government photo

Supply dump in area of Rapido River.
—U.S. government photo

Field Marshall Albert Kesselring, commander of German forces in Italy 1943–45

Snow and white sheets camouflage machine gun emplacement of the 142nd Infantry at Bischwiller on January 24, 1945.
—Signal Corps photo SC 364324

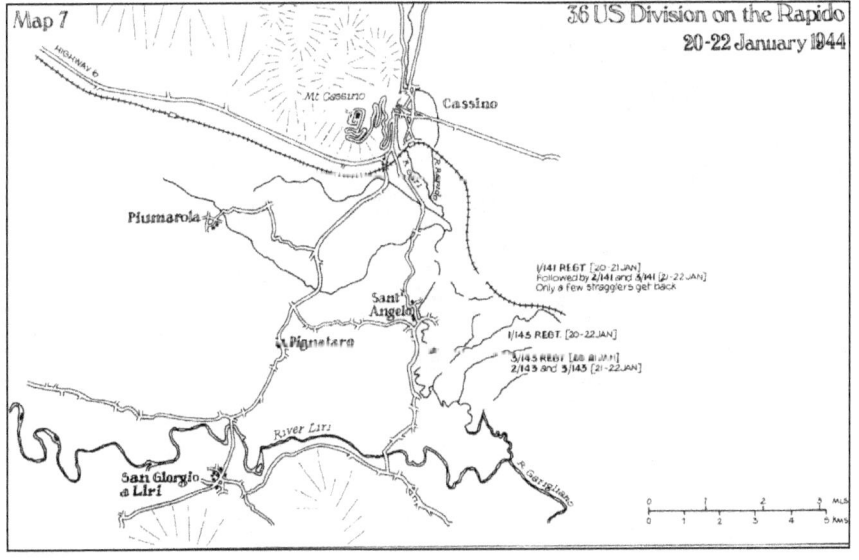

Helmet and rifle mark the initial grave of a member of the 36th Division. This was the way bodies were marked for graves registration teams to come by and pick up. An appropriate religious burial was given. Bitschoffen, March 15, 1945.

—Signal Corps photo
SC 202219-S

World War I, 1914-1918. "Bringing in wounded under the Red Cross flag. The Somme, September, 1916." From a photograph in the Public Archives of Canada.

Lone Star flag flies over the customhouse at the border town of Schweigen, Germany, on March 22, 1945.
—Signal Corps photo SC 203262

Lt. Gen. Mark W. Clark, commander, United States 5th Army.

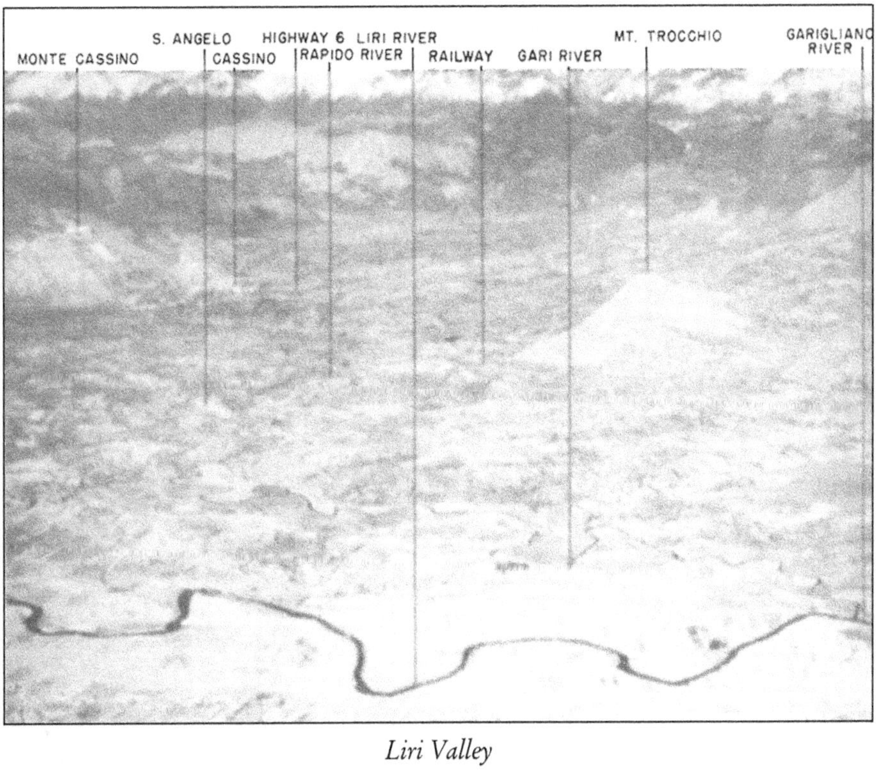

Liri Valley

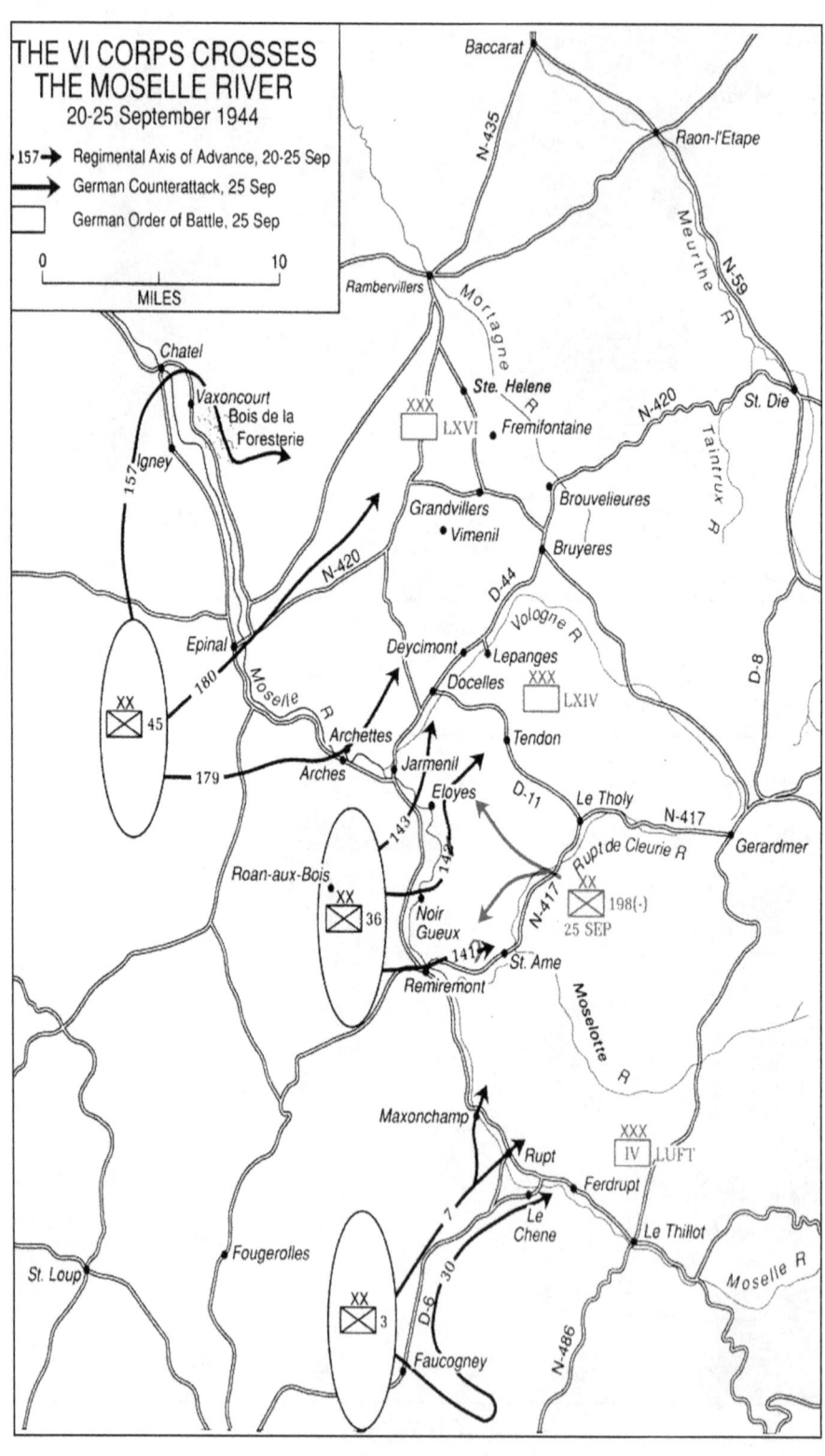

TO VALENCE
Livron
Allex
Drome R
Loriol
Grane
Crest
TO DIE
N-7
La Coucourde
Condillac
Pass
Marsanne
Rhone R
Hill
430
Hill
300
Hill
294
Puy St. Martin
Sauzet
D-6
Bonlieu
Roubion R
Montelimar
Jasron R
MONTELIMAR
BATTLE SQUARE
Highground
ELEVATIONS IN METERS
0
3
MILES

XIII SS
31 DEC
17SS
36
Sarreguemines
103
44
44
100
XC
1 JAN
LXXXIX
1 JAN
Landau
Lauter R
Bundenthal
Rimling
Hottviller
100
Bitche
559
257
256
Nieder Schlettenbach
Nothweiler
45
Bobenthal
Singling
Holbach
Camp de Bitche
36
Rohrbach
Siersthal
Wissembourg
Schweighofen
Sarre R
Wingen
Rott
Lemberg
TF HUDELSON
Neunhoffen
Climbach
79
Schleithal
Salmbach
Scheibenhardt
Sarralbe
6 Mtn
SS
2 JAN
Baerenthal
Dambach
Lembach
Riedseltz
Lauterbourg
Montbronn
Philippsbourg
XXXIX
PZ
16 JAN
Scheibenhard
Sarreinsberg
Mouterhouse
Nehwiller
Sarre-Union
Ratzwiller
Wildenguth
Niederbronn-les-Bains
6 Mtn SS
Soultz-sous-Forets
21
25
Eberbach
Reipertswiller
24 JAN
Woerth
Surbourg
Niederroedern
Rhine R
Wingen-sur-Moder
Puberg
45
Rothbach
Gundershoffen
Hatten
Seltz
Tieffenbach
Frohmuhl
Wimmenau
36
Rittershoffen
Durstel
Moder R
24 JAN
47
Walbourg
10 SS
16 JAN
Wolfskirchen
Eywiller
Ingwiller
Rothbach R
Mertzwiller
Haguenau Forest
Rastatt
La Petite-Pierre
7 Pchr
Soufflenheim
Fenetrange
103
Haguenau
10 SS
Bouxviller
Schweighausen
24 JAN
Schirrheim
Fort Louis
79
Kaltenhouse
Camp d'Oberhoffen
Phalsbourg
36
Bischwiller
21
24 JAN
Gries
Saverne
Sarrebourg
Zorn R
Rhine-Marne Canal
Weyersheim
Herrlisheim
Offendorf
Gambsheim
THE LAST GERMAN OFFENSIVE
31 December 1944 - 25 January 1945
Seventh Army Frontline, 31 December
German Attacks, Beginning Dates Indicated
Seventh Army Frontline, 25 January
0
10
MILES
SEVENTH
XXXX
FIRST FR
3 Alg
La Wantzenau
553
Wasselonne
XIV SS
5 JAN
Ill R
Strasbourg
Kehl
NINETEENTH
ARMY ATTACK
5-11 Jan 1945
Strasbourg
Kehl
Molsheim
Bruche R
Erstein
Canal
Benfeld
Rhinau
Ill R
Rhine R
Selestat
0
10
MILES

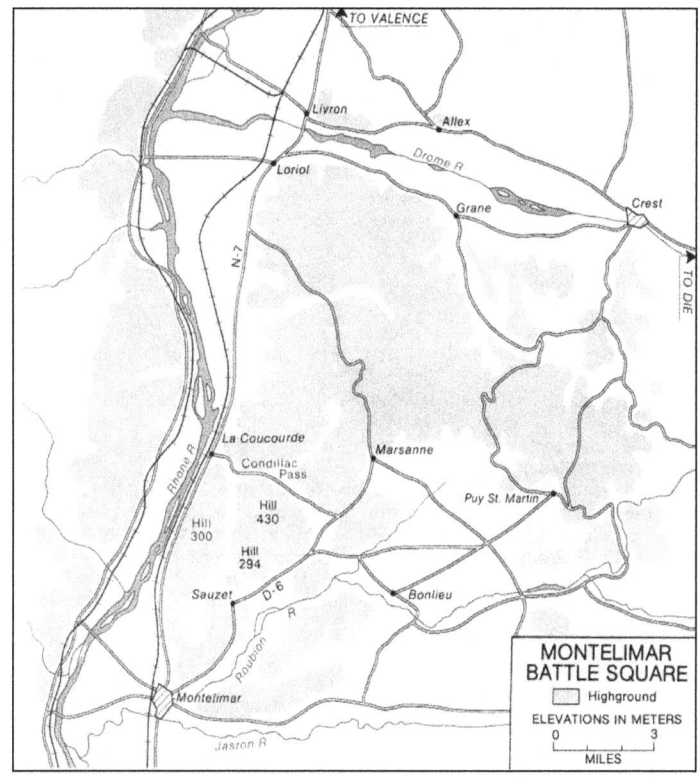

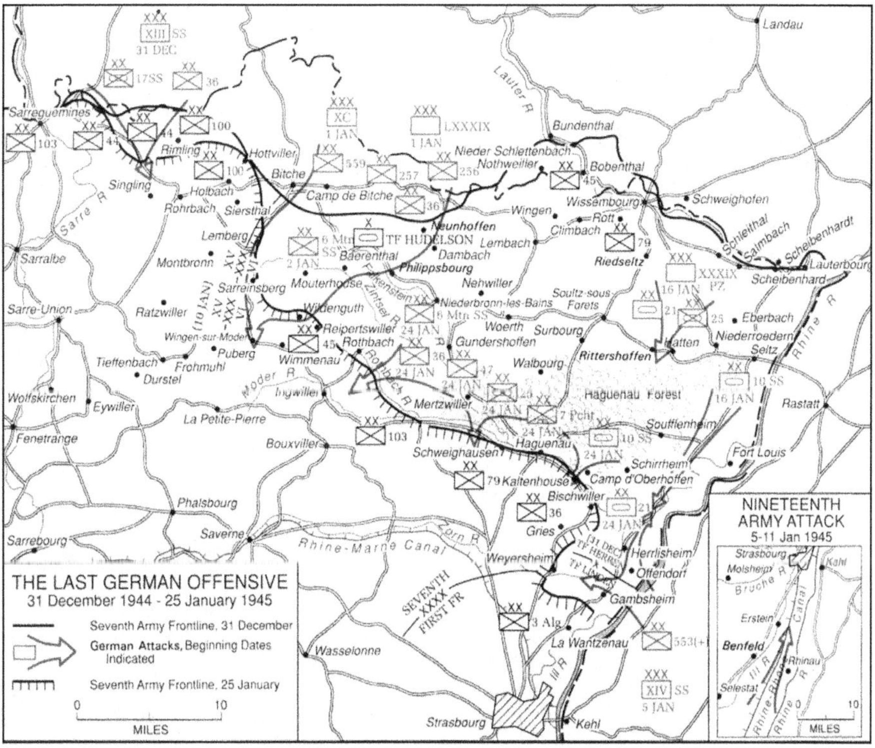

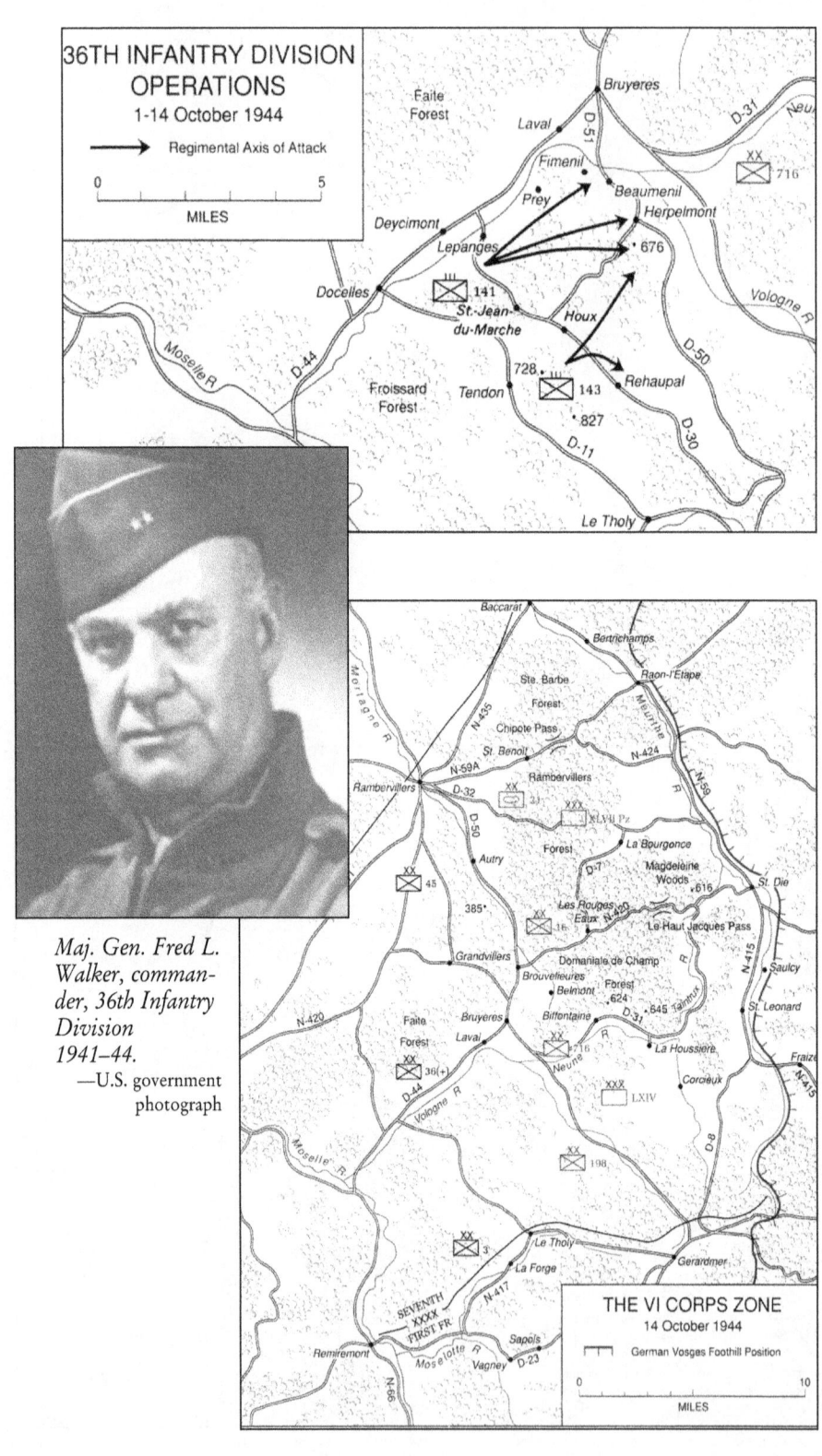

Maj. Gen. Fred L. Walker, commander, 36th Infantry Division 1941–44.
—U.S. government photograph

THE VI CORPS ADVANCE
26-30 September 1944
Forwardmost Division Location, Evening, 25 September
Approximate Front, 30 September
German Order of Battle, 30 September
0
10
MILES
Baccarat
21
N-435
Raon-l'Etape
OELSNER
XLVI
Meurthe R
N-59
157
Rambervillers
Mortagne R
180
16
Ste. Helene
Fremifontaine
N-420
St. Die
Taintrux R
Chatel
Vaxoncourt
Bois de la Foresterie
Igney
45
Grandvillers
Brouvelieures
179
Vimenil
Bruyeres
N-420
D-44
143
Lepanges
Deycimont
Epinal
716
LXIV
D-8
Docelles
141
Moselle R
36
Tendon
Archettes
142
Vologne R
Jarmenil
Arches
D-11
Le Tholy
N-417
Gerardmer
Eloyes
Rupt de Cleurie R
Roan-aux-Bois
Noir Gueux
30
N-417
198(-)
St. Ame
15
Remiremont
Moselotte R
LUFT
IV
3
338
Maxonchamp
7
Rupt
Ferdrupt
308
198
Le Chene
Le Thillot
KIFFLER
Fougerolles
St. Loup
D-6
Moselle R
N-486
Faucogney

St. Quirin
Vezouse R
Cirey
Sarre Blanche R
Sarre Rouge R
XV
XXX
VI
N-393
Strasbourg
Bruche R
25
N-392
Mutzig
26
Molsheim
N-59
Badonviller
N-392
Plaine R
25
25
Vacqueville
D-8
117
14
21 NOV
708
10 NOV
Neufmaisons
25
Schirmeck
Baccarat
Bertrichamps
D-9
19
24
Hantz Pass
23
N-420
Meurthe R
18
La Trouche
100
Raon-l'Etape
N-424
Saulxures
St. Blaise-la-Roche
Barr
21
23
N-424
Clairefontaine
Etival-Clairefontaine
20
716
10 NOV
Saales
F R A N C E
Rambervillers
3
21
N-59
22
Saales Pass
D-19
Ville
26
St. Michel
20
24
Provencheres
Mortagne R
N-420
22
St. Die
N-420
18
22
16
10 NOV
Wisembach
N-59
Ste. Croix-aux-Mines
26
Selestat
103
Taintrux R
Saulcy
24
Ste. Marie-aux-Mines
25
Brouvelieures
Ban-de-Laveline
Ste. Marie Pass
St. Leonard
198
10 NOV
Bruyeres
19
Neune R
Anould
Fraize
26
N-415
Vologne R
36
24
Clefcy
Le Bonhomme Pass
Rhine R
GERMANY
19
SEVENTH
XXXX
FIRST FR
Le Tholy
Gerardmer
Colmar
3 Alg
19 NOV
VI CORPS ADVANCE
12 - 26 NOVEMBER 1944
26
Axis of Infantry Attack
25
Axis of Armored Attack
0
10
MILES

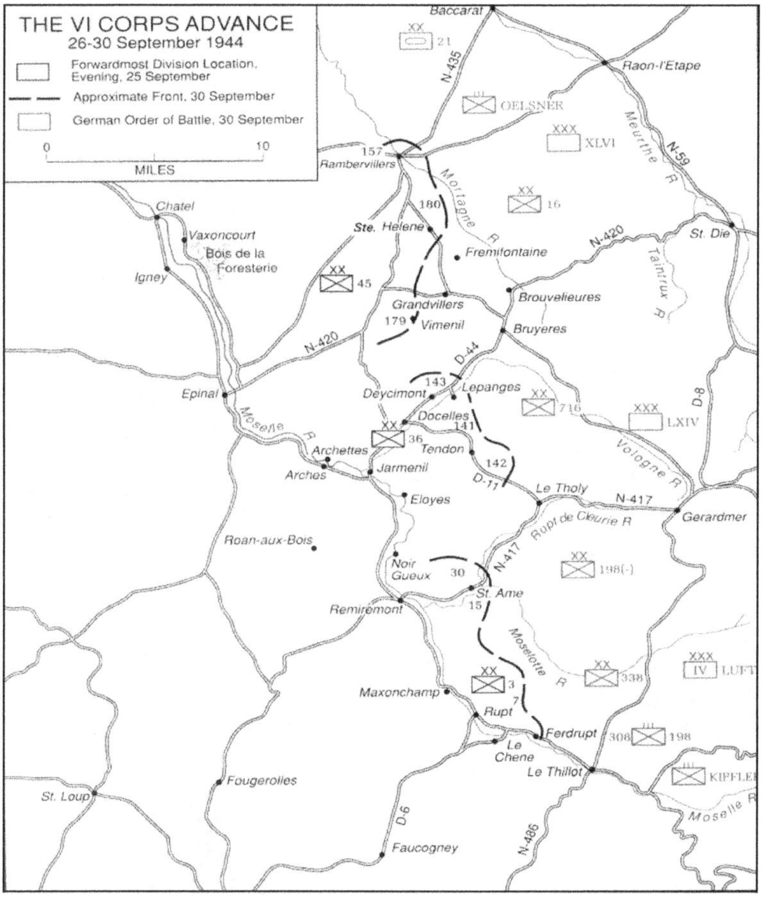

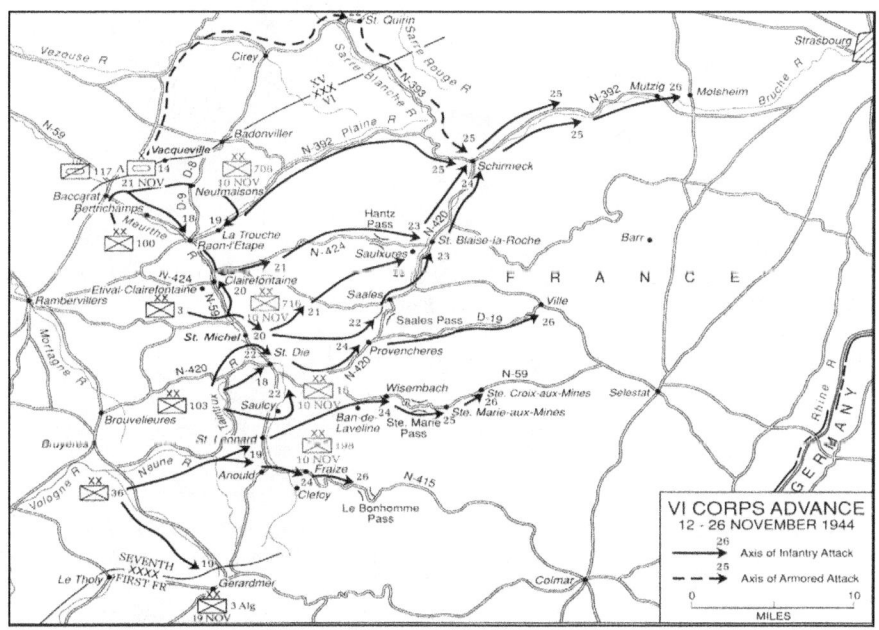

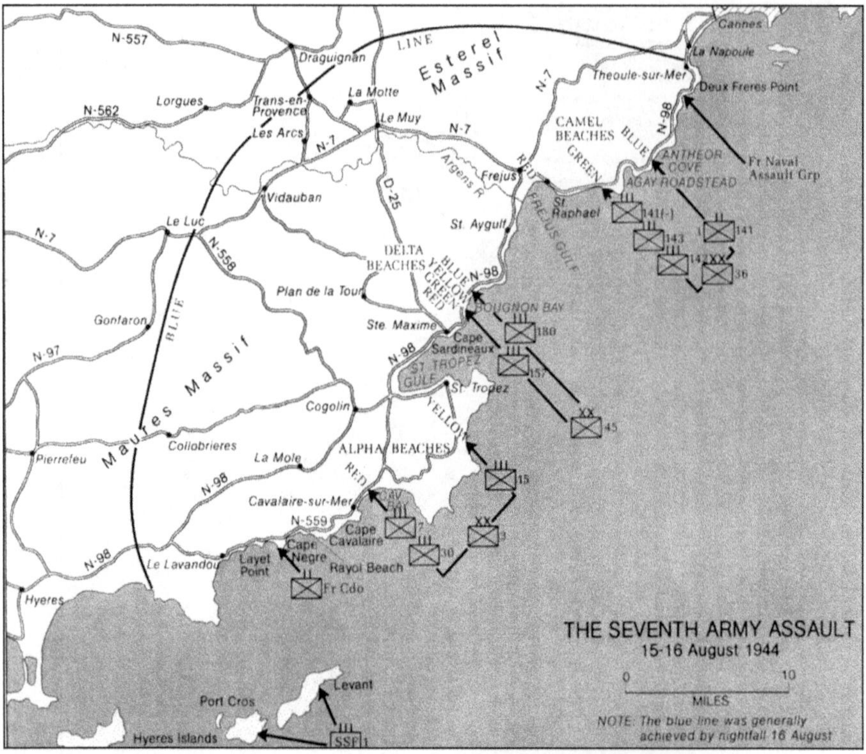

Hermann Goering, number-two Nazi, is interrogated by Maj. Gen. John E. Dahlquist (back to camera) and Brig. Gen. Robert I. Stack (right), who effected the capture of the field marshal and took him to the Grand Hotel in Kitzbuhel, Austria.
—Signal Corps photo SC 231807 taken May 9, 1945

CHAPTER 12

Anvil/Dragoon

> "Now we were 2000 yards offshore and the great rocket ships began to send their screeching cargo into the air. The sea was rolling lightly and the increased speed threw a fine salt spray into our faces. At 1000 yards the din of thousands of rockets and the shells crashing into the beach ahead become a steady roar in which the concussion caused by no single shell or group of shells could be heard."
> —*Five Years–Five Countries–Five Campaigns*, 1945[1]

The debate over invading southern France continued after the June 1944 Normandy Invasion. The debate continued after the fall of Rome, when the Allied high command was uncertain as to whether the forces in Italy should continue trying to advance northward or should stop at the first good defensive position. The Green Book series volume on the Riviera campaign states that "Although ultimately proving to be one of the most important Allied operations of World War II, the invasion of southern France has also remained one of the most controversial."[2]

Active campaigning in Italy had momentum on its side, including Mark Clark's desire to stay with a major campaign.[3] The Allies wanted to complete the liberation of Italy and to maintain the option—however unlikely its exercise—to continue into the Balkans. German commander Albert Kesselring, however, remained as skillful as ever. Not long after the 36th, 45th, and 3rd divisions left the lines to prepare for Anvil/Dragoon, the slow slog returned.

The Allied command felt that invading the French Riviera would divert German troops from Normandy. The Allies could form a complete line of advance across France. Significant numbers of French soldiers could help liberate their country. Capture of several major ports, particularly Marseille (the largest port in France), would help supply the Allied advance. Capture of English Channel ports was going more slowly than expected, and the ports were greatly damaged. On August 11, 1944, just four days before the scheduled date for what had been changed from Operation Anvil to Operation Dragoon, the final "go" decision was made.

★ ★ ★ ★ ★

Few British soldiers would participate in the Anvil/Dragoon landings. American units would provide the main invasion force, with French units brought in as soon as possible. The American units were the 7th Army, now under the command of Alexander M. Patch, though the actual invasion forces would be Lucian K. Truscott, Jr.'s VI Corps. Maj. Gen. John E. O'Daniel, who performed good work as temporary sector command under Fred Walker at Salerno, commanded the regular army 3rd Division. Maj. Gen. William W. Eagles commanded the 45th Division, originally consisting primarily of Oklahoma National Guard. Maj. Gen. John E. Dahlquist had been brought in from outside (the 70th Division) to replace Walker as commander of the 36th Division.

The VI Corps did not control the other units in the invasion. Airborne assault troops, including a British parachute brigade and light artillery battalion, would be commanded by Robert T. Frederick, now a major general. His old 1st Special Service Force was now under the command of Col. Edwin A. Walker. Two French commando units would also take part in the landing. Additional assistance would probably come from the French Forces of the Interior, FFI, better known as the Maquis. The Maquis had 75,000 men and women in the field, but only about one-third were armed.

The main body of French troops would come in after the first wave. Gen. Jean de Lattre de Tassigny, a full general, would temporarily serve as commander of the first French corps brought into southern France. Operations would be upgraded to army level, under LTG Alexander Patch. De Lattre was waiving rank so French units could get into action. When the second French corps arrived, de Lattre would command the 1st French Army. It would remain under Patch until Jacob Devers, also a lieutenant general, took command of the 6th Army Group.

Vice Adm. Henry K. Hewitt, United States Navy, would command naval forces for the invasion. He would have charge of the ground troops until Patch could establish his headquarters onshore. Hewitt had commanded naval forces at Salerno. Rear Adm. Spencer S. Lewis would have charge of naval units in the 36th Division area. Direct air support would be provided by the XII Tactical Air Command. The 15th Air Force had major responsibility for area strategic bombing.

As to the enemy, Army Group G, under Gen. Johannes Blaskowitz, had charge of the south of France, under the overall direction of OB West. The German 1st Army, in southwestern France, had withdrawn most of its men to the Seine River in early August. The LXIV Corps was the only remaining element in the area. The 19th Army had seven divisions in the area but only one reserve division near the invasion beaches. The strong 11th Panzer Division was spotted on its way to southern France, however. The Germans had about 30,000 men in the assault area, 200,000 within a few days' march, perhaps another 50,000 also available.

German evaluation of Allied intentions in southern France varied as much as Allied plans. They finally came to expect an invasion after Normandy, in roughly the same location and time as the actual August 15 assault. An early-August German counterattack in northern France had failed, and the Germans were desperately trying to get as many men as possible out of the so-called Falaise Pocket. The Germans had switched commanders in northern France twice, plus were suffering from the confusion caused by the July 20 plot to kill Adolf Hitler. Their basic plan in the French Riviera would eventually end up somewhat similar to Kesselring's plan in Italy—delaying the Allies as much as possible while withdrawing to stronger, mountain-based defense positions.

According to the Green Book, "The choice of assault sites along the Mediterranean coast of France was in large measure dictated by the ANVIL operational concept—to land in southern France, seize and develop a major port, and exploit northward up the Rhone valley."[4] Three major mountain masses, with natural corridors to the north, rise close to the coast. Directly assaulting Marseille and Toulon would hit beaches both poor in quality and heavily defended. Beaches to the west of Marseille were first-rate, but soon after landing, forces would run into land the Germans could easily flood. The best area was the coastline between Toulon and Cannes. Toulon was the first major Allied target, but several smaller ports, including the famous St. Tropez, made this area attractive. The invasion area was eventually narrowed down to one between Cape Negre on the west and Theoule-sur-Mer, roughly fifty miles to the east. The three infantry divisions would assault the inner thirty or so miles of this area, with commandos on both flanks.

The 1st Special Service Force would begin the attack at about midnight on August 15, landing on two islands at the left flank of the beach. These islands were supposed to hold extensive German artillery positions. Soon after this, French commando units would attack both beach flanks. About the same time, the paratroopers were scheduled to drop in the Le Muy area, about ten miles inland, clear it for glider landings, and block German attacks from that general direction.

The main landings would begin at the relatively late hour of 8:00 A.M., in the Cape Cavalaire–Antheor Cove beaches. Daylight was required for accurate bombardment, and the landing forces needed as much daylight as possible to get sufficient troops and supplies ashore, secure the landing—and, having learned the lessons of Salerno and Anzio, capture nearby hills.

The 3rd Division would land on the left, initially with two regiments, clear its area, capture St. Tropez, and head toward Toulon. The two assault regiments of the 45th would land in the center, take the nearby hills, and capture Ste. Maxime. The 36th would play a basically defensive role on the right. The division would advance to the nearby hills, make contact with the paratroopers, and seize the small port of St. Raphael. Only one regiment was scheduled to land at H-hour, about four miles east of St. Raphael. A battalion would land at Antheor Cove, three miles away. The second regiment would land at the same beach, with the third landing at 2:00 P.M. between the mouth of the Argens River and St. Raphael. The invasion planners expected that the third regiment would be supported by elements of the 45th. If necessary, however, the third regiment could land at the main landing beaches to the east.

The convoys left as scheduled, and things seemed likely to go well. However, the same had been true at other landings, including Salerno, as veteran members of the 36th had to remember. The Green Book notes:

> Although Allied intelligence had pointed out the disabilities of the German defenders, there was always the chance that ULTRA or the other intelligence sources had missed some critical last-minute German troop deployment and that the assaulting force might be in for surprise.[5]

The poetically written "Operations in France for the Month of August 1944," for the 142nd Infantry, describes the landing:

> A hundred small craft bobbed up and down in the blue Mediterranean within a thousand yards offshore, waiting. Hundreds of other vessels, large and small, were in the vicinity, each with an assigned job to perform . . . Looking toward shore from the sea, . . . spectacular white streamers of smoke fanned out in clusters as phosphorous smoke shells burst in the air. Billows of smoke poured from floating pots laid to form a protective screen . . .
> The three thousand men in the hundred small craft watched all this with anxious interest . . . Southern FRANCE [sic] was being invaded.[6]

Every previous landing on the continent was met by the Germans with heavy resistance, either on the beach itself, as in Normandy, or with heavy counterattacks soon after landing, as happened at Salerno. There was every reason to believe the same would happen on the Riviera. The Allies sought, by every means available, to make it as difficult as possible for Army Group G and its 19th Army to reinforce the beachhead, or to counterattack the landing troops.

The French resistance, the FFI, actually struck the first blow of Anvil/Dragoon. Increased sabotage efforts in southern France after August 1 made it difficult for Army Group G to maintain communications. The Germans had to use combat troops to keep supply routes open to the Riviera. Rail lines and phone lines were regularly cut. Army Group G had a great deal of trouble communicating with OB West near Paris, and with its units on the Atlantic coast. The mountains made it hard to communicate by radio. The FFI had become so aggressive that the Germans could only move large, well-protected convoys along the highways and railroads in southern France. By August 7, General Blaskowitz considered the FFI as being an effective army in his rear.[7]

The commander of Task Force Butler, an ad hoc pursuit force General Truscott had created, wrote of the FFI:

> The German dread of the Maquis came to the surface continuously during our race into the interior. Really, some of our adventurous young officers became quite persuasive salesmen. Many and many a garrison was taken after a few shots—an American advanced under a white flag and a parley. If the German commander could be convinced that he and his force would become American prisoners, and not be turned over to the French, surrender usually was accomplished forthwith.[8]

By the time of the invasion, in southern France the Germans had effective control over only the Rhone Valley, the Carcassone Gap, and strips along the Atlantic and the Mediterranean. Extensive strategic Allied bombing contributed to the German confusion by destroying most bridges over the major rivers. Tactical bombing and shelling, however, did not heavily damage German defenses in the landing area.

The actual invasion began with paratroop landings in the early-morning hours, the same technique used in Normandy. Aided by cooperating French resistance forces and commandoes landed by sea, the paratroopers succeeded in cutting off most of the invasion beaches from reinforcing German forces. Task Force Butler's commander later wrote that "The airborne operation was highly effective and so disrupted communications in paralyzing the [German] 62nd Corps headquarters (the enemy was forced to destroy his radio the first day) that Corps lost track of what was happening on the beaches."[9]

A recent historian states that "Other Allied boats roamed ostentatiously along a considerable stretch of coast, listening to the screams of German sentries being knifed by the successful commandoes, and trying to create still further misconceptions about where the main landings would put down."[10]

Indications are that German commanders were not fooled by these and other deception measures, including dummy "paratroopers" similar to those used in Normandy.

The attack on the beaches themselves opened with an artillery barrage from naval guns. Two of the three assault divisions, the 3rd and 45th, experienced little trouble or significant resistance to their landings. By the time August 15 was over, the 3rd Division had entered St. Tropez, but found it already had been liberated by FFI and paratroopers dropped by mistake in the area. By the end of the day, elements of the 3rd had linked with elements of the 45th Division. The 45th Division did not get as far inland, as German resistance was slightly stronger. However, it did not require the services of its 3rd regiment, the 179th Infantry, being kept in reserve. A platoon of the 45th Reconnaissance Troop met up with paratroop units south of Le Muy.

The 36th Division did not have as easy a time. The division occupied the right flank of the landing. Four initial beaches had been selected, near the French city of St. Raphael, from east to west, Red (east of St. Raphael), Green, Yellow, and Blue.

Yellow Beach would have been an excellent landing site. Unfortunately, landing on this beach would have required going into an estuary, hitting the shore with land, and likely German fire, behind them as well as in front. Regimental records for the 141st Infantry report that

> the area around the bay was so heavily fortified and its waters so extensively mined that no attempt was to be made to assault the beach.[11] This beach would be flanked and attacked from the rear. When the beach was taken, engineers would prepare the beach for landing supplies.

There was little resistance on most beaches that day, unlike the case in Normandy. Still, an amphibious landing was an unforgettable experience. A history of the 141st Regiment, prepared by its veterans, describes their landing:

> Now we were 2000 yards offshore and the great rocket ships began to send their screeching cargo into the air. The sea was rolling lightly and the increased speed threw a fine salt spray into our faces. At 1000 yards the din of thousands of rockets and the shells crashing into the beach ahead become a steady

roar in which the concussion caused by no single shell or group of shells could be heard. Now the water became rough and the boat lurched violently from side to side.[12]

The first battalion of the 141st Regiment was the only one to use Blue Beach, the farthest to the east. From the same history:

> The regimental combat team had been split up to assault two separate beaches. The 2d and 3d Battalions assaulted GREEN beach while the 1st Battalion assaulted BLUE beach. Neither GREEN nor BLUE beaches could be called beaches in the technical sense of the word. An extremely narrow strip of rocky shale separated the water from steep embankments directly to the rear of both beaches. Beyond the coastal highway and railroad which paralleled the shoreline along the bluff above the beaches, the terrain ascended in heighth [sic] to high hills, which in turn arose into a mountainous sector to the north. The entire shore area was dotted with villas, hotels and resorts of the famed Riviera "Cote d'Azur" . . .[13]

The area was also dotted with German defenses, including "pillboxes . . . in advantageous defensive positions flanked by trenches."[14] "Other than the narrow beach, all portions of the land bounding the narrow inlet [the 1st Battalion assaulted] comprised rocky cliffs."[15]

The 1st Battalion met some resistance but managed to quickly seize high ground overlooking the beach and protecting the division's right flank, and to capture 1,200 German prisoners. The battalion won a Presidential Unit Citation for these actions. The other two battalions in the 141st landed on Green Beach just after 8:00 A.M., with little problem and light resistance. "The beach line was a sight to behold. The Navy had done a superb job. The only remaining signs of the barb [sic] wire entanglements was a picket here and there," reported the 2nd Battalion of the 141st.[16] These battalions quickly took the high ground near this beach.[17] The postwar regimental history continues,

> For most of us [in the 141st Regiment] the memory of D Day in southern France is not too unpleasant. It has been a spectacular show. We had suffered few casualties and our regiment had taken the beach, over which [almost] the entire division had landed, before midnight on D Day. Almost to a man we

liked the first impression of France. D Day was warm and sunny and the reddish firm soil was a welcome contrast to the powdered ankle-deep dust of Italy.[18]

The 143rd Regiment started to land on Green Beach at 9.45, followed soon after by division commander John Dahlquist and his staff. They planned to watch, and assist if needed, the 142nd landing on Red Beach, several miles to the west. The 2nd Battalion headed directly toward St. Raphael but encountered stubborn resistance from several German strongpoints. Its lead elements had not reached St. Raphael by 2:00 P.M.

The Red Beach landing was planned to take place several hours after that on Green Beach, to give the 143rd time to head west and take the heights dominating Red Beach. The division was eventually supposed to occupy a line on the right flank of the corps, with a twelve- to fifteen-mile depth inland.

Red Beach had the heaviest German defenses, including underwater obstacles (which had proved such a hazard in the Normandy Invasion). The 142nd later found out that "The defenses of GREEN BEACH were almost entirely pulled over to reinforce those of RED BEACH the day before our landing."[19] Heavy bombardment, and the use of drone boats filled with explosives, failed to sufficiently clear obstacles from the beach. At one point, one of the radio-controlled drone boats, loaded with explosives, reversed course and headed straight at the convoy—to the distress of everyone watching. Fortunately, it turned again and ran aground on an inlet near Red Beach, though without exploding.

The 142nd's records report:

> RED BEACH was the obvious place to land in this coastal sector and its defenses had been stoutly organized by the enemy. A broad sandy beach was near the small port of ST. RAPHAEL and the town of PREJUS. Access to it was essential for the quick follow-up of supplies and equipment which must accompany any invasion force, and in this particular facilitate the intended drive of the Army Northwest in the ARGENS river valley to cut off the ports of TOULON and MARSEILLES...
>
> The alternate plan anticipated the possibility of the 142nd Infantry landing behind the 141st and 143rd on GREEN BEACH. An apparent disadvantage lay in the fact that it

hardly seemed feasible to land a whole Division with the necessary armor and vehicles over the one rock-enclosed beach...."[20]

It was 2:00 P.M., 1400 hours, the time that was finally set for the 142nd RCT to land on Red Beach. Col. George E. Lynch, commanding the 142nd, radioed his leading battalion to find out the reason for the delay and learned of the poor performance of the drone boats.

At 3:15, the navy commander of the assault on Red Beach, Rear Adm. Spencer Lewis, ordered the landing craft to head to Green Beach. General Dahlquist ordered the 142nd to land according to the alternate plan, to use Green Beach. "During the course of D-day, the alternate plan was put in effect for the 142nd Infantry as the naval demolition teams were unable to reduce the underwater obstacles on Red Beach."[21] Reports from shore indicated that Red Beach could be captured from overland, while Green Beach, where the bulk of the 36th had landed in the morning, was proving to be a better unloading beach than expected. St. Raphael and Red Beach were captured from the rear within two days.

Lt. Gen. Lucian Truscott was watching the planned Red Beach landing. He thought the bombardment was going well and was shocked to see the landing flotilla first stop and then head out back out to sea. His boss, overall invasion and 7th Army commander Lt. Gen. Alexander Patch, and overall navy commander Vice Adm. H. K. Hewitt, shared Truscott's feelings. Truscott later wrote, "Then, while we watched, helplessly, to our profound astonishment the whole flotilla turned about and headed out to sea again. Hewitt, Patch and I were furious."[22] Truscott, in fact, threatened to relieve the 36th's commander, Maj. Gen. John Dahlquist, or court-martial Col. George Lynch, the commander of the 142nd Infantry, if either of them had ordered the diversion.[23] Truscott, in his 1954 memoirs,[24] wrote that the diversion and the apparent delay in capturing Red Beach and St. Raphael set back the schedule for the whole operation:

> It was in fact almost the only flaw in an otherwise perfect landing. Failure to carry out this delayed landing as arranged was to hold up the clearing of Beach 264A by more than a day ... It was in my opinion a grave error which merited reprimand at least, and most certainly no congratulations. Except for the

Chapter 12: Anvil/Dragoon

otherwise astounding success of the assault, it might have had even greater consequences.[25]

Others disagreed. A recent history of the invasion concluded that "Truscott's criticism appears unjustified."[26] The 36th Division G3, Lt. Col. Fred W. Sladen, wrote in his dairy:

> August 17, 1944: . . . Red Beach is finally opened this evening. Good thing we didn't land there, mines by the thousands and casements and underwater mines and obstacles. We would have taken heavy casualties.[27]

A member of the division, who later wrote a history of the 36th Division in the last year of the war, also quoted the commander of the French forces in the invasion, Gen. Jean de Lattre de Tassigny, as supporting the decision to switch beaches. The same division veteran offered his own opinion that "The diversion was correct, the decision to do so was correct, and the results were more than satisfactory."[28]

George Lynch, commander of the 142nd Infantry, later said he felt his regiment would have been so damaged by a Red Beach landing that it would have been unable to meet most of its later objectives. The diversion did not hurt Lynch's career, as he eventually retired from the army as a major general.

The diversion came as a surprise to the Germans. The 142nd's record of operations reports that "A [German] CORPS staff officer admitted that they had never imagined that we could or would attempt to land a whole division on GREEN BEACH."[29]

During the Normandy Invasion, June 6, 1944, the forces on Omaha Beach came in where they had planned to come; Utah Beach was a mile west of target. The end results of Utah Beach were probably more valuable, thanks to the capture of causeways leading inland. There were less than 200 casualties at Utah Beach, 2,000 at Omaha.[30] The accidental diversion at Utah Beach and the intentional diversion to Green Beach seem to have met the only bottom-line justification, the one criterion for variance from superior orders—they worked, and they did not damage the overall plan. "Beaches secured and positions consolidated, as Division continued advance . . . against sporadic enemy opposition," the 36th Division narrative of operations for August reported.[31]

By the end of the second day, as the poetic clerk of the 142nd phrased it, "The beachhead was now secure and our forces were rapidly fanning out beyond. The invasion success was won with lightning suddenness and meager casualties."[32] The controversial diversion had worked and left all regiments of the 36th Division in excellent shape for the work that followed. "Now the amphibious aspect of the operation was satisfied. What lay ahead was land fighting, that which the Army is used to doing."[33] The "blue line" of the initial bridgehead was reached, and Truscott issued new attack orders before dark on August 16.

The two primary French Riviera ports, Marseilles (the largest in France) and Toulon, had been designated "fortresses" by Adolf Hitler and ordered to fight to the death. They were captured by French divisions serving with the 7th Army on August 28—far sooner than expected—repaired, and put into service as major supply ports for the entire Allied effort.

On August 16, Hitler reluctantly agreed with his generals. Withdrawal orders were transmitted the next day. Army Group B was ordered to withdraw from the Falaise Pocket in Normandy to more defensible positions closer to Germany. Army Group G was ordered to pull back to the tough Vosges Mountains.[34] The Allies knew about these orders almost immediately.[35]

August, and a good portion of September, consisted of a giant race, sometimes hindered by the need to overcome supply problems—"Our vehicles are too few in number for what we have to haul,"[36] one company reported—to try and stop the withdrawing German forces from Army Group G, under Gen. Johannes Blaskowitz, from reaching stronger positions in the mountains along the German and Swiss borders.

The coast of southern France was undermanned, as many troops had been withdrawn to fight the Allied effort in Normandy, when the Germans finally realized it was the main invasion. Hilly terrain focused this withdrawal to the roads running up the Rhone River valley:

> With the disorganization of the Germans and their attempt to withdraw up the RHONE valley the campaign developed into a fight for the valley and control of road nets. With great mobility, units were maneuvered on the flanks of the enemy, de-

stroying their convoys and large quantities of materiel and personnel.³⁷

Apparently the shortage of vehicles was overcome before August 22, when Task Force Butler and elements of the 141st Infantry arrived at Montelimar.

CHAPTER 13

The Battle of Montelimar

> "As the Germans attacked over our positions, we called the artillery down on our own location. There was nothing else we could do."[1]
>
> —LTC Theodore Andrews,
> 143d Regiment, 36th Division, 1981

The Germans were withdrawing their troops in southern France back to the Vosges Mountains. Intelligence reports indicated that they were not going to move troops from Italy to southern France, but would continue to mount a strong defense north of Rome. Lt. Gen. Alexander Patch, commander of the 7th Army, could plan accordingly. With no concern about his flanks, Patch could use virtually all of his forces in as rapid an advance as possible up the Rhone River. Patch wanted to get ahead of the Germans, block their path, and force them to attack at places the Americans chose. One such place was just north of the city of Montelimar, 120 miles from the invasion beaches, eighty miles inland from the coast.

The Rhone River valley, and its two major highways, run nearly north-south. It was the quickest, most logical route for a retreating German force—especially one protected by the powerful 11th Panzer Division, which was heading to the area. The valley has one particularly significant characteristic. It varies in width. Usually a broad plain, with a lot of space along its two main highways, it frequently narrows. These bottlenecks are obvious points for defending the valley or stopping a retreating army. Montelimar itself originated as such a defensive position, a Roman fortress at a particularly narrow spot, with rugged hills and cliffs near both banks of the river. "The Montelimar Gate," the area came to be called.

Task Force Butler was created as a rapid-movement force to try to block the German retreat. In order to leave him with an exploitation force if French armor was taken from his command, about a week before the Anvil/Dragoon invasion VI Corps Commander Lucian K. Truscott "planned to constitute a Provisional Armored Group from elements of the corps."[2] Placed under the command of Brig. Gen. Frederick B. Butler, VI Corps deputy commander, the force consisted of a motorized infantry battalion, thirty medium tanks, twelve tank destroyers, twelve self-propelled artillery pieces, and a light cavalry squadron with armored cars, light tanks, and trucks. Task Force Butler has been described as a "balanced, mobile offensive force."[3] With a strength equivalent to somewhere between one and two regiments, however, it was not particularly strong.

By August 19, Butler's force (to which had been added a battalion of the 143rd Infantry) was already speeding northwest, preparing to swing around in front of the Germans. The force reached Sisteron, about halfway to Montelimar, by noon. About this time, Patch told Truscott to have a division ready to advance northward, in the general direction of Grenoble. Truscott ordered the 36th Division's commanding general, Maj. Gen. John Dahlquist, to have the 36th Division ready to move early the next day. Truscott expected that at least one regiment of the 36th would reach Sisteron by the end of that day. Butler was told to await Dahlquist, but also to continue patrols westward to seize the high ground north of Montelimar.

Butler never got this message. The mountains, which had interfered with Army Group G and 19th Army radio transmissions, were neutral. The only instructions Butler received, on the night of the 19th–20th, stated that the mission of the task force was unchanged—reconnaissance northward. Shortly before midnight, Butler reported that he intended to continue in the morning. He also reported a shortage of fuel and asked for further instructions, particularly whether he should head north to Grenoble, or west to Montelimar. Butler was uneasy remaining stationary at Sisteron, within enemy territory. The French Forces of the Interior (FFI) had reported a strong German force at Grenoble, and one within thirty miles of Task Force Butler. Butler was oriented toward a northward advance but sent his operations officer by liaison plane to corps headquarters for more specific orders.

The evening of the 20th, Brig. Gen. Robert I. Stack, assistant division commander of the 36th, arrived with part of the division headquarters and the remaining two battalions of the 143rd Infantry. The operations officer had returned with the news that further orders would be coming that night. Stack reported that the 36th Division was headed north, with the 142nd now thirty-five miles away. The 141st would follow the next morning. Stack warned Butler, however, that the division was being hindered by shortages of fuel and trucks, and he did not know when it would arrive. The 143rd was to head for Grenoble the next morning. Stack thought this was the most logical move for Butler.

Radioing Dahlquist that evening, Stack relayed instructions for Butler to stay in the area of Sisternon until most of the 36th had arrived. Several hours later, after having spoken with Truscott, Dahlquist told Butler to stay in the area and cancel the move to Grenoble. Finally, at about 8:00 P.M. on August 20, Truscott finally ordered Butler to move to Montelimar at dawn the next morning, as rapidly as possible, to seize the town, and to block the German route of withdrawal. The 36th Division would follow as quickly as possible. A slight change in the written orders, received by Butler early the next morning, told him to seize the high ground north of Montelimar, but not the city, before dark the next day. Two battalions of corps artillery were being sent, but only one infantry regiment, with supporting artillery and other units, the 141st, was specifically ordered to immediately join Butler. The rest of the division would not be ordered to the Montelimar area until twenty-four hours later. Dahlquist would take command as soon as he arrived on the scene.

Butler's force, which had been spread out and oriented north to Grenoble, was regrouped the morning of the 21st. Some of his force had to be left to hold gap in his rear, and Croix Haute Pass until the 36th Division elements could arrive. By the end of the day, however, the bulk of Butler's force was within thirteen miles of the Rhone River, at the town of Crest.

Crest was the northeast corner of what came to be known as

the Montelimar Battle Square.⁴ Boundaries for this area were the Drome River on the north, the Rhone on the west, the Roubien River on the south, and the implied line north–south from Crest to the Roubien on the east. The area alternates from flat farmland to rugged, hilly country. Route N-7 runs through Montelimar, two miles from the Rhone, and then continues up the valley. A secondary road, D-6, passes through Montelimar, then cuts into mountainous country before rejoining N-7 about fifteen miles north of the city. Railroad tracks run on both sides of the Rhone, with another road paralleling N-7 to the west of the Rhone.

Butler's men reached Marsanne, in the center of the "square" and blocking D-6, late on the 21st. The advance party, under Lt. Col. Joseph G. Felber, recognized that a hill called Hill 300, immediately over N-7 and with a clear view of the other Rhone arteries, was the key terrain feature. Unable to occupy the whole ridge, Felber set up outposts, road blocks, and other positions. Germans were already passing through on the road. Guns opened fire as soon as they could unlimber. A light armor and infantry roadblock was chased back into the hills by a German attack at dusk. The Americans also destroyed the N-7 bridge over the Drome, in the northwest section of the disputed area, with no resistance from the Germans.⁵

On the north bank of the Drome, light armor sent west from Crest spotted a German truck column fording a stream. Fifty of these trucks were destroyed before the troop was ordered back to Crest, to protect the roads to Puy St. Martin, in the middle of the "square."

Just before midnight on August 21, Butler radioed Truscott to report the task force's arrival at the objective area. Butler reported that his forces were thinly spread out. With the reinforcing artillery and a 36th Division regiment, he thought he could hold against determined German attacks, and attack Montelimar successfully the next afternoon. These reinforcements, and supplies, had not arrived by the next morning.

The Germans moved first. Units attacked north from Montelimar, taking Sauzet and forcing an American outpost back into the hills. This was just a feint, however. Larger German force attacked from south of the Roubion River, advancing on Puy St. Martin and Marsanne, behind Butler's defenses. Puy was occupied that afternoon, cutting Butler's supply line back to Crest and Sisteron. Fortunately, Butler's unit left at Gap had been relieved by the 36th and arrived in time to successfully counterattack into Puy.

Butler thought the German attacks were still only probes, to determine his strength. He expected a far larger attack the next morning, August 23. No part of the 36th arrived during most of the 22, just Butler's own men from Gap and the Croix de Haute pass, and the two battalions of corps artillery. The task force was also beginning to run out of tank and artillery ammunition.

About 10:00 P.M. the evening of the 22nd, the 2nd Battalion of the 141st regiment arrived. The regimental operations report later stated that "The many miles covered by the regiment in the previous two days had placed a severe strain on the Service Company. The drivers received very little rest and the supplying of gasoline, rations and water for all units of the combat team proved to be a most difficult and arduous task."[6]

General Truscott and General Butler would be critical of the 36th Division and its delayed arrival at Montelimar.[7] Butler later wrote that, at Montelimar, "Even had the 36th Division swung in behind me rather than continue north to Grenoble the effectiveness of my position would have been enhanced," and referred to the "slowness of the relief at Gap"[8] by the 36th. An aide to General Truscott even wrote in his journal for August 21, 1944: "36th fouled up."[9]

The orders to reinforce Butler were late. The Germans had contributed to the delay, with a possible accidental deception. Ultra intercepts had already informed the Allied high command of their plans to withdraw, but by the 21st, Patch was not sure if the withdrawal had started. Truscott, that morning, had ordered the 179th Infantry, of the 45th Division, to Sisternon when word came that elements of the 11th Panzer Division had crossed the Rhone and were approaching. Truscott cancelled the move, though it turned out that only a few tanks had crossed the Rhone. It is unclear whether the message was misinterpreted or the Germans had guessed it might be intercepted and phrased it accordingly. The ruse, accidental or not, worked.

The Green Book on the Riviera campaign states that:

> The lack of reinforcements reflected American indecision.

Throughout the day and evening of 21 August neither Patch nor Truscott had been willing to make Montelimar the major effort. They were still unable to predict when Toulon and Marseille would fall, or confirm the beginning of a complete German withdrawal up the Rhone valley.[10]

Truscott finally ordered Dahlquist to start moving toward Montelimar at 11:00 P.M. on August 21. That evening at 9:00 P.M., the 141st was in Digne, eighty "crow" miles southwest (more by road) from Montelimar. Most of the division was still swinging north, toward Grenoble, in accordance with its latest orders. Division orders instructed the 141st to move towards Aspres-sur-Beach, on the route to both Grenoble and Montelimar, and still less than halfway toward Montelimar. The 141st itself had just completed, not long before midnight on August 19, a sharp action in capturing the town of Callian, only sixteen miles from the beach.

The next morning, Truscott flew up (by command Piper Cub), to find out what was happening to cause the delay in movement. Truscott remembered flying in to the 36th command post just before noon on August 22.[11] A staff officer of the 141st, present when General Truscott arrived, later recalled that the VI Corps commander's plane landed on the highway between Aspres and Sisternon, where the 141st was moving mostly by foot, toward Division HQ at Aspres:

> General Truscott got out of the plane and introduced himself. I was travelling in a jeep with [141st Regiment commander] Colonel Harmony. We told General Truscott we were on route to the Division CP to get instructions on our next employment. He got into the jeep with us and we hastened to the Division CP in Aspres.
>
> We were met by Colonel Stewart T. Vincent, the Chief of Staff, who said that General Dahlquist was "up front somewhere." He had no means of communicating with him.
>
> Truscott asked Vincent if he had received the message which he, Truscott, had sent the night before, telling Dahlquist to get the division to Montelimar as quickly as possible, with the mission of blocking the highway. He said that was the key to stopping the German Nineteenth Army.
>
> Vincent had no knowledge of the message, but Truscott demanded that the Message Center chief check. The sergeant

found that the message had been delivered late the night before to Lieutenant Colonel Fred W. Sladen, the Division G-3.[12]

Apparently Sladen had not thought to awaken Dahlquist. When Sladen got up the next morning, Dahlquist was gone. Sladen then went to Gap and Grenoble with Stack, the assistant division commander. He appears not to have mentioned the orders to Stack. There is no mention that Truscott had asked to have his orders confirmed by Dahlquist:

> Truscott was bitterly disappointed, and he turned to Harmony and told him to get his regiment to the west, vicinity of Crest, as soon as possible. He then turned to Vincent and told him to get in touch with General Dahlquist by whatever means possible and inform him to get his forces to Montelimar as soon as possible.[13]

Truscott returned to his command post and sent Dahlquist written orders. He expressed disappointment with the 36th Division's deployments. According to Truscott's memoirs, he emphasized that the

> primary mission of the 36th is to block the Rhone Valley in the gap immediately north of Montelimar. For this purpose you must be prepared to employ the bulk of your Division. If this operation develops as seems probable, all of your Division will be none too much in the Rhone valley area.[14]

The 19th Regiment was to replace the 143rd, at Grenoble, with the 45th Division eventually assuming responsibility for that area. Truscott also sent a convoy of fuel trucks to the 36th, enabling the rest of the 141st to join Butler. Telephone conversations with Dahlquist—on a line finally opened—stressed the basic orders, though making a few changes in details. The confusion with the 36th finally seems to have been straightened out on August 23, when "Division began shifting of forces to the West to block the RHONE VALLEY."[15]

The bulk of the 11th Panzer Division had actually gotten south of Montelimar. The division's reconnaissance battalion had fought with Butler on the 22nd, while most of its elements were blocking roads leading to the Rhone from the east and the landing beaches. The division was then ordered to move northward, and to take the high ground around Montelimar and secure Highway N-7 running northward. The 198th Infantry Division was also ordered to the area. Fortunately for Butler, Dahlquist, and Truscott, the Germans had the same supply problems as the Allies. The first elements reached the Montelimar area only on August 23, with the rest not expected until the next day. The Germans still intended to try and seize the initiative.

Terrain provided one final obstacle to the immediate execution of Truscott's orders. Records of the 141st Infantry describe the terrain.

> The country through which the regiment travelled was mountainous, comprising portions of the French Alps. A good highway was available most of the way but the Germans had made enough demolitions to cause numerous detours. The enemy had evidently left the sector very hurriedly as there were many places where effective demolitions could have appreciably held up the motor columns. This mountainous region was a stronghold of the FFI (French Forces of the Interior) and the information and assistance furnished by this force was invaluable. It was extremely improbable that the Allied thrust so far north in a few days could have been accomplished without their aid.[16]

Fighting started again with the arrival of the lead elements of the 141st. "For us," as a regimental history of the 141st put it, "Montelimar and the week from August 24th to the 30th was one of the fullest 168 hour weeks of the war."[17] Colonel Harmony was with the 2nd Battalion and received his regimental assignments from General Butler. The 2nd Battalion was assigned to attack Montelimar itself, with a jump-off point about five miles from the city.

Three German probes were fought off during the day. Fortunately for Task Force Butler (formally abolished on August 23 but still together) and the 36th, the Germans also did not have full strength in the Montelimar area. Company A of the 1st Battalion was able to seize the north slopes of Hill 300, about five miles north of Montelimar, without opposition.

The "main" American attack of that day came down Highway D-6, through Sauzet, aimed at Montelimar. Troop B of the 117th

Cavalry Reconnaissance Squadron moved out in Company F, the leading element of the 2nd Battalion, led by Company F. Tanks and tank destroyers had not arrived in time for the attack. As the battalion commander later remembered, "We reached a point a little more than a kilometer from Montelimar when we got into a heavy fire fight with the Germans . . . Our attack was actually brought to a halt by a combination of an intense grass and brush fire and by the Germans."[18]

Butler had yet not blocked Highway N-7, though on the afternoon of the 23rd, Truscott ordered all Rhone Valley roads blocked.

Both battalions were counterattacked that evening and night, but held their positions. However, the Germans began to infiltrate around the units, on the maze of area roads. The 2nd Battalion mission changed from an attack on the town to withdrawing to presumably better defensive positions north of the road east from Montelimar. Both units were alerted to expect a German attack the next morning. The rest of the 36th Division was on its way, but still not quickly. Increasingly strong German counterattacks started the next morning. One was against 2nd Battalion. The battalion's commander, Lt. Col. James H. Critchfield, later recalled:

> At dawn the next morning, we again attacked, with Company F in the lead. We were quite successful and they actually reached and cut Highway 7 just north of Montelimar . . . After a while they launched an attack out of the village; the situation stabilized, and turned into a fight that raged throughout the day. I can only describe it as intense . . . During the day, we had a series of attacks against us of varying intensity . . . Gradually we were forced back into a perimeter defense. Our connections to the rear were very tenuous—we really didn't have a rear area except for those up near Marsanne. There were no stable lines at this time.[19]

American units were holding on in the general area, aided by artillery and other supporting elements. However, they were pushed back from blocking N-7, their primary goal in the area. Dahlquist seemed more concerned with holding his position than with any active offense against the German escape route. The 142nd and 143rd, still not arrived at the area, received confusing orders from division headquarters. Truscott spoke with Dahlquist, again reminding the division commander to block N-7. One can only imagine what Truscott would have said if he had learned about a copy of Dahlquist's operational instructions for the next day falling into German hands. A liaison officer left them in a jeep, fleeing a German roadblock.[20] Fortunately, all of the 36th Division arrived in the battle square that night.

The complex German battle plan for August 25 called for attacks from six different directions. If all went well for them, they would be able to surround and virtually destroy the 36th Division and Task Force Butler. Coordination proved impossible, and the attacks had different degrees of success. One task force from the 11th Panzers, based on a panzer grenadier battalion, attacked eastward along the Drome River from N-7, in the north of the combat area. The attack started out just before noon, reaching Grane before 2:00 P.M., about halfway to Crest at the northeast corner of the "battle square." Other German units seized the town of Allex, north of the Drome, about the same time. Dahlquist sent what remained of Task Force Butler, a weak battalion with supporting elements, north from Puy St. Martin halfway to Crest to cover his supply route. Butler sent a tank platoon toward Grane. It was not able to retake the town, but set up a blocking position just short of the town. No further German attacks in this area followed. The Germans in Grane seemed content to defend their accomplishments. German attacks in the southern part of the box accomplished little.

The main American offensive of that day was an attack directly at Highway N-7. The Hill 300–La Coucourde area, about five miles north of Montelimar, was left unprotected by the Germans for most of the day. However, defending a six-mile front made it impossible for Colonel Harmony to put together a 141st Regimental attacking force until late in the day. About 4:00 P.M., units of the 2nd Battalion, 143rd Infantry, secured the northern portion of Hill 300. The 1st Battalion, 141st Infantry, moved directly at the highway. Elements of the German task force supposed to be in the area, Group Wilde, arrived but were fought off. By 7:00 P.M. an American rifle company (later reinforced by a second company), four tanks, and seven tank destroyers had set up a roadblock. German traffic was again piling up south of Montelimar.

American artillery kept the Germans from assembling for a counterattack until nightfall, but Colonel Harmony was concerned about whether the road could be held. German pressure against the rest of his regiment was making it impossible to reinforce the blocking force. He was having trouble keeping it supplied. The colonel suggested the force retire into the nearby pass at Hill 300 for the night, blowing up several bridges before leaving. They could return in the morning. Dahlquist vetoed this idea, in compliance with Truscott's orders.

By this time, Maj. Gen. Wend von Wietersheim, commander of the 11th Panzers and acting commander of a temporary corps combining the 11th and 198th infantry divisions, had become annoyed with the failure of his plans, and of his men to keep the road open. He organized an attacking force from what he could scrape together near Montelimar and personally led a midnight attack on the roadblock. By 1:00 A.M., three tanks and six tank destroyers had been knocked out, with the rest of the force dispersed. Wietersheim sent some of his forces to seize the high ground over the highway, to prevent an easy American attack in the morning.

When the fighting on the 25th (and early on the 26th) ended, neither side had done particularly well. Dahlquist had still used little of his strength, with most of the 142nd and 143rd having seen little action. The road was open. Opening the road, however, was all the Germans had accomplished. Their forces were spread out, vulnerable to attack at weak points.

On August 26, Truscott ordered the 45th Division to send the 157th Regimental Combat Team (infantry regiment and attached supporting elements) and the 191st Tank Battalion north to the Montelimar Battle Square area. Part of the force arrived on the 25th, to be used as 36th Division reserve. The rest, arriving the next day, stayed at Crest as corps reserve. The 3rd Division was also on its way to the area. No major German units arrived, though part of the LXXXV Corps was still heading toward Montelimar from the south. However, von Wietersheim asked to be relieved of responsibility for the acting corps so he could just handle the 11th Panzers. Lt. Gen. Baptist Kneiss, commander of the LXXXV Corps, was given control of the German troops in the Montelimar area.

The only offensive action Dahlquist planned for the 26th was an attack by Task Force Butler, through the Condilac Pass, to restore the roadblock. The major portion of this attack started in the early afternoon. Butler sent two rifle companies of the 3rd

Battalion, 143rd Infantry, over the northern slope of Hill 300. As they neared the highway, they were reinforced by a platoon of medium tanks and some tank destroyers moving on the road directly out of the pass. The attacking Americans ran into Germans moving up from the south and attacking down from the north, and the best they could do was hang on to Hill 300 after again being pushed off the road. The tanks and tank destroyers withdrew. Efforts by the 143rd in the next few days were also unsuccessful.

The commander of the 3rd Battalion later remembered the battle:

> We didn't have much trouble taking the high ground, but we had trouble keeping it . . . Fortunately, we had good artillery communications, and they were blasting the valley.
>
> As the Germans attacked over our positions, we called the artillery down on our own location. There was nothing else we could do.[21]

The Presidential Unit Citation the battalion won described the rest of their battle:

> The following day, when attacked by an enemy force of battalion strength, units of the battalion fought valiantly to repel the attackers and inflict upon them an estimated 30 percent dead, while other units of the battalion courageously beat off successive enemy tank and infantry attacks from the north.
>
> The members of the battalion directed thousands of rounds of mortar fire into the enemy, blocking the highway with the debris of destroyed vehicles and trucks.
>
> On 29 August 1944, the enemy in overwhelming numbers desperately attacked the battalion and succeeded in infiltrating through and dividing it into small units. Although completely isolated from other units and faced with possible annihilation, the members of the 3rd Battalion fought furiously to hold their positions and by mid-morning had completely beaten the hostile forces, who suffered tremendous losses in personnel and equipment.
>
> During this action, the 3rd Battalion captured more than 600 prisoners, including the Commanding General of the German 198th Infantry Division.[22]

Blocking of the main German escape routes through the Rhone was intermittent and was mainly accomplished by artillery.

American artillery was short of ammunition, not able to do as much damage as it would have liked to the steady stream of targets. The 3rd Division was hampered by supply and transportation shortages in getting to the combat area. Units of the 36th Division could not move as fast as hoped in the battle square itself.

Truscott, however, was able to travel. He arrived at Dahlquist's headquarters the morning of August 26, intending to remove Dahlquist. Truscott complained that Dahlquist had failed to carry out instructions to block the highway and was sending in faulty situation reports.[23] Truscott wrote that he told Dahlquist,

> John, I have come here with the full intention of relieving you from your command. You have reported to me that you held the high ground north of Montelimar and that you had blocked Highway 7. You have not done so. You have failed to carry out my orders. You have just five minutes in which to convince me that you are not at fault.[24]

Dahlquist blamed much of this on the confusion of battle, including one case where the wrong hill was taken.[25] German attacks in the northeast corner of the square, at Crest and Puy, made their defense necessary. Supply and transportation shortages made it impossible to concentrate sufficient combat power on Hill 300 or to establish a permanent physical block across Highway N-7. Truscott saw the terrain difficulties but ordered further attacks the next day. Though Truscott remained unhappy, Dahlquist kept his job.

The author of the 141st's regimental history later wrote:

> In the fighting around Montelimar, operations were often in a confused state. Near Crest a fleeing Jerry motorcyclist approached a crossroads near the Division Command Post. Standing there was one of the faithful Division MP's attempting to keep straight the mobile affairs of the 36th. Seeing the Jerry speeding toward him in frantic flight, the MP, following the dictates of his habits, helpfully waived him on.[26]

Tactical operations remained basically stalemated through the next few days. Neither side did as well as they might have. Both failed to capitalize on opportunities. Supply problems made it impossible for the Americans to close the road and trap the German troops. Most made it out. But they took heavy materiel damage, and high casualties in numbers and percentages of strength engaged. German units passed into the Vosges far weaker than they might otherwise have.

The 36th Division was primarily responsible for this damage, though Dahlquist probably was not as aggressive as he could have been. In a letter written midbattle, Dahlquist blamed himself for letting the Germans escape.[27] Truscott would agree. However, a recent book[28] points out that Dahlquist was new to commanding a combat division and had never before worked with Truscott. He probably could have used closer guidance in his first major engagement. He definitely could have used more supplies, transportation, gasoline, and ammunition. The rapid advance inland of the VI Corps, and the fact that Marseille and Toulon were not captured until the end of the battle, not repaired as ports until sometime later, made this unlikely.

There was room for improvement, but objectively the 36th Division passed its first real test in France:

> Our operations in Southern France were completed. Ahead of us, behind misty grey curtains of rain and fog, rose the formidable Vosges Mountains, which no army had ever before managed to penetrate in force.[29]

CHAPTER 14

Routine Combat

"The fighting during the month of October was comparable to jungle fighting . . ."[1]
—36th Division after-action report for October 1944

The Battle of Montelimar was still underway when Lt. Gen. Alexander Patch, 7th Army commander, began planning and issuing orders for the next phase of the American advance. Patch told Truscott that his VI Corps would continue its advance up the Rhone. The first goal, to be reached as rapidly as possible, was Lyon, 75 miles to the north of Montelimar. The next, 110 miles away, was Dijon. Further advances, subject to coordination with Theater Supreme Commander Gen. Dwight D. Eisenhower, would be 160 miles to Strasbourg on the Rhine. Eisenhower was to assume operational command over the 7th Army on September 15, about the time it was expected to link up with George Patton's 3rd Army, on the right of the forces coming from Normandy. The French 1st Army would be established, and Jacob Devers would take command of the newly activated 6th Army Group, including the 7th Army.

"The character of combat in which the Seventh Army soldiers engaged during the late summer of 1944 differed dramatically from what they had experienced in Italy or elsewhere,"[2] writes one modern historian. In the two weeks since Montelimar, they made excel-

222

lent progress. Lyon, for example, was evacuated by the Germans on September 2. The retreating Germans continued to suffer heavy materiel destruction.

On September 8, Truscott sent a memo to his subordinate commanders describing his "ideal" pursuit operation:

> The purpose of this operation is to destroy by killing or capturing the maximum number of enemy formations.
> Therefore the following should be observed:
> a. Make every effort to entrap enemy formations regardless of size. Long range fires, especially artillery, will merely warn and cause a change in direction.
> b. All units, but especially battalions and lower units, must be kept well in hand. Commanders of all ranks must avoid wide dispersion and consequent lack of control.
> c. Tanks must accompany leading infantry elements and tank destroyers must accompany leading tanks. All must be supported by artillery emplaced well forward.
> d. Reconnaissance must be continuous and thorough—foot elements to a distance of five miles, motor elements to contact the enemy.
> e. Contact once gained must be maintained. The enemy must not be allowed to escape.
> f. Every attack must be pressed with the utmost vigor. Be vicious. Seek to kill and destroy.[3]

For a while, at least some of Truscott's units were able to carry out his wishes. The 2nd Battalion of the 141st Infantry was one such unit. A regimental history tells of the advance and the first major resistance at the Moselle River:

> From the 3rd of September until we reached the Moselle River line on the 21st of September, the war moved rapidly and we met Jerry in one sharp clash after another. Only once did the enemy stop his retreat long enough to fight. After passing the 3rd Division at Besancon, we stopped at Vesoul while the 1st and 3rd Battalions joined forces with the 3rd Battalion of the 15th Infantry to attack a strong German force that stood and slugged it out for six hours before it turned and fled leaving burning trucks, 88's and dead behind. But mostly the towns and rivers and villages we took are very vague in our minds...[4]

At least one member of the 36th had a introduction to a side of the Germans that would become all too clear at the end of the war. Cpl. Zeb Sunday, for a minor disciplinary infraction, was assigned to help some paratroopers in Dijon locate and help rebury thirty-two Frenchmen murdered by the Gestapo. He "persuaded" a German captain to disclose the location of the secret graves by threatening to shoot the captain and bury him secretly. The German POWs exhumed the bodies so the French could rebury them properly. Sunday and the other Americans returned "cold sober."[5]

The rapid advance came to a halt at the beginning of October, at what was known as the Belfort Gap, when the German 19th Army turned to fight in more defensible mountainous terrain.

The Vosges Mountains, particularly the section known as the High Vosges, provided an obstacle on the path of the 7th Army to Strasbourg and Alsace, and then into Germany. The mountains were a defensive obstacle from both directions. No modern army had ever pushed through opposition in the Vosges. French revolutionary armies had gone around. During World War I, French and German forces had stalemated in trenches in the Vosges western foothills. German armies, in the blitzkrieg of 1940, also went around the Vosges and the connecting Maginot line, plunging through the equally feared but far less formidable Ardennes Forest. Despite the great success of this attack, when the Germans turned their attention to the second line and garrison troops in the Maginot line, some fortifications gave the Germans great trouble.

An army attacking from the south, as was the 7th Army, would encounter the highest part of the Vosges, appropriately called the High Vosges. The Americans in 1944 faced more than just mountains, formidable enough alone. The Germans had reinforced the natural defensive strength of the area. Additionally, due to Eisenhower's "broadfront" strategy of attacking across all of Western Europe—arguably to prevent Germans from being able to concentrate against a single drive—the 7th Army had to pass through the entire chain, not just the Low Vosges.

The easily defensible mountains, and the expected bad weather starting in late September, would provide an excellent place for the retreating Germans to make a stand. However, if the Vosges could be reached and attacked before the Germans could fully "dig in," and before the early rains turned to snow, the 7th Army's task would be far easier.

Chapter 14: Routine Combat

The VI Corps, under Lt. Gen. Lucian K. Truscott, would remain in the forefront of the 7th Army attack. In addition to the 36th Division, the corps still included the 3rd and 45th divisions. All were veterans of the tough mountain fighting in Italy and had appreciated the relatively fast-moving combat of the first few weeks in southern France. The XV Corps, with the 79th Division (relieved in late October 1944 by the 44th Division), would support the main attack. The main German opponent was Army Group G, now under Lt. Gen. Herman Balck, particularly the 19th Army, under Lt. Gen. Friedrich Wiese.

Intense fighting early in the Vosges campaign had already earned the following description, from a 36th Division after-action report for October:

> The fighting during the month of October was comparable to jungle fighting, where ... maintenance of direction was most difficult because of the dense forests. This alone resulted in many erroneous reports as to locations of units and enemy positions. Difficulties arose [from] orders based on the best available information, which was frequently inaccurate, miscarried, and at times resulted in bitter and unexpected fighting ... All commanders must report actual conditions carefully, avoiding all possibilities of errors in locations of units and omitting entirely reports based on optimism rather than fact. Forest areas must be mopped up thoroughly. Small, well-dug-in enemy detachments if not mopped up will harass supply columns, and present difficult problems of liquidation because of our inability to use our supporting weapons inside our lines ... Sometimes the enemy deliberately lets us get as close as seventy-five or one hundred yards to him before disclosing his presence with fire, and on occasion lets the leading elements pass by. This reduced the fight to a small arms fight with the enemy enjoying the advantage of good cover ... Holding the top of a hill or even what is normally termed the military crest of a wooded hill does not necessarily give us control of the surrounding terrain. We must require all units engaged in capturing a hill covered with forests to continue down the forward slopes until the open country is under small arms fire and artillery observation.[6]

The 36th Division was steadily, though slowly, progressing toward Germany. However, the increasingly tough going for the 36th

Division, and its two fellow divisions in the VI Corps, was showing on the men. As early as late September, John O'Daniel, commanding the 3rd Division, noted a decline in his units' aggressiveness. Dahlquist noted an increase in "desertions," probably AWOL (absent without leave, defined as short-term, unauthorized absences), and straggling. He thought at least part of the problem might be due to heavy officer and NCO casualties, with the need to promote men —who might not be ready—from the ranks to command positions. Frequent changes of command would result in some inconsistency of leadership. Most units were recognizing the necessity of training subordinates to take over from commanders, should this prove necessary, with a minimum of disruption.

Combat exhaustion was also coming to be accepted as a factor in declining combat effectiveness. "One of the outgrowths of extended combat," a division veteran remembered, "was that the human body and the human mind would simply give out. Exhaustion set in."[7] Exhaustion was also cumulative. British soldiers, generally given a few days' rest after two weeks or so of combat, could last a total of about 400 days. American troops, without this rotation, tended to be able to stand only about 200 days of combat. A study of American combat effectiveness, including combat morale, notes that

> With little unit rotation, men faced prolonged periods of danger, physical exertion, emotional anxiety, and mental stress without rest and relaxation . . . The price of such extended combat service was the loss of memory, perception, and judgement abilities and a lowering of morale. Casualty rates rose as soldiers became increasingly fatigued . . . Even men who refused to break under the strain suffered from deteriorating combat skills, fatalistic attitudes about surviving the war, and a demoralization that adversely affected incoming replacements.[8]

Enabling replacements to learn from veterans was a double-edged sword. Replacements could learn combat survival skills from veterans—such as the easiest way to judge how close you were to danger was to see how quickly the truck drivers drove back to the rear after dropping you off[9]—and be aided in developing the combat instincts virtually impossible to teach. However, contact with veterans of long-term combat could also "teach" replacements negative morale. Perhaps one can postulate the existence of a combat-effectiveness "window," starting about the sixth day of combat (replacement life expectancy in combat was five days) and lasting about a month, when exhaustion started to show.

Col. Paul D. Adams, commander of the 143rd Infantry, reported "an almost alarming mental and physical lethargy among the troops of his regiment."[10] In one incident, his best battalion commander, soon promoted to regimental executive officer, feel asleep during a jeep ride and conversation with Adams.[11] Writing in the monthly narrative of action, Adams elaborated, stating that:

> Future operations must be scheduled with some planned rest periods wherein officers and men are able to recuperate from the strain of combat. While little shooting occurred during many days of this pursuit, there always was present the possibility of immediate engagement which places a strain on those concerned.[12]

Short-term rest camps had been set up in Italy.[13] Dahlquist approved their establishment in France. Some officers felt the 36th did better in combat once division headquarters became more sensitive to fatigue.

One such camp was established at Plombieres-Les-Bains, where the 36th Division rear headquarters had recently been moved. The small town was undamaged, with six modern hotels and several smaller hotels. It had been a resort before the war, a famous hot springs attracting tourists. Postcards of the town were even available.[14]

Each unit had a number of rest slots allocated, first for two and then for three days, after Dahlquist approved the program on October 2. The division band helped staff the rest hotels, as well as providing musical entertainment. Dentists and medics were available, as were laundries. New uniforms were available. One of the theaters was operating to show films borrowed from army headquarters. Red Cross volunteers operated centers at the rest areas,[15] complete with much-appreciated female volunteers. The idea, at which the Plombieres-Les-Bains and a second camp at Bains-Les-Bains succeeded, was to make the men forget the war. One division veteran recalled,

> this was wintertime, and bitter cold, and most of [the men at the rest camp] had been in the line a month to six weeks or

more without any relief, without changing clothes, very poor food, muddy and dirty and no place to sleep. They were physically and mentally exhausted...

... some of them were in a stupor, actually, and really didn't know where they were. Their buddies were leading them along. They were so exhausted, so listless, with so little interest in what was going on around them.

[After processing, a shower, a hot meal, and new clothes] most of them were so tired and exhausted... Once they got to the new hotel, most of them would sleep a straight 24 hours. They seldom got out of their bed...

At the time I didn't realize how important this was to these people. As I look back over the years, I realize that the... break provided by the rest center probably saved the lives of a lot of people.[16]

"Gee do I feel good,"[17] this writer's uncle (and namesake) wrote his family on October 16 from a rest camp. The next day he wrote, "Boy, what a life. Definitely not G.I."[18] The rest camp seemed to be having its desired effect, helping the letter writer rest from whatever had delayed him writing since his previous letter some time earlier. Regrettably, wartime censorship prevented details from being sent home until a certain amount of time had passed. These particular details were never sent.

This writer's uncle, Bruce Kouser, was a replacement rifleman, a member of the 2nd Platoon, Company A, 142nd Infantry. The destruction of a sizable portion of World War II service records in a fire makes it impossible to trace the exact days of his service. However, he joined the division in mid-September, about the end of the "chase" portion of the pursuit north. His unit was apparently considered tired enough to have men with a month or so in action merit a rest. Allowing three days at the rest camp, it means he probably arrived October 13 or 14 and left the afternoon of Tuesday, October 17, to return to action.

Private Kouser appears to have missed most of the attack on Bruyeres. Private Kouser rejoined the 142nd Regiment just about the time it went into defensive positions in the lower Vosges Mountains, covering the right flank of the VI Corps and facing the German 198th Division. The regiment later reported that "The enemy to our front, themselves in a desperate circumstance and

Chapter 14: Routine Combat 229

weak in numbers, presented no great threat to our defense, and only on occasion attempted to take limited objectives."[19] However, despite reinforcement by the 442nd, the 36th Division was understrength, and not all men had been able to be rotated back for short rests.

The division was becoming involved with preliminary moves to a major corps attack code named Operation Dogface, a twenty-mile advance to high ground near the town of St. Die. Dogface itself was a prelude to a major 6th Army Group offensive, designed to get over the Vosges and through the Belfort Gap.

The preliminaries were quite effective, with the 143rd and the 442nd making good progress, ten miles into the German lines in the direction of Belmont. The 142nd and 143rd were holding most of the ten-mile front between Biffontaine and Le Tholy, near Rehaupal, Champdray, and Jussarupt. Most of this line would eventually be held by the 142nd, including a 4,000-yard addition the night of October 22–23. Truscott was not worried, given the reduction in German strength from two months of rapid flight and heavy combat. Besides, what proved to be a successful attack just left of the line was taking German attention from the 142nd.

The 142nd, however, had to send out patrols. The regimental operations report for October records that:

> Patrolling to the front became a daily and systematized routine. It constituted the main combat action during the period and afforded invaluable training and experience particularly to the new men, but also for old-timers. By aggressive patrolling the enemy was kept guessing . . .
>
> Patrolling is a risky and sometimes eerie business. Once out in front of our positions a sudden burst of fire may come from any quarter. A Company "K" combat patrol on 18 October went to search out a certain house on its sector. The patrol got to within fifty (50) yards of the house where an enemy sentry challenged. A brisk fire fight then followed. . . yet the patrol got out without a casualty.[20]

Other patrols from the 142nd were not so lucky. Second Platoon, Company A, sent out a patrol on Sunday, October 22, presumably connected with the extension of the regimental defensive line. This patrol included Pvt. Bruce Kouser. Several letters to his family give a brief outline of what happened next. This patrol was a combat patrol to find the strength of the enemy. In a wooded area between Jussarupt and Chamdray, France, his platoon was attacking a strongly entrenched enemy across an open field. It soon became apparent that the enemy had superior fire power and by counter-attacking were surrounding the platoon. Our artillery fire was immediately directed against the German attacks and the platoon was ordered to withdraw.[21]

Another letter continued:

After this action was over a concerted effort was made to account for the men of the Second Platoon. All men were accounted for except Bruce. It was therefore necessary to report him as missing in action.

It was not until the enemy had been driven from the area that it was learned that Bruce had been killed during the attack and had been buried by the Germans.[22]

This writer remembers being told that his uncle was "on point," leading the patrol. The telegram reporting Bruce Kouser missing did not arrive until December 1, 1944. His death was reported by telegram of January 22, 1945, and letter of January 27, 1945.

The patrol did not rate a specific mention in regimental journals. Casualty reports even neglected to mention the MIA for October 22. It was just a "routine" patrol.

CHAPTER 15

Lost Battalion

"We used to talk about food, mostly . . . We talked about chocolate cakes and bacon and eggs and everything our mothers and wives used to make for us back home. I remember once we spent a whole afternoon just talking about flapjacks . . ."

—*A Pictorial History of the 36th Division*, 1946[1]

Operation Dogface, the continuing VI Corps push into the High Vosges, would continue under a new corps commander. Lt. Gen. Lucian K. Truscott, Jr., turned over command to Maj. Gen. Edward H. Brooks on October 25. (Brooks had actually arrived a few days early, working with Truscott on planning the operation and ensuring continuity.) Truscott would return to Italy to assume command of the 5th Army in December, when Clark moved up to take over 15th Army Group.

The difference in rank between Truscott and Brooks was a result of Eisenhower's policy of having his corps commanders serve as major generals. Truscott had assumed corps command, and a third star, in Italy.

Truscott may not have always gotten along with General Dahlquist, but he had become known as a hard-driving, effective, and dominating commander. The Green Book summarized his command of the VI Corps in France:

For over two months he feverishly kept his three American infantry divisions on the move, forever harassing the retreating Germans and allowing them little time to rest and reorganize. His new offensive was yet another attempt to destroy the German defenses before they could solidify—an attempt that, like those before, was to be undertaken in the face of severe Allied logistical difficulties. However, the effort was also marked by much tactical and operational imagination . . .[2]

The 36th Division was helped by having a driver such as Truscott as its corps commander. Dahlquist seems to have been a competent commander, but not as aggressive and innovative as Fred Walker. Truscott, though occasionally overestimating how fast an infantry division could move, provided a useful push. Brooks was not as flamboyant as Truscott, but was also a first-rate commander. He would, however, have to adapt to commanding infantry in his first corps assignment after coming from an armored division. The mountainous terrain was more difficult than that in the north, where Brooks had served under Omar Bradley. The supply problem, however, was on its way to being alleviated.

This may seem an odd problem to be faced by an army which had the reputation of being lavishly supplied. United States production figures for World War II can be staggering. The value of materiel to a war effort, the value of sheer economic power, was shown by the effort the Allies put into destroying the German and Japanese economies. Albert Speer, the brilliant German minister of armaments in the last three years of the war, was able to counteract much of the effects of the Allied bombing. German production on several key weapons, including airplanes and tanks, actually increased in 1944. Speer was, however, never able to fully mobilize or direct the German economy to the degree this was done in the United States and the Soviet Union. The Germans never overcame the tendency to look to a few great weapons over a lot of good weapons. "No war was more industrialized than the Second World War. The Allies made better use of their industry than their enemy."[3]

The problem for the Americans, and for the British, was in getting war materiel to where it was needed. This problem began in northern France, barely two weeks after the Anvil/Dragoon landings. On September 2, for example, tanks from the 743rd Tank Battalion, 2nd Armored Division, sat on the crest of a hill overlooking

Tournai, Belgium. They were out of fuel. Historian Stephen Ambrose, in his recent work *Citizen Soldiers,* writes:

> The crisis was inevitable. It had been foreseen. It could not have been avoided. Too many vehicles were driving too far away from the ports and beaches... the front line continued to move east and north, and the system just couldn't keep up.[4]

The 7th Army, in which the 36th Division served, had the same supply problems. The supply lines from the Riviera coast were also long and difficult. A modern study of this campaign points out that "By the first week of October, in fact, the American supply situation had been deemed 'critical.'"[5] Use of artillery ammunition was limited. The situation improved by mid-October. The VI Corps now would be able to attack well-prepared German emplacements—which the Germans had been given two additional weeks to improve. Truscott's last two weeks in command of the VI Corps would not be dull.

The first step would be to capture the French village of Bruyeres. Bruyeres lies in a small valley, surrounded by several major hills which dominate the village. Buildings were constructed of stone and masonry, characteristic of the Alsace area of France. "Any house can be turned into a pillbox, practically impervious to small-arms fire, within a few hours, let alone the two weeks the Americans' logistical problems had afforded the defenders,"[6] a military officer/historian noted after a visit. The defending Germans, from the 16th and 716th volksgrenadier divisions, made good use of their time.

The 36th Division conducted the attack, using the 143rd Infantry (without the 2nd Battalion in division reserve) and the attached 442nd Regimental Combat Team, the Nisei Regiment, for the main assault. The 141st and 142nd regiments were deployed to the south of Bruyeres, to cut off the town from the south and conduct supporting attacks toward the Vologne River. The 45th Division carried out much the same functions in the north of the village.

The 143rd Regiment advanced up the Laval-Bruyeres road, from the southwest and the south. The regiment would advance through open, easily defended territory. The men would be seen by the defenders soon after the advance began and could be engaged by the Germans at long range. The Germans would be using landmarks as registration points, to give them the range, for indirect artillery fire.

The main attack unit, however, was the 442nd. This regiment would envelop the defenders' right flank by advancing quietly through thick woods to the west of the town. The 442nd would be advancing through thick woods, which would conceal them—and, unfortunately, defense positions—until relatively close range.

Early phases of the operation began on the morning of October 15. The advancing regiments first went after the villages of Laval-sur-Vologne, about two miles to the southwest of Bruyeres, and Champ-le-Due, about a mile and half directly south of the eventual target. Elements of the 45th Division were already attacking to the immediate north of the 36th. German resistance took good advantage of the terrain. One veteran of the 36th Division remembers,

> the enemy had withdrawn to key defensive positions in dense woods, had minded buildings along roads, and felled trees across them. Such roadblocks were, with German military efficiency, well guarded and kept under artillery and small arms fire. The mountainous terrain often provided excellent observation for the enemy. The climb up slopes was slow and difficult torture.[7]

Despite strong German resistance, the 143rd and 442nd made reasonably good progress. On the night of the 15th–16th, the 143rd was even able to send a patrol to within 200 yards of Bruyeres.

By October 18, both the 143rd and 442nd were pushing into Bruyeres. Their assignments had been slightly altered, with the 143rd given responsibility for the town, the 442nd for the area to the north. By the next day, the Germans had been forced out of the town.

Another problem had arisen. The cooperating 45th Division was making slower progress, and a gap was opening between it and 36th. Truscott did want to slow the 36th's momentum, so he moved a regiment of the 3rd Division into the gap.

The 45th Division's attacking regiments began crossing the Mortagne River on October 23. The 180th and 197th infantry regiments had taken casualties crossing but were in good shape and advancing into the Rambervillers Forest by midday. Patrols from the 45th Division met patrols from the 3rd, attacking to its south, that day. By October 26, Brooks' first full day in command of the VI Corps, the 45th and 3rd divisions had made substantial progress in

Chapter 15: Lost Battalion

the forest. In combination with 36th Division operations, they had almost isolated the German 16th Volksgrenadier Division.

The right flank of the 3rd Division ran into unexpected problems. It was advancing along a road, N-420, supposed to be the easiest axis of advance. However, the American unit ran into opposition from a regiment of the 338th German division. This was the first major German response to the VI Corps offensive. The 19th Army was reinforcing a gap opened by the 36th, hitting the dividing line between two German corps. The gap continued to open for a day or so, but the German regiment slowed the 3rd Division advance. A newly arrived panzer brigade was held back as army reserve, against an expected American attack into the Meurthe River valley. For a few days, at least, this seemed a good idea, as the 16th Division seemed likely to be forced back. Reinforcements would have wasted fresh troops. The 16th Division did not do well, but the other German units in the area had stalled the American advance by October 28. The American 7th Infantry regiment was still stuck on N-420 and needed help from the 36th, on its right flank.

★ ★ ★ ★ ★

Dahlquist's division had its own problems. Simultaneously with the 45th and 3rd division movements to its north, the 36th pushed into the Domaniale de Champ forest. On October 23, the attached 442nd Infantry on the division's left moved several miles into the forest. One battalion reached Biffontaine. The next day, the 442nd was relieved by the 141st and 143rd, which continued the drive to the east. The 1st Battalion occupied Hill 624, northeast, and Hill 645, southeast of Biffontaine. The Germans, backed by artillery and sensitive to potential dangers in their intercorps boundary, immediately counterattacked with fresh units of the 933rd grenadier regiment. Supply lines were cut to Hill 645, and a relief column of the 2nd battalion was fought off. During the night of October 24–25, Germans overran the command post of the 1st Battalion, completely blocking all routes to Hill 645. That morning, the 36th discovered that 241 men were trapped near Hill 645, 237 from the battalion, four from other units.

The Germans were not fully aware of the pocket. The dense forest and hilly terrain made it as hard for them to find out what was going on as it was for the Americans. Rescue parties from the 2nd and 3rd battalions of the 141st were unable to reach Hill 645. That afternoon, a thirty-six-man patrol was sent out by the trapped elements. Five men made it back to hill, and a sixth man reached 36th Division lines five days later.

The "Lost Battalion" was lost.

One historian, a veteran of the 36th Division, makes comparisons to the Alamo.[8] Two key differences are in the enemy not knowing the battalion was lost, and in the different ending of the World War II incident. Another historian points out that the term "Lost Battalion" was a media comparison to elements of the 308th Infantry, 77th Division, in World War I, whose location actually was unknown for several days. Many battalions got "lost" in World War II, when attacking units advanced faster than their flanking units and got cut off by infiltrating enemy units.[9] The division veteran/historian notes:

> Much has been written and even a movie has been made, focusing on the action during the last week of October 1944 . . . but the isolated group were never really "lost" and they were not a complete battalion.[10]

By October 26, the trapped men were running desperately short of ammunition, medical supplies, and food. The 2nd Battalion of the 442nd came back into line that day, after a two-day rest, to enable the non-trapped elements of the 141st to concentrate on rescuing its comrades. The next day, Dahlquist brought the remainder of the 442nd back into line on the 141st's left. However, the Nisei Regiment was immediately subjected to a sharp German counterattack. The 141st was also making limited progress. Dahlquist replaced its commander, only on duty since October 7, with Col. Charles H. Owens, divisional chief of staff. This did not rescue the Lost Battalion, however. October 28 and 29 saw the 442nd continuing its push but still not reaching the trapped unit.

A *New York Times* article later reported that within the Lost Battalion itself,

> The isolated doughboys huddled in cold foxholes and fought off numerous attacks but apparently the enemy did not realize the Americans' desperate plight . . .
>
> The Americans were without food for five days. Rations

and medical supplies finally were parachuted to them from low-flying Thunderbolts and fired in by artillery shells.[11]

Division records show that priority order for supplies was medical supplies first, batteries (for radios) second, ammunition third, and rations fourth.[12] The regimental operations journal for the 142st reported an interesting "friendly fire" danger that the relief effort presented:

> the [artillery] is preparing 100 rds to be filled with medical supplies and fired as time bursts over the 1st bn area the same manner propaganda leaflets are fired. These will probably be read at 1700 [5:00 P.M., two hours from the time of notation] and Col Lynch will notify the fwd CP in time for the 1st BN to receive notice of the plan and to take cover since the shells will contain sufficient explosives to crack them open and the flying pieces will be a little dangerous.[13]

The *New York Times* quoted one officer with the battalion on the unusual method of supply delivery:

> We just crouched in our foxholes and sweated out the 'attack' like we would have done if the Germans had been shelling us. Those shells may have had chocolate in them, but they could have killed you when they landed.[14]

The Germans were not fully aware that they had cut off the better portion of an American unit, apparently thinking it was just a salient. The first night saw heavy German shelling, but no attack.

The 1st Battalion, 141st Regiment, was cut off on the treeless top of a heavily wooded hill, in an area roughly 300 by 350 yards. Officers on the hill, apparently with no one over the grade of lieutenant remaining,[15] had the men form a circular defense, with riflemen and carefully located machine guns covering all angles. A surprise German attack was no longer possible, but with at least 700 Germans in the immediate area, it might not be necessary. The men made it a point to stay as quiet as possible, trying to avoid attack until promised rescue arrived. When cutting trees, with pocket knives, to improve defenses, they wrapped the trees in blankets to cushion the noise when they fell. Food and fuel were pooled from all packs. But this food quickly ran out. Water was also a problem. A veteran remembered, the water situation became critical, more critical than the food shortage. Finally they found a mud puddle out of their area. It was dirty-stagnant, but it was water. They could boil some of it for the medicos—not much. Even the smallest fires caused smoke, which might give away their position.

They couldn't get this water whenever they wanted it. They had to crawl quietly during the blackest part of the night, with their fingers itchy on their triggers. The Germans were using the same water hole.[16]

Food was not sent through to the battalion until October 28. The first efforts missed. When they got supplies through, unfortunately, they also confirmed for the Germans the unit's location. Fortunately, it was not until October 29 that the disorganized Germans were able to mount a serious attack. "Fortunately," a division historian remembered, "the Germans picked the one spot [to focus their attack] where the battalion had concentrated most of its heavy machine guns."[18] The attack was beaten off, with heavy casualties to the enemy.

The 442nd Infantry continued fight to its way towards the battalion on October 29 and 30 against heavy resistance. In the words of their history:

> Progress was slow on the 27th. The terrain was next to impossible, heavily forested and carpeted with a dense growth of underbrush. Fighting went back to the days of America's Indian wars; every tree and every bush was carefully investigated before the troops passed on . . .
> The following day, both battalions continued the drive forward in the teeth of stubborn resistance and heavy artillery and mortar fire. Casualties went up and up . . . At the end of the day, the regiment was 1,500 yards closer to the 'lost battalion,' but only at terrible cost in men and materiel. During the night, biting cold and rain kept the men from resting.
> On the 29th, the regiment jumped off again, cleared one knoll, and ran into the enemy's main defensive position thrown astride the ridge which was so narrow that maneuver was impossible. Any attack was hopelessly canalized into a direct frontal assault. In the meantime, word had come. . . that the situation of the "lost battalion" was becoming desperate . . .[19]

Heavy fighting on the 29th was won by the 442nd, combining

Chapter 15: Lost Battalion 239

a rare bayonet attack with the capture of a hill controlling the German ridge position. On the afternoon of October 30, the 442nd Regimental Combat Team's history continues,

> a 1st Battalion rifleman on outpost saw a figure approaching through the trees. Fearing it might be a Jerry, the rifleman remained in place and strained his eyes to obtain a better view. Then like a crazed person, he ran down the hill laughing, crying and yelling at the approaching figure. Private First Class Mutt Sakamoto was at a loss for words as the rifleman ran towards him and hugged him and shouted words of thankfulness. About all Sakamoto could think to say was: "Could you use a Cigarette?"[20]

The "Lost Battalion" had been found.

Brooks and Dahlquist were upset by the whole incident. Relief of the battalion had consumed the energies of two regiments, plus much of the division's supporting units. The 36th was unable to continue its advance, or to start toward St. Leonard, its final objective under Operation Dogface. The Green Book continues,

> The VI Corps' attack, which had seemed so promising at first, appeared to be stalled in the Vosges forests at the end of October. Troop fatigue, poor weather, difficult terrain, ammunition shortages, and fresh German reinforcements all conspired against any rapid advance . . . time seemed to be working against the American attackers . . . But the Germans also had severe problems. The broad, grinding American advance was chewing up their infantry units bit by bit, never allowing them to build up their reserve for anything more than localized counterattacks . . . The fighting . . . seemed to be coming down to a matter of will and stamina—a war of attrition that, without assistance from the outside, only the more determined opponent would win.[21]

While the rescue of the "lost battalion" was underway, the 111th Combat Engineer Battalion was building a road quite reminiscent of the difficult project it had undertaken at Velletri. The road, with a plank floor on part of it—called "corderoying" in the American Civil War—took nineteen days to build and was operated for another week until the Germans were cleared from the area. It aided the 442nd in its rescue efforts and helped the division in its offensive until mid-November. A picture of the road could show something out of the Civil War—until you see the tank, which appropriately appears to be a Sherman. Once again, old techniques aided in a new war.

CHAPTER 16

Winter Warfare

"Dawn found the situation critical in many sectors . . ."
—141st Infantry regimental
history, 1946[1]

The 442nd RCT, or what was left of it, came out of the lines November 9, leaving the "workhorse"[2] 36th Division. The regiment had taken heavy casualties in very hard fighting, and heavy non-battle casualties from mostly Hawaiian-born and -raised infantrymen having medical problems in the cold, wet winter. Two battalions went back into the line on November 13 but came out again November 18 to join the other battalion on the French Riviera. The 442nd would be put back into combat in Italy in late March of 1945.[3]

The XV Corps, to the VI Corps' left, and the II French Corps, to the right, had succeeded in their semi-diversionary attacks in October. The American line of advance had been evened out by the time the "Lost Battalion" was found in late October.

Slow but steady advances continued throughout November— costly in casualties and equipment. This was nothing like the early-fall "chase" northward, having gained only about twenty miles since mid-September, but also nothing like the virtual stalemate the 36th Division remembered from the Italian mountains. The 36th Division, after three months fighting, was assigned a supporting role in the VI Corps late November offensives. The 3rd Division would lead the corps attack across the Meurthe, to within fifteen

241

miles of the major Alsatian city of Strasbourg. The 36th would take over positions vacated by the new 103rd Division which had just joined the VI Corps, west of the Meurthe, move forward to blocking positions and maintain contact with the French II Corps to its right. When ordered to do so, the 36th Division would be ready to attack across the Vosges. VI Corps' Operations Instructions Number 9, issued on November 19, 1944, directed the 3rd Infantry Division to attack across the Meurthe River just before 7:00 the next morning.[4]

Despite a heavy current, with some flooding, the two assault elements of the division were able to infiltrate across the river at night. The assault began from across the Meurthe. Tactical air support was possible for the first time in the Vosges campaign, and for one of the few times in that campaign. Division and corps artillery added additional weight to the attack. VI Corps commander Brooks, when he saw how well the attack was going, ordered two regiments from the 103rd Division to cross the Meurthe at the same location and join in the offensive. Germans from the 716th Volksgrenadier Division, withdrawing in the face of the strong American attacks, burned most of the city of St. Die. Though most Americans did not actually enter the city, German actions were known. In the words of a recent study of the Vosges campaign, "Furthermore, since the attack pushed immediately beyond St. Die, the destruction of the structures in no way deprived the Americans of shelter. It was pointless, counterproductive savagery."[5]

Just before midnight on November 21, Brooks ordered pursuit of the Germans. Ad hoc combined-arms task forces were to advance to the Vosges passes as quickly as possible. Units were instructed to bypass German strongpoints but cut off German lines of retreat. Defenders would be destroyed by American infantry formations following the initial advance.[6]

Supporting did not mean "inactive" during this period for the 36th. On November 19, the 36th Division was able to reach its original Operation Dogface objective, the high ground west of the Meurthe near St. Leonard. It also controlled forested high ground to the south near the Meurthe River. The division boundaries and mission were then changed. When Brooks ordered the 103rd to cross the Meurthe over bridges built by the 3rd Division, he moved the boundary between the 103rd and the 36th north by three miles.

The role of the 36th was also changed from support to active pursuit. Brooks ordered the division to advance along its front to the east and northeast, keeping abreast of the rest of the VI Corps.

The first divisional objective was Ste. Marie-aux-Mines, on Route N-59, twelve miles east of St. Die, two miles east of the Ste. Marie pass. After reaching the pass, the division would continue twelve miles east to the city of Selestat, on the Alsatian plains, between Colmar and Strasbourg. Dahlquist was also instructed to launch a secondary attack, generally along Route N-415, through Le Bonhomme Pass to Colmar.

The 36th began crossing the Meurthe River on November 21. Opposition primarily came from artillery and mortar fire, with heavily mined areas and flooded banks also creating problems. The same day, the division seized a large portion of N-415. However, more resistance was met at Route N-59. German defenses in the 36th's area were the best in the VI Corps zone. Route N-59 narrowed to a twisting two-lane road running through mountain passes. It was an easy area to defend. The German 198th Division, one of the best in the area,[7] had probably set up the defenses. However, its replacement, the 16th Division, lacked sufficient manpower to hold all roadblocks and defensive works.

German resistance on November 22 and 23 temporarily held up the 36th near the Meurthe, but the division moved faster the next day. Progress was also held up at Le Bonhomme Pass for two days. On November 25 the division took Ste. Marie-aux-Mines and the Ste. Marie Pass. Thirty years later, George Lynch, the former commander of the 142nd Infantry, wrote about the capture of the pass, assigned to his regiment:

> The offensive moved slowly and then gathered speed as the 142nd entered Wisembach, the last town before ascending to the high Ste. Marie Pass. The pass itself, at 2,547 feet . . . was simply a narrow, hard-surfaced road cut into the side of the precipitous mountain whose peak rose 3,214 feet . . . Heavily wooded as well as precipitous, travel off the road was a matter of climbing and clambering on foot. The pass, with timber barricades, was defended by antitank guns and about 100 infantrymen.
>
> By this stage of our advance, it was clear that the Germans were no longer defending, just delaying our advance wherever

the terrain offered an advantage. So there was no question of linear defenses. Road blocks and delaying along principal arteries became the pattern of German tactics at this time.

A map study showed a dirt cart trail (probably an old lumber road) leading northwest out of Wisembach, crossing the high saddle of the mountain north of the pass.

A plan was devised to have the 3rd Battalion divide into two forces. One element was made up of Company L, reinforced by a platoon from Company B, 753rd Tank Battalion, and commanded by Major Ross Young... This force was to attack the pass frontally, up the road approach, as a diversion.

The remainder of the battalion, commanded by Lieutenant Colonel A. Ward Gillette... the battalion commander, was to mount on trucks, follow the dirt trail northeast out of Wisembach. They were to go as far as possible mounted, and after crossing the saddle, dismount and attack the town of Ste. Marie aux Mines on foot...

At daylight on the morning of 25 November, the men of Gillette's 3rd Battalion (less Company L) mounted their trucks and moved out of Wisembach up the dirt trail and disappeared into the forest without incident or interruption.

The attack of Young's diversionary force against the fortified pass itself was delayed until the 132nd Field Artillery Battalion could complete its move to firing positions. Into place, they laid a generous amount of artillery fire on the fortifications while Young's force advanced up the road.

While the fortifications were still under our artillery fires, the tank platoon attempted a frontal attack against the pass. With this approach denied to us, Company L deployed into a shallow envelopment around the north (uphill) side of the pass. After many hours of tortuous climbing, this force descended on the pass and at 1815 hours completed its capture, with 28 prisoners and 30 dead.

The 3rd Battalion, meanwhile, had passed over the saddle, two miles north of the pass, and had continued mounted on trucks toward the town of Ste. Marie. At 1320, having dismounted, the battalion entered the town and captured a stunned German garrison of 150, while suffering only two casualties—both just light wounds.[8]

The 1st Battalion was ordered to follow the 2nd Battalion and attack St. Croix. The town was taken, but not without more resistance from German troops alerted by the capture of Ste. Marie. George Lynch, who retired from the army as a major general, called the capture a "truly remarkable achievement for which the 3rd Battalion and its attached troops received the Presidential Unit Citation."[9]

Though failing to destroy retreating German units, the VI Corps advance in this phase of the campaign has to be considered successful. The 2nd French Armored Division captured Strasbourg on November 25. Selestat was captured by elements of the 36th and 3rd divisions, and Combat Command A of the 14th Armored Division, on December 4. German resistance in the 7th Army area of the High Vosges was at an end.

As November came to an end, the German 19th Army was gradually being destroyed by Patch's 7th Army and Jean de Lattre's 1st French Army. With the exception of an area known as the Colmar Pocket, the 19th Army was close to being unable to offer resistance west of the Rhine River. The 19th Army commander was considering withdrawal of most of his units east of the Rhine. Jacob Devers, commanding the 6th Army Group, believed the pocket would soon be withdrawn. German estimates were that without major reinforcements, the Colmar Pocket could be held for three weeks.

Devers thought the French would have little problem getting rid of additional German resistance west of the Rhine. The Panzer Lehr armored division was making some progress in a southward attack, however—including nearly chasing Eisenhower and Omar Bradley away from XV Corps headquarters[10]—which added an extra element to Allied planning.

Devers wanted to have the 7th Army cross the Rhine and then head north. This would outflank German defenses against the northern portions of the Allied drive. One motive for Anvil/Dragoon, perhaps the chief motive, had been to help what were considered the main Allied drives to the north. At the end of November, Devers would learn just how high a priority this was.

On November 24, Eisenhower and Bradley left for a tour of the southern portion of the Allied front. Their first stop was George Patton's 3rd Army headquarters. Two months earlier, Patton had been told to go on the defensive, as supplies and fuel had to be diverted to Montgomery's British forces in the north. Patton interpreted "defensive" rather loosely and had been sending out continuous "patrols" to "adjust" his lines. By November he was formally back on the offensive. However, by the end of the month, bad weather and supply problems, among other factors, were slowing down the 3rd Army.

Bradley, the 12th Army Group commander, considered Patton's offensive against Germany's Saar Basin his main effort. Something had to be done to assist this effort. Bradley thought shifting forces, such as transferring the XV Corps back to Patton's control, was too time-consuming—though Patton favored this course of action. Bradley wanted to leave command arrangements as they were but shift some of Patton's front to Patch's 7th Army. Eisenhower also seems to have made up his mind to help Patton.

After joining Devers and Patch, Eisenhower and Bradley proceeded to XV Corps headquarters. They then visited Brooks at VI Corps headquarters. Both were making arrangements for crossing the Rhine. At XV Corps, Eisenhower issued verbal orders to stop these preparations. He wanted the 7th Army to switch directions to a northward attack, against the flank of the German 1st Army, in support of Patton.

Devers was upset. That night, after a formal dinner, he met with Eisenhower and Bradley. The volume of the Green Book history series on the Riviera campaign records that this was not a pleasant evening. "The ensuing discussing lasted until the early hours of the following day and saw a heated argument between Eisenhower, Bradley and Devers."[11] Eisenhower continued to insist that Devers stop preparations for a Rhine crossing and have the 7th Army turn to the north to assist Patton. Overall SHAEF directives provided that all forces would seize bridgeheads over the Rhine if the opportunity presented itself. However, Eisenhower ordered Devers not to make any such crossing. He and Bradley rejected Devers' arguments that the Germans were sufficiently weak, that crossing the Rhine, even against the West Wall defenses, would not be a problem, and that Patton would best be aided by a second force enveloping the Saar from the south. Eisenhower ordered Devers to clear up Germans west of the Rhine as soon as possible—presumably including the Colmar Pocket—and attack to the north. Devers had no choice but to comply.[12]

Eisenhower's decision did not result in Patton's quick crossing of the Rhine and can be questioned in the 20-20 hindsight available nearly fifty years after the event. However, it might not be fair to evaluate historical decisions with evidence other than that available at the time. The Colmar Pocket gave Devers problems. Things might have gone easier if more than the 36th Division had been available to help the French. However, the Germans might have been quicker to withdraw if a Rhine crossing by the 7th Army, especially in conjunction with that of the 3rd Army, had endangered the German 19th Army supply lines.

Devers thought Eisenhower should have been willing to exploit the opportunity the 6th Army Group's rapid advance presented the Allied command. Arguments exist in favor of Eisenhower's "broad front" strategy, in particular that this strategy prevented the Germans from concentrating against one thrust and that it avoided danger to the flanks of the advance. However, Eisenhower may not have been sufficiently flexible in responding to changing circumstances.

Politics, an unavoidable (and even necessary) part of every war seems to have played a role in Eisenhower's decision-making. Devers, not known as the most reflective of commanders, may not have considered that higher level Allied political decisions put the focus of effort on the northern thrust by the 12st Army Group, under Bernard Montgomery. Montgomery was not always the most cooperative of commanders under the best of circumstances, as Clark[13] (and Patton) might have told Devers. Initial American policy had been to pay perhaps excessive attention to British feelings. Eisenhower probably concluded that Montgomery was enough of a problem as he was. After having already rejected Montgomery's idea for a total concentration in the north, under Montgomery, Eisenhower did not want the additional problems of convincing him, and Churchill, of a major change in political plans for the war.[14]

The "ground level" view remained that Eisenhower might have made a mistake. Patton later wrote that were he Eisenhower, he would have told Patch to send the VI Corps across the Rhine.[15] The bulk of the XV Corps would have remained west of the Rhine to protect Patton's right flank. The 3rd Army could also have been used to exploit the XV Corps breakthrough. Political factors are hard to estimate, as the prospect of upsetting Montgomery, if it went no further, might have had a sort of perverse appeal to Eisenhower and Bradley. From the American point of view, stopping the 7th Army probably showed a lack of imagination and flexibility.[16] This decision, which the Germans would quickly guess—if not learn from intelligence sources—fit in with Hitler's plans. The December surprise counterattack in the Ardennes was already in the works. There was no need to worry about a southern breakthrough.

★ ★ ★ ★ ★

The 7th Army attacked in two directions during the last week of November. The XV Corps attacked to the north, supporting Patton. The VI Corps attacked to the east in the general direction of Colmar, with the 1st French Army. Devers was not satisfied with the progress either attack was making. Effective December 5, 1944, army borders were changed, giving the French full responsibility for the Colmar Pocket. Additionally, operational control of the 36th Division and the French 2nd Armored Division was transferred to de Lattre's 1st French Army.

The main 7th Army effort was aimed to the north. The Germans had adapted the captured Maginot line to face the other direction. Noted historian Russell Wiegley writes that "Against these defensive resources, the new drive of the Seventh Army in early December became a plodding affair similar to the Third Army's recent and concurrent attacks, or to the September to mid-November experiences of Patch's own troops."[17] The attacks would continue until mid-December, with progress but with no overwhelming success, until new developments forced a rapid change in plans.

Devers and de Lattre were surprised that the German 19th Army continued to hold the Colmar Pocket. Hitler, however, made an unusual appointment, but one reflective of his determination to hold the pocket. On December 10, Hitler appointed SS chief Reichfuhrer Heinrich Himmler as commander of Army Group Oberrhein, a new headquarters controlling the 19th Army (whose commander Himmler quickly replaced) and some units east of the Rhine. Army Group Oberrhein was semi-independent, formally a theater command reporting directly to OKW, German military high command. Actually, Himmler reported directly to Hitler.

Himmler's previous military command experience had been

taking over as head of the replacement army after the July 20, 1944, attempt on Hitler's life. The Green Book states:

> For the next few weeks the command changes proved effective. However questionable his military abilities, Himmler was able to accelerate the infusion of replacements into both the Colmar area and the east bank defenses by having the immediate German interior scoured more thoroughly for supplies, equipment, and manpower.[18]

Himmler's presence ensured that no local German commanders withdrew without permission or otherwise failed to fight more vigorously.

The full-scale French 1st Army offensive would start on December 11 but would be temporarily cancelled on December 21. The 36th Division had been seeing some action, however, almost since it was formally assigned to French operational command. Yet several weeks of hard fighting failed to eliminate the Colmar Pocket. German counterattacks were fought off, including one in the Selestat area in the northwest of the pocket, on December 10. As described in the 142nd regimental narrative for December:

> At Selestat, a heavy artillery concentration alerted the 1st Battalion at about 0630 hours on 10 December 1944. A platoon of Company B at a road junction was the first to report the attack. On both flanks, approximately 100 enemy struck but in time were driven off and 30 prisoners taken. The main weight of the German attack was then realized to be further to the north. Before daylight, the enemy infiltrated into the Vienweg area, quickly seized and controlled a half-mile stretch of houses along the first north-south street, and cut off another platoon of Company B, at another road junction. Another route of infiltration was along the stream bed and railroad tracks on the north. One column may have gone as far west, on the north bank of the railroad crossing, as Scherwiller . . . crossed a footbridge, then attacked east to gain control of a factory area. When a friendly unit holding one of the northern posts saw the enemy in strength early in the morning, it withdrew, leaving a wide gap through which the Germans could have infiltrated in large numbers. This gap was not known until nearly noon, and posed a serious threat . . . Artillery in the area was seriously threatened. Although Company B was isolated in small units, some only squad size,

with some attached antitank platoons, they held their positions and even took prisoners and kept them under guard.[19]

The 142nd Regiment commander attached Company I to the 1st Battalion and placed the rest of the 3rd Battalion in the town of Chatenois, west of Selestat on the Ste. Marie road, to block further penetration in that area. Two light tanks made it into Selestat that morning, two medium tanks and a Company C platoon by noon. It took most of the afternoon to clear Selestat, and defeat a dusk attack. The 1st Battalion later won a Presidential Unit Citation for this action.

However, general progress had not been as good as expected. The overall attack was cancelled. The 36th Division was not involved when the attack on the Colmar Pocket resumed in January. It was shifted to reserve on December 25.

On February 8, 1945, what was left of the German 19th Army withdrew back across the Rhine.

★ ★ ★ ★ ★

The first priority when the 36th left the line was the surprise heavy German offensive in the Ardennes, the Battle of the Bulge. When Eisenhower sent part of the 3rd Army to attack the German salient, the "bulge," the 7th Army had to take over the vacated line. Devers was asked for two divisions to go into SHAEF (overall theater) reserve. The 36th Division and the 12th Armored Division got the honor. Offense was impossible for the 6th Army Group until the emergency in the north had passed, but the removal of these two divisions gave Devers even fewer troops to hold a longer line. Devers does not seem to have said anything to Eisenhower, as he recognized the necessity for reinforcing the Ardennes. He noted in his diary, however, that he still thought the high command should have taken advantage of the success in his sector.[20]

The 36th Division enjoyed the brief rest behind the front lines. Early in January it would be back in the line. The Battle of the Bulge was winding down, but the Allied commanders were expecting a second wave of attacks in the south. Both Hitler and Field Marshall Gerd von Rundstedt, overall German commander in the west, realized that the Allies had to greatly weaken their forces in the south to stop the German offensive in the north. At the very least, they expected the attack to relieve pressure on German forces in the north.

Operation Northwind, *Nordwind* in German, would start on New Years's Eve 1944. Part of the German 1st Army would attack in the Sarre River valley, part through the Vosges. Two panzer divisions would remain in reserve to exploit a breakthrough. Army Group Oberrhein would launch supporting attacks north and south of Strasbourg, but only after success in the Sarre valley.

Intelligence indicated a German buildup in the area of the 7th Army, and an attack was expected. Eisenhower wanted Devers to withdraw from the Lauterbourg salient, and from Strasbourg, to better positions in the rear. Devers thought Strasbourg could be defended. He told Patch to prepare three intermediate withdrawal positions, to be occupied only in the case of heavy attack. A fourth emergency defensive position was back at the Vosges. There would be military problems defending the area, and Eisenhower continued to insist that Devers withdraw. Charles de Gaulle and the French added a complicating political factor in the case of Strasbourg, insisting that it be defended for moral and emotional reasons. For a very brief period, the only Allied troops in the city may have been two staff members from *Stars and Stripes*, the famous newspaper published for soldiers in Europe.[21] Strasbourg would eventually be defended by an Algerian division.

Most 6th Army Group activities during the last two weeks of December, aside from the attacks on the Colmar Pocket, were spent switching over to defense and preparing for the expected attack. Allied defense lines were thin. Three new infantry divisions were immediately brought into the lines. Devers authorized Patch to use the 36th Division and 12th Armored Division SHAEF reserves to form a secondary line of defense behind the XV Corps. Ultra intercepts of German communications were far less valuable with the Germans on their own soil, where they could send instructions by far more secure telephone wires. Devers and Patch were guessing that the main German attack would come in the Sarre River valley, against XV Corps.

The Northwind assault's first wave, as predicted in the Sarre valley, proved to be weaker than expected. "The German attack barely made a dent in the beefed-up Allied line,"[22] according to the

Green Book. Small German advances were pushed back by counterattacks from the 44th, 63rd, and 100th divisions. This attack was finally called off on January 4.

The second attack, launched at about the same time, was more of a surprise. It started around the area of the city of Bitche and headed south through the Low Vosges. The American commanders had been more concerned with the area west of the mountains near the Sarre River, and with the Strasbourg area. They had not expected the Germans to come through the rough terrain. The Germans did not open with an artillery bombardment, and the overcast skies hindered Allied air observation.

The western edge of the German advance hit the 100th Division, aided by an additional regiment from the 63rd Division and the 141st Regiment, which arrived late on January 1, 1945. In the words of the 141st regimental history:

> Those first days in the Bitche area were for many the worst of the war. The fact that the Germans were really on the offensive was fully substantiated. Snow lay deep on the hard frozen ground, and the ominous, non-directional sounds of armor echoed through the splintered woods and hollows. Dawn found the situation critical in many sectors ... Our mission, we learned, was to protect the weak right flank of the 100th Division and to stem an attack that had the impetus of a three-day start. Even who the enemy were was in doubt, for the treacherous Krauts were found to be using American tanks as well as dressing in American and British uniforms.
>
> When three or four machine pistols are powdering the snow around your neck and 88's are chopping firewood overhead, you aren't too greatly concerned about such a general thing as straightening out a defense line, but that was actually our job ... Along our line there were gaps to be filled, isolated squads and platoons to save, tanks to be met and terrorizing reports to be checked and dismissed as fables, or met with hurried strength if found to be true. It was evident that Jerry had his mind made up to go someplace, and though we might change his mind at noon, there was nothing to assure us that nightfall wouldn't find him with the same idea ...
>
> By noon of the second day we were pretty well aware of what was up and the prospects for heavy fighting were much too good. Company I, just north of Lemberg, had one platoon completely cut off, and enemy armor and infantry had worked

into the area of a second platoon. Further south . . . attacks were coming from all directions . . .

Meanwhile the 2nd Battalion had turned the tables on the aggressive enemy with an attack east from the towns of Goetzenbruck and Sarriensburg. So unexpected was this uprising in the face of their offensive, the Germans become completely disorganized for a time and were never able to renew their attack in this sector. . . There wasn't even so much as an enemy patrol around Goetzenbruck for four days following the encounter.

By the third day the line was slowly and painfully beginning to straighten, but in order to save our positions . . . a platoon had to be moved here, then there, to meet the numerous threats, a small herd of Germans had to be cleaned out from behind our lines, and the abundant artillery had to be placed quickly and accurately . . . the offensive subsided and reverted to the old familiar 'stagnant warfare.'[23]

German attacks continued for several weeks after the formal end of Northwind. Many of the attacks were not particularly well planned, but with the tired state of the 7th Army they created major problems. One attack, on January 5, hit the VI Corps ten miles north of Strasbourg. A second, two days later, struck at the northern edge of the Colmar Pocket, also not far from Strasbourg. A fourth attack also began on January 7, somewhat to the north of Strasbourg. The Americans and the French stopped the German attacks, barely.

A fourth attack started on January 7, based on a panzer division, a panzer grenadier division, and several assorted smaller units. The attacking force hit in the area of the town of Herrlisheim, being attacked by the 12th Armored. The 36th Division and the 3rd Algerian Division were moving up. However, until they arrived, the new American 12th Armored Division was the only unit trying to stop the attack. After several days' fighting, on the night of January 16–17, the 12th Armored sent its 43rd Tank Battalion and 66th Armored Infantry Battalion to attack the town. They ran into advanced elements of the 10th Panzers. A few American infantrymen made it back to American lines.

Brooks pulled back his lines north of the Haguenau Forest, on the night of January 20–21. The movement took the attacking Germans by surprise. The VI Corps was able to establish more defensible positions along the Zion, Moder, and Rothbach rivers.

Early in the morning of January 20, 1945, Germans attacked the 143rd Regiment in the area of the Bowden Woods, south of the town of Bischwiller. A small German group of about 150 men attacked positions held by Companies E and G. Twenty-seven Germans were killed, four captured, in this unsuccessful attack. In the words of the unit journal of this regiment, "Operations in France," for January 1945:

> The enemy, expecting to find cover in the woods occupied by the Second Battalion, found only strong positions and grazing fire bearing down on them . . . It was the scene of the Second Battalion's determination to hold the enemy off at all costs.
>
> The remainder of the early morning of 20 January 1945 was quiet. . . . It was believed that within the next forty-eight hours the sector would be the recipient of a large enemy attack. Our Observation Posts were reporting great enemy activity with tanks and track vehicles. Prisoners of war, when questioned, stated that a large attack was planned and would occur shortly.[24]

"The morale of the troops was high"[25] is the entire text of the next sentence. One presumes it was American morale which was high.

The weather was a problem, the same conditions which bothered troops in the northern part of the western front. The regimental narrative continues, "The snow was heavy on the ground, and at times visibility was very limited. Again our troops were fighting under the most difficult conditions with the elements a detriment to the operation."[26]

Intelligence was correct about pending attacks. A second smaller attack occurred the afternoon of January 20, a third and larger attack, supported by armor, the next morning. The woods were not cleared of Germans until that afternoon. Elsewhere on its front, the VI Corps was able to fight off further German attacks, the night of January 24–25, with little difficulty.

There was fighting to come, including final efforts to reduce the Colmar Pocket, and the tough advance into the German defensive system known as the Siegfried line. Evaluating the way the battle was run is hard. It is possible that the lack of imagination which led the high command to stop Devers' advance allowed both the

Ardennes offensive (Battle of the Bulge) as well as Operation Northwind and the follow-on attacks to take place.

There were still errors, but the initial decisions were made, and the necessary operational and tactical moves were carried out. There was no way to avoid sending Patton to rescue the American 1st Army attacked in the Battle of the Bulge. This made it necessary to weaken the 6th Army Group, which invited German attack. The attack was repulsed. Devers, Patch, and Brooks seemed to have handled their troops well. Dahlquist and the 36th Division did not accomplish all they set out to in the Colmar Pocket, but they were a very tired division and probably did all they could. They handled themselves well in January, particularly providing valuable assistance to the 100th in fighting off the second attack.

The 36th Division took part in the rest of the fighting in France and then entered Germany. The division was relieved at the end of March 1945 but would be back in the line in late April, for the "kill." It would see the end of the war and be among the American troops to experience a most unpleasant surprise when they saw some of the things the Germans had been doing behind the lines: "Inside... a sight revealed itself that daunted even a battle experienced man like myself."[27]

CHAPTER 17

The End of the War

". . . it was much harder to accept the fact that a screeching shell in the last minute of the war has the same lethal qualities it had during the most bitter of the early days in combat."

—141st Infantry regimental history, 1945[1]

The 36th Division staged a symbolic ceremony to mark its passage of the Siegfried line, the German defensive system. Division commander John Dahlquist was not available, so staff officer Armin Puck was given the task of raising the Lone Star flag over the Siegfried line as the division passed through. On March 26 he went to a customs building in the town of Schweigen, on the border between France and Germany. "I raised the Texas flag over the Siegfried line to the accompaniment of machine gun fire and mortar fire, coming in from the Germans and our machine gun fire and artillery fire and mortar fire from our Division. Needless to say, that was a real quick flag raising."[2]

The final major battles in the west had taken place in January and early February, with the crushing of the major German offensives and the reduction of the Colmar Pocket. Fighting remained, but the defeat of Germany was a question of weeks, even days, no longer years. German resistance was still determined, though. There were actually more American battle casualties in March than in February, though fewer than during the major German offensives of December and January—53,200 in March; 39,400 in February; 69,100 in January; 77,000 in December.[3] Retrospect shows that on April 1 the war was almost down to mopping up—and some horrible discoveries. Allied commanders "did not dare predict what sort of desperate defense Nazi fanaticism might yet muster."[4] They could not know that the plans the German government actually had for guerrilla resistance and withdrawal to the so-called "national redoubt" would remain just plans.

The push to the Siegfried line still met some tough German resistance. On March 21, 1944, elements of the 143rd Infantry Regiment assaulted the German town of Dorrenbach. The regimental report on "Operations in France and Germany" for March 1945 describes the engagement:

> Heavy resistance, consisting of machine gun fire from well concealed pillboxes, self-propelled fire and pre-arranged artillery and nebelwerfer concentrations, made the 1st Battalion's advance slow and treacherous. Despite the heavy resistance imposed on the 1st Battalion from the cleverly prepared Siegfried defenses, the Battalion continued to wedge its way forward, by-passing and isolating some of the enemy's positions and destroying those that impeded its advance.[5]

That night, the 3rd Battalion attacked a hill east of town. The operations report continues, "From the beginning of the attack, the enemy offered strong resistance firing machine guns from concrete pillboxes, with cleared fields of fire through the forest."[6]

The new German commander in the west, the 36th Division's old "friend" Field Marshall Albert Kesselring, was still a skillful and tough commander. However, at least west of the Rhine, commanders were authorized to withdraw if destruction was imminent.[7] As had been the pattern in Italy, the Germans fell back under pressure, but usually with fighting withdrawal. Some German soldiers surrendered, but there was little general loss of cohesion.

The Rhine itself was crossed in various places. Eisenhower still favored a broad-front approach, but some emphasis was given to a crossing in the north by Field Marshall Montgomery and his British and Canadian troops. Montgomery was interested in public credit for his men and for himself. He was quite happy to accept American help and had in fact been nagging Eisenhower for another ten American divisions—to Omar Bradley's horror. Montgomery did not get his Americans. Montgomery's March 23 crossing had been announced in advance ("21 Army Group Will Now Cross the Rhine," a message to troops was titled) and was turning out to be a rather ponderous effort.

Montgomery even asked Winston Churchill, with the army to watch the crossing, which was the last British unit to fight in Germany. Churchill, an expert military historian, knew it had been the Rocket Brigade, Royal Horse Artillery, attached to the Swedish Army in 1813. Their descendants were in Italy. So the 51st Highland Division had to lead the successful effort that night.[8] Unfortunately for Montgomery's ego, on March 7, American troops from the 9th Armored Division in the American 1st Army had captured the relatively intact Ludendorff railroad bridge across the Rhine at Remagen. Patton also beat Montgomery, but by only one day.[9] The 7th Army got into the Rhine crossing act on March 26.

The 36th Division took part in the difficult push through the Siegfried line to get to the Rhine at the end of March. It did not actually cross the river, being pulled out of the line on March 30 for occupation duty. "Division was completely relieved from combat at midnight on this date,"[10] according to division records. One officer of the division later wrote about a strange nighttime adventure at about this time. He was being driven at night, apparently just with his driver and some men in a second vehicle. Stepping out of his jeep for a few minutes waiting for traffic to move, he was approached by a figure who spoke to him in English, but with a German accent. The German wanted to surrender his just under 200 men:

> My driver got out of the vehicle and I saw his cigarette drop to the ground. "Take it easy!" I said to him. "Tell the men in the vehicle behind you to stay put."
>
> I turned to the German again and said, "Tell your men to make as little noise as possible but to place their weapons on the ground now." I could hear some low voices and then some scraping sounds of metal on rocks but, over all, the noise level was far lower than I had expected.
>
> "We will walk down the middle of the road in a column of twos," I said to the disembodied voice. "Tell your men there is

Chapter 17: The End of the War

to be absolutely no talking by them. I'll answer all questions that may be asked."

Again, there was a low relaying of orders. I cautioned my driver to keep up with the column when it moved and to be on lookout for me. I had no idea where I would be.

I told the German whom I believed to be an officer, to move his men quietly and without cadence, onto the road. We walked at a fairly slow pace beside the vehicles of our division headquarters, receiving occasional inquires as to who we were and what we doing. I answered each time saying we were needed up forward. One wise guy managed to come close to guessing my mission correctly by saying, "Thought we weren't taking any prisoners," followed by a laugh.

I didn't even want to think about the great likelihood that some of the surrenderees had retained possession of pistols and other weapons against the possibility of a fight breaking out in the darkness.

We proceeded in silence except for the shuffling of feet and the idling of engines along the roadside until we met a couple of MPs in a jeep a mile or so from where I'd taken the Germans in tow. The MPs challenged us and I spoke to them quietly telling them the circumstances of our being there.

"How do you know they laid down their weapons," one of them wanted to know.

"We've come more than a mile without any trouble so far," I said. "I've got to find some place to leave them under guard."

"The Provost Marshall has taken over a large building about a quarter of a mile up the road," the other MP said. "OK," I agreed, "but you men ride behind our column in your jeep. Don't say a word about the surrenderees. It could cause trouble. There should be no need for any rough stuff."

We started moving again and it occurred to me to ask the German spokesman why he had come up to me to make his request. "You were der only vun standing outside your vehicle like a man von a fighting unit," he said.

We reached the Provost Marshal's building in short order and the MPs who had stopped us earlier drove up and made arrangements for the surrenderees to be housed in the basement of the building. I never saw, except as shadows, any of the Germans who had walked with me for well over a mile at night, nor did I ever learn whether any weapons were found on them after they were incarcerated.

Later that night . . . I kicked something in the dust of a country road. On picking up the object I thought it an appropriate memento of the time and the place since it was a small lead soldier that some child had lost.[11]

Some of the incidents of the period include the adventures of the chief division chaplain, Lt. Col. Herbert E. MacCombie, first in finding food and places to sleep for 175 rescued displaced persons, and then in organizing a Passover Seder for Jewish members of the division.[12] Summarizing his successful Seder in Herxheim, MacCombie commented:

> It was the first Seder Feast there since Hitler had come to power. Perhaps it was the first in all of Germany. Everything went well... When the party was over several men came to me and offered to help clean the place. I told them "That will not be necessary. This time the Germans will clean up after you." The Germans did.[13]

On April 24, 1945, the 36th Division was moved back to the front for the final phases of the war. Any fighting tended to be quick skirmishes with diehard SS troops, some of the few Germans actually determined to fight to the last man. Most Germans surrendered en masse. Individually, as one division historian commented, "German doughboys, like Americans, aren't the glamorous type. Prisoners we had taken had always been the muddy, weary GI Joe of the German army with a weeks growth of beard... it had been a long war for the German doughfoot, too."[14]

A different feeling was developing for the SS troops they encountered. According to the regimental history of the 141st:

> Our casualties in the Bavarian and Tyrolian Alps were by comparison to earlier campaigns negligible, but it was much harder to accept the fact that a screeching shell in the last minute of the war has the same lethal qualities it had during the most bitter of the early days in combat. American soldiers do not hate easily and they forget quickly, although there is no more deadly and determined animal than the GI at the climax of a battle. Yet during these last days there developed a new and acute loathing and hatred for the stupidity of the SS who fought on as American armored columns drove into the very heart of the Reich, making their cause utterly hopeless.[15]

And there was more to come. A division veteran, author of a memoir of the Riviera campaign, wrote:

If the battles were not sustained and furious, the gruesome evidence of Nazi horrors brought the inhumanity of war strongly to the sight and minds of the T-Patchers. This was the area of Dachau, Hurlack and Landsberg. The advancing Americans were appalled at the stench of bodies, the roadside litter of dead, and the walking dead who were little more than skin stretched over bones. There was clear evidence that the SS Guards, before being forced to flee, had machinegunned their victims.[16]

The captor of the 200 Germans discussed above also advanced into this area. The 36th was not the first,[17] but arrived soon enough to view the camps relatively unchanged. The above officer saw Landsberg am Lech, near the prison where Adolf Hitler had written *Mein Kampf.* Landsberg, a subcamp of Dachau, was liberated, two days before the main camp, by elements of three United States divisions, the 103rd Infantry and 10th and 12th armored.[18] When the Americans entered the town, they found German civilians hanged from lampposts. The Wehrmacht had withdrawn from the town. Civilians hung white sheets out of windows. The Waffen SS (SS combat troops), before they withdrew, dragged as many civilians as they could find outside and hanged them.

Some prisoners from Landsberg had already been transferred to the main camp, to share the fate of the prisoners at Dachau itself. Others were locked into their barracks and then burned alive.

The main camp of Dachau was almost liberated by accident, not being on the Allied attack plans. On April 26, some escaped prisoners ran into American troops at the town of Pfaffenhofer and told them prisoners in Dachau were in danger of murder by withdrawing SS guards.[19] The 157th Regiment of the 45th Division was engaged in an attack on the town of Dachau (a suburb of Munich) when orders came to immediately take the camp.[20]

The 36th Division veteran, the one who had 200 Germans surrender to him a few nights before, told of his first view of Landsberg:

> My first view of the camp at close range was of a bundle of rags near one corner of the main enclosure. As I drew nearer I realized the rags contained a body and protruding from them was an arm stretching through the barbed wire fence grasping at a

> tuft of grass. A bullet through the head had been the price paid by some prisoner for reaching for something to eat . . .
>
> On looking into one of the hutments, I saw dozens of bodies on the platforms and signs of charring of their clothing and the surrounding woodwork and roof where fires had been set. The prisoners, emaciated to the point of extinction in any event, had been suffocated and burned to death . . .
>
> The worst scene lay along a railroad track to the rear of the camp where the most heinous treatment imaginable had been accorded nearly a thousand such walking skeletons. I learned later from the survivor that the Germans had told the inmates that since the war was almost over and that Germany was losing it, many of the prisoners were to be taken south to be rehabilitated before the Allies saw them. In order to help the prisoners survive the trip a woman doctor whom they could trust would inject them with sedatives to alleviate their suffering and make the trip easier for them . . .
>
> . . . I learned that the woman doctor had injected poison into the right arm of each person selected to be "rehabilitated." Once the poison took effect, the diabolical step was taken of placing the bodies on the railroad track and running over them with a train so that there could be no way to identify them. The effect was enough to make anyone observing it ashamed of being a member of the human race . . .
>
> . . . I saw what looked like eight of the haunted prisoners from the Stalag, standing half concealed in fox holes. Closer inspection showed that we were looking at the upper portions only of mens' bodies that had been cut in half by machine gun fire and propped up for pistol practice.[21]

The writer notes that the decision was made to make the surviving German adults of Landsberg visit the camp, a frequent practice by the Allies as the camps were liberated. Some of the Germans were heard to say that they had no idea what was going on. The writer did not believe them. Further:

> Our troops, well aware that the war was drawing to a close, had been inclined on re-entering combat to lay back a bit—until they saw Stalag 3. From then on few prisoners were taken. Our troops moved swiftly to eradicate anyone standing in their way.[22]

This, also, was not uncommon among advancing troops encountering the true unprecedented horror of the Nazi regime. One can totally identify with the 36th Division staff officer who described what he saw at the camps, and a German "research" hospital, who said, "This experience was the worst I had in my life."[23]

The 36th Division captured Hermann Goering, chief of the Luftwaffe and at one time the second-most-powerful man in Nazi Germany. Goering had gotten himself into trouble with Hitler in late April. With Berlin surrounded by Soviets, Goering, at Hitler's old retreat of Berchtesgaden, thought he should take over. Hitler disagreed, and only agreed not to have Goering shot if Goering resigned all his offices. Even after doing so, Goering was arrested by the SS. A few days later, orders came down to shoot Goering and his family. The SS commander wanted to confirm the order, but an answer was never received. After being moved to Austria, Goering was rescued by Luftwaffe troops. He still thought he would be safer in Allied hands. According to an attachment to the 36th Division narrative for May 1945:

> On the morning of 8 May 1945, Colonel Brauschitz of the German Luftwaffe, was brought into the C.P. of the C.G. 36th US Division at Kufstein, Austria. Thru an interpreter, he explained to General Dahlquist he had two letters from Goering, one for General Devers, Commanding Sixth Army Group and one for General Eisenhower. General Dahlquist had the interpreter read the letters.
> In them, Goering offered to surrender . . .[24]

Goering actually surrendered himself, and his family, to assistant division commander Brig. Gen. Robert I. Stack. Stack took his aide and half a platoon from the 142nd, commanded by 1st Lt. Jerome N. Shapiro, of New York City, to get the Reichmarshall.

The 36th also captured Field Marshall Gerd von Rundstedt, Air Marshall Ritter von Griem (Goering's successor), and SS Gen. Sepp Dietrich. They liberated some important French prisoners being held in Itter Castle, including two former prime ministers and Charles de Gaulle's sister.

The 36th Division suffered a total of 26,718 casualties in

World War II, from Salerno to Austria,[25] including 5,957 killed.[26] The division took 175,806 prisoners directly, counted in its cages.[27] Others were undoubtedly taken prisoner by other units as a result of 36th Division actions. There is also no way of counting casualties inflicted by the division.

★ ★ ★ ★ ★

On April 29, 1945, German forces in Italy, Army Group C, were surrendered—to take effect May 2, 1945. As befitted the complexities of the Italian surrender and the start of fighting in Italy, the ending had many twists and turns. They included replacement of General von Vietinghoff for negotiating with the enemy, his chief of staff not only refusing to accept a similar firing but working with an SS general to arrest their replacements and see that the surrender was carried out.[28] It was an appropriate end for the Italian campaign, a "cruel, bitter campaign that all too often seemed to be going nowhere,"[29] in which the 36th Division had played so key a part.

On May 4, with the surrender finally assured, advance elements of the 5th Army, commanded since December 1944 by Lucian Truscott, contacted elements of the 7th Army.[30]

On May 5, 1945, Army Group G surrendered to Jacob Devers. On May 7, Alfred Jodl surrendered all German forces to the Allies, effective May 8, 1945.

Japan remained to be beaten, but this was a matter of just a few months. Eisenhower could send the following message to Washington: "The mission of this Allied Force was fulfilled at 0241, local time, May 7th, 1945."

Postlude and Meaning

> "The whole nature of war and conflict resolution has changed, and I think the United States—in order to be an effective leader in the world—has to maintain maximum flexibility with its troops."
> —Governor George W. Bush of Texas, 1996[1]

The story of the 36th Division did not begin in 1917, either with its formal formation or its first commitment to combat in the trenches of France. Through its component parts, the division formed in 1917 already had a long and distinguished history. Texas tradition claims for the division an even longer and even more distinguished history.

The history of the 36th Division did not end with the end of World War II, though that was the last combat the division experienced. The division existed until 1968, though it was not mobilized for service in Korea or Vietnam. The 141st Regiment component still exists, part of the 36th Brigade, itself part of the 49th Armored Division. It was mobilized, but not deployed, for the 1990 Persian Gulf War against Iraq—though guard support units were used in that war. In a decision still controversial, an offshoot of the citizen soldier versus standing army debate older than this country, the

Defense Department felt that National Guard units could not be made ready for combat in the five months between the Iraqi invasion of Kuwait and the start of Desert Storm. With the striking success of the military aspects of the Gulf conflict, one might not want to second-guess Secretary of Defense Dick Cheney and Gen. Colin Powell. Future wars may not be so quick and easy.

After World War II, the Texas National Guard was initially reorganized under the command of Fred Walker. Walker, who had retired from the United States Army in 1946, that same year was appointed lieutenant general in the Texas National Guard. The guard got back into disaster assistance that very year. A combined unit of National Guard and State Guard troops was sent to Texas City, to help cope with the aftereffects of the disastrous dust explosion in the harbor. The State Guard, which had served the local militia function while the National Guard was in federal service during World War II, was officially abolished in August 1947. (It has since been re-created.)

In 1950 Audie Murphy, perhaps the most famous American soldier in World War II below the rank of general, joined the 36th Division. He saw no further combat, but was used as a kind of "poster boy" for the National Guard. Murphy served until 1957. A Texas historian notes that "Audie did much to help promote not only the 36th Infantry Division and the Texas Guard but also the National Guard program across the nation when it badly needed support."[2]

Sizable portions of the National Guard were mobilized for service in the Korean War. The 36th Division, and the 49th Armored Division (also from Texas), were not among them. Virtually all the Texas Air National Guard, however, saw active service in Korea. The next major Texas Guard mobilization was September 10, 1961, when Hurricane Carl hit the Texas coast near Port Lavaca. Guard members helped in the recovery from that storm. The 49th Armored Division was mobilized in October, with increased tensions with the Soviets over Berlin.

The Vietnam War, for political reasons, marked the first time that the National Guard was not immediately mobilized for a major conflict. The only significant mobilization for that war occurred in May 1968. Just over 20,000 troops were called up:

President Johnson refused to declare a national emergency, to seek congressional legislation for a mobilization, or to seek a declaration of war. During the period 1965-67, the President rejected all recommendations for a mobilization. The principal reason for these decisions was his overriding concern for the domestic political arena. Never before in US history had a President declined to use in war military forces whose very purpose was for such activity.[3]

In early 1968, the 36th Division and the 49th Armored Division were abolished, to be replaced by three brigades. The 49th was later reactivated.

The Defense Department continues to reassess the potential role of the guard. State governors continue to be a strong lobby supportive of the guard. Typical of their comments are two from the February 1996 National Governors Association Conference in Washington, D.C. Governor George W. Bush of Texas, a Republican, spoke of the guard's role in national defense:

> The whole nature of war and conflict resolution has changed, and I think the United States—in order to be an effective leader in the world—has to maintain maximum flexibility with its troops. That means having people in ready reserve, and the Guard is just that . . .[4]

Maryland governor Parris Glendening, a Democrat, spoke of the guard's domestic role:

> I don't think you can as easily take the regular army and conveniently backfill traditional domestic National Guard activities . . .[5]

Guard support units were deployed in Bosnia in late 1995. However, the immediate military response to the Iraqi moves in September 1996 was with regular troops. The role the National Guard will play in the American military of the future, not to mention the question of the responsibility of men, and women, to serve their country will continue to be debated.

★ ★ ★ ★ ★

On first glance, looking at the 36th Division proper (1917–45), one might conclude that the division did not equal the extraordinarily motivated group of fighting men known as Hood's Brigade. The world, and war, however, are different. The division could not have been, nor could more modern units totally duplicate the Stonewall Brigade, the 20th Maine or the 1st Minnesota, Hood's Texas Brigade's colleagues and opponents in the American Civil War. These exceedingly motivated groups of neighbors fought not just for their homes, or their friends, or their countries, but for their ideals. Unlike the Confederate units, however, the 36th was on the winning side, and the right side (forgive the writer's Yankee arrogance) of the two wars in which it fought.

Hood's Texas Brigade was small. The 36th Division was large in World War II, huge in World War I. The Texas Brigade was from Texas and neighboring Arkansas. Though based on the Texas National Guard, both national incarnations of the 36th Division were diverse and national, heavily draftee dependent. Drafting initial and replacement troops as individuals, though enabling the unit to stay close to "book" strength, hindered cohesiveness, not to mention what it did to the life span of replacement riflemen (five days' life expectancy in combat).

The men of the 36th, and their bosses, usually performed with the steady competence and professionalism expected of American soldiers. The division could be a bit slow, as at Montelimar. The division could be unable to help a bad plan, as at the Rapido. Divisional tenacity, or luck, could save a close call—Salerno and the "Lost Battalion." The 36th Division could achieve excellence, as at Forest Farm or at the Ste. Marie pass. The division could be brilliant. One only has to look at the superbly conceived, planned, and executed infiltration at Velletri, leading directly to the fall of Rome.

If there is a lesson from how the 36th Division conducted its operations, it can be found looking at the divisions's two most dramatic battles—Rapido and Velletri. At Rapido, the division took part in a battle plan set in stone. It was told to cross the Rapido River in an effort to divert German troops from opposition to the Anzio landing. The division was given a poor, and a dangerous, plan—a frontal assault on strong German positions—and told to "just do it." The army commander ignored the fact that a second diversionary effort was working, that his ostensive goal was being achieved. "Facts on the ground" seemed to have had little impact on higher-level command—a not uncommon complaint during World War II.

Though not quite a useless bloodbath on the scale of some of the Civil War battles in which Hood's Brigade participated, winning and losing, the Rapido crossing was useless and damaging.

Things were different at Velletri. Fred Walker anticipated problems with the plan he had been assigned. This time he found an alternative plan, to exploit a gap he had discovered in German lines. Walker was able to act quickly and convince his superiors that his plan was better. This time commanders were flexible and able to adapt to a changed situation.

The men themselves *seem* to have performed better at Velletri than at the Rapido. At Rapido they showed many of the negative characteristics stereotypical of American combat soldiers, in particular dislike of night fighting. Four months later, at Velletri, everything changed. Did they suddenly decide they liked night fighting? Did men who are supposed to have been reluctant to kill from a distance with guns (Do we really want Americans to like killing with guns?) suddenly become quite willing to kill with knives? Or was the improved performance the subconscious result of believing in a good plan? Do positive expectations provide just that extra little bit that might make the difference between success and failure?

Perhaps the ultimate lesson is to make maximum use of resources, men (and now women), supplies, and, perhaps most importantly, information. Go with your strength. When your army is the best in artillery and air support, only an idiot would fail to maximize use of these arms. When you can read enemy messages, read them. When man-to-man fighting is necessary, be sure your men are able to prevail. Objectively looking at the record of the United States Army in World War II shows a fair number of errors in policy and implementation. However, it also shows more skill than some postwar experts chose to recognize, perhaps due to the unfortunate modern tendency of seeking to "debunk" rather than to find the truth. The United States Army paid in lives for its errors in World War II. We must have been doing something right. We won. The Germans, with the "genius for war," lost.

Finally . . .

Any dull narrative in this story might be explained away as reflecting the reality of war—short periods of terror punctuated by long periods of boredom. Fortunately, nothing interesting happened to the division after World War II—in war, as in traditional

Chinese culture, "interesting" is a curse. We can blame the arsonist who destroyed Texas military records after the Civil War for making it impossible to start the book with a discussion of Hood's Texas Brigade, if not the Alamo, rather than post–Civil War law enforcement. This is the legendary tradition of the 36th Division; the proven tradition begins during the relatively quiet Reconstruction period.

Hopefully, any boring transition periods are balanced by the excitement of the combat narrative. Hopefully, the job of the writer has been done, producing a work balanced, factually accurate, and most importantly, a good read. And hopefully it has shown a military unit with excellent days, disastrous days, but mostly very good days. This makes the 36th Division reflective of the American divisions which won World War I and World War II. I hope it makes this book worth reading. It made the book worth writing.

APPENDIX

The Cost

Perhaps the best way to give an idea of the human cost the 36th Division paid in World War II (and its other wars) is to reprint some material from the files of the 36th Division Association (apparently a rough draft of material for an early newsletter).

T-PATCH MATERIAL

PERSONALS

Mrs. G. E. Cameron is very anxious to hear from some of the buddies of her husband and to learn some of the details surrounding his death. He was S/Sgt Gerard E. Cameron, 31030719, of B. Co. 141st Infantry. He was first reported missing in action on 9 December 1944 and later reported as having been killed that day. If any of the old members of that company can furnish any information please write to her at Tasker St., Saco, Minn.

Mr. Arthur Hilty, Box 642, Gary, Indiana, is very anxious to hear from some of the former members of L Co. 143rd Infantry who might have known his son, Sgt. Harold Hilty. Harold made the supreme sacrifice for his country at the Rapido River on 1 January 1944. If any of you who knew Sgt Hilty would write his father it would be greatly appreciated.

Mrs. J. Franchi is seeking any information that anyone may have concerning the death of Sgt. David Franchi formerly of L. Co 143rd Infantry. He was killed in the invasion of France in September 1944. Mrs. Franchi is his sister and would like very much to hear from some of you who knew David.

John Kotkovetz former T/Sgt of the Anti-Tank Co., 143rd Infantry is improving steadily from wounds received in Action. You former members of this unit will remember that he was wounded 12 March 1945. He is very anxious to hear from some of you former members of his unit. His address is Fitzimmons General Hospital, Denver, Colo.

Mrs. Sophia Griesmer would like to hear from some of her son's buddies in C Co. 143rd Infantry. Her son was killed on the move from Southern France to Chenimenil on 25 September 1944. Please write to Mrs. Griesmer if you know any of the details surrounding the death of her son.

Mr. & Mrs. Charles McLagan, 413 Houghton St, North Adams, Mass., have written the following plea: "We would like very much if you could give us some information concerning our son who was wounded at Cassin, Italy, 25 Feb 1944 and was confined to the hospital in Italy and near Naples. His last letter to us was on 21 May, and he said he was well and was being moved, possibly home. The War Department said he died from his wounds but his buddies who have returned home and who were with him overseas all have different stories. The last story we heard and the one that sounds most logical was that he died en route home and was buried at sea. We would like to know if this is true or if he was buried in a grave in Italy. He was Herbert H. McLagan, Hq 3rd Bn., 142nd Infantry. We would like to know if someone who was with him when he died would communicate with his parents."

Wesley M. Garrison, 2355 Vanderbilt Rd., Birmingham, Alabama, has performed a notable deed for the parents of Pvt. William E. Boos Jr., who was killed in action at Montelimar, France. Wesley was the only survivor of the fight in which William was killed. He has made two trips to see Bill's parents and they appreciated his visits very much. He was able to give them some account of Bill's death. They have ordered a History to send Wesley to repay in part his thoughtful deeds.

Endnotes

INTRODUCTION

1. Jefferson Davis to twenty companies of Texas volunteers on their arrival in Richmond in the fall of 1861. Quoted in A. V. Winkler, *The Confederate Capital and Hood's Texas Brigade*, Austin: Von Boekmann, 1894, page 33.

2. From a letter written by John Marshall from Richmond, Virginia, on June 7, 1862. Quoted in Col. Harold B. Simpson, *Hood's Texas Brigade: Lee's Grenadier Guard*, Waco, Texas: Texian Press, 1970, page 41.

3. Interview with Theodore H. Andrews, then lieutenant colonel commanding 3rd Battalion, 143rd Regiment, quoted in Vincent M. Lockhart, *T-Patch to Victory*, Canyon, Texas: Staked Plains Press, 1981, pages 45-46.

4. Quoted Lockhart pages 46-47.

CHAPTER 1: PRELUDE

1. Excerpt of the speech of Capt. William K. Martin on the mustering of his company (Henderson Guards, later Company K, 4th Texas Infantry Regiment), May 1861, Fincastle, Texas. Quoted J. J. Faulk, *History of Henderson Country, Texas*, Athens, Texas: Athens Review Publishing Company, 1929, page 129.

2. Louis Cantor, *The Creation of the Modern National Guard: The Dick Militia Act of 1903*, Unpublished Ph.D. dissertation, Duke University, Graduate School of Arts and Sciences, 1963, page 44.

3. Allan Robert Purcell, *The History of the Texas Militia, 1835-1903*, unpublished Ph.D. dissertation, The University of Texas at Austin, 1981, pages 1-2.

4. Walter Millis, *Arms and Men: A Study in American Military History*, New York: Putnam, 1956, pages 34-35.

5. Constitution of the United States, edition published by Commission on the Bicentennial of the United States Constitution, undated but circa 1987.

6. Russell F. Weigley, *History of the United States Army*, New York: The MacMillan Company, 1967, page 81.

7. H. P. N. Gammel, *The Laws of Texas, 1822-1897*, Austin: Gammel's Book Store, 1902, volume I, page 6.

8. Ibid., vol. I, page 13.

9. Purcell, page 65.
10. Gammel, vol. I, page 1114.
11. Stephen L. Hardin, *Texian Iliad,* Austin: University of Texas Press, 1994, page 250.
12. Hardin, page 5.
13. Mark E. Nackman, "The Making of the Texas Citizen Soldier," *Southwestern Historical Quarterly,* LXXVIII (1974), page 243.
14. Weatherford Whiteman, September 13, 1860, quoted in David Paul Smith, *Frontier Defense in the Civil War: Texas' Rangers and Rebels,* College Station: Texas A&M University Press, 1992, page 19.
15. Ibid.
16. Archer Jones, *Civil War Command and Strategy,* New York: The Free Press, 1992, pages 3-4.
17. Stephen E. Ambrose, *Upton and the Army,* Baton Rouge: Louisiana State University Press, 1964, page 171.
18. Quoted A. M. Williams, *Sam Houston and the War of Independence,* Boston: Houghton Mifflin Co., 1893, page 353.
19. Ralph A. Wooster, "Texas in the Southern War for Independence," pages 71-85, Joseph P. Dawson, editor, *The Texas Military Experience,* College Station, Texas: Texas A&M University Press, 1995, cited page 74.
20. Col. Harold B. Simpson, *Hood's Texas Brigade: Lee's Grenadier Guard,* Waco, Texas: Texian Press, 1970, page 9.
21. D. H. Hamilton, *History of Company M, First Texas Volunteer Infantry,* Waco, Texas: W. M. Morrison, 1962, pages 9-10.
22. D. H. Hamilton, Sergeant, Company M, 1st Texas, V.I.C.S.A., *History of Company M, First Texas Volunteer Infantry, Hood's Brigade, Longstreet's Corps, Army of the Confederate States of America* (no publisher or city listed), 1925, page 20. This is likely an earlier edition of the work cited in endnote 4.
23. D. H. Hamilton, 1925, page 69.
24. Robert James Lowry, *Diary,* edited by John E. Hamer, Alexandria, Virginia: privately printed, 1965, page 32. Lowry quotes cited in Simpson, *Lee's Grenadier Guard.*
25. Simpson, page 37. See also Harold B. Simpson, "Foraging with Hood's Brigade," *Texana,* vol. 1, no. 3, summer 1963, pages 258-276.

CHAPTER 2: RECONSTRUCTION AND BEYOND

1. Sheridan to Rawlins, November 14, 1866. Quoted in William L. Richter, *The Army in Texas During Reconstruction,* College Station, Texas: Texas A&M University Press, 1989, page 9.
2. Richter, page 13.
3. W. D. Wood, *Reminiscences of Reconstruction in Texas and Reminiscences of Texas and Texans Fifty Years Ago* [No city or publisher listed], 1902, page 5.
4. David Paul Smith, *Frontier Defense in the Civil War,* College Station: Texas A&M University Press, 1992, page 171.
5. Ibid., page 168.
6. John H. Reagan, "To The People of Texas," in prison, Fort Warren,

Boston Harbor, August 11, 1865, typescript in John H. Reagan papers, Center for American History, University of Texas, Austin; and John H. Reagan, *Memoirs, with Special Reference to Secession and the Civil War*, New York and Washington: The Neale Publishing Company, 1906, pages 286-295.

    7. Reagan, Fort Monroe letter.
    8. Reagan, *Memoirs*, page 227.
    9. Charles William Ramsdell, *Reconstruction in Texas*, New York: Columbia University Press, 1910, page 89.
    10. Resolution of March 15, 1866, in *Journal of the State Convention, 1866*, Austin: Southern Intelligencer, 1866.
    11. Ibid., pages 94-105; and Richter, pages 49-50.
    12. Richter, page 57.
    13. Mary Whatley Clarke, *David G. Burnet*, Austin: Pemberton Press, 1969, page 241.
    14. Richter, page 61.
    15. Edgar P. Sneed, "A Historiography of Reconstruction in Texas: Some Myths and Problems," *Southwestern Historical Quarterly*, volume LXXII, no. 4, April 1969, pages 444 and 445.
    16. Jim Dan Hill, *The Minuteman in Peace and War*, Harrisburg: The Stackpole Company, 1964, page 104.
    17. United States Congress, House of Representatives, "Communication from Governor Pease of Texas Relative to the Troubles in that State," 40th Congress, 2nd Session, 1868, H Misc. Doc 127.
    18. Robert W. Shook, "The Federal Military in Texas, 1865-1870," *Texas Military History*, vol. VI, no. 1, Spring 1967, page 41.
    19. House Executive Documents, 40th Congress, 3rd Session, no. 1, xvi-xvii.
    20. Christian G. Nelson, "Rebirth, Growth and Expansion of the Texas Militia, 1868-1898," *Texas Military History*, vol. 2, no. 1, February 1962, page 5.
    21. Hans Peter Nielson Gammel (editor), *The Laws of Texas 1882-1897*, Austin: H. P. N. Gammel, 1898, volume VI, page 190.
    22. A. G. Malloy to J. Davidson, October 22, 1871, quoted in Otis A. Singletary, "The Texas Militia During Reconstruction," *Southwestern Historical Quarterly*, volume LX, no. 1, July 1956, page 27.
    23. *Historical and Pictorial Review, National Guard of the State of Texas*, Baton Rouge, Louisiana: Army and Navy Publishing Company, 1940, page xxx.
    24. Ann Patton Baenziger, "The Texas State Police During Reconstruction: A Reexamination," *Southwestern Historical Quarterly*, volume LXXII, no. 4, April 1969, page 470.
    25. Otis A. Singletary, "The Texas Militia During Reconstruction," *Southwestern Historical Quarterly*, July 1956, volume LX, #1, pages 23-35, quote from page 28.
    26. Ibid., page 29.
    27. *Daily State Journal* (Austin), February 10, 1871. Singletary, pages 29 and 30. Bruce A. Olson, "The Houston Light Guards: A Case Study of the Texas Militia, 1873-1903," thesis, presented to the faculty of the Graduate School of the

University of Houston, in partial fulfillment of the requirements for the degree of Master of Arts, May 1985, pages 35-36.

28. *Daily State Journal,* April 9, 1872.
29. Singletary, page 30. *Daily State Journal,* August 13, 1871.
30. Singletary, pages 32-33.
31. George E. Shelley, "The Semicolon Court of Texas," *Southwestern Historical Quarterly,* XLVIII, pages 449-468, specific discussion of incident on pages 463-466.
32. Ibid., page 468.
33. *Daily Democratic Statesman,* January 16, 1874.
34. *Houston Daily Telegraph,* October 1, 1874.
35. Clarence P. Denman, "The Office of the Adjutant General in Texas: 1835-1881," *Southwestern Historical Quarterly,* XXVII, pages 319-320.
36. Report of the Adjutant-General of the State of Texas for the Fiscal Year Ending August 31, 1878, Galveston: The Book and Job Office of the Galveston News, 1878.
37. Nelson, page 9.
38. Ibid.
39. Ibid., page 10.
40. "Department of the Army, Lineage and Honors, 141st Infantry (First Texas)," undated sheet in the files, Center for Military History, United States Army, Washington, D.C.
41. "Souvenir: Houston Light Guard," commemorative brochure, undated but circa 1902, Texas State Archives.
42. "Souvenir: Houston Light Guard," page 1.
43. Ibid., page 17.
44. Ibid., page 15.
45. Olson, pages 139-154 and *Houston Daily Post,* September 27-30, 1887.
46. Ibid.
47. *Houston Daily Post,* September 27, 1887.
48. Olson, page 142.
49. W. A. Hickey to Gov. Lawrence Sol Ross, September 25, 1887. Quoted Olson, page 144.
50. John Hope Franklin, *The Militant South: 1800-1861,* Cambridge: Belknap Press, 1970, page 76; Olson, page 144; and general analysis of this writer.
51. Ross to F. A. Reichardt, September 26, 1887. Quoted Olson pages 145-154.
52. Olson, page 149.
53. September 28, 1887.
54. Christian G. Nelson, "Organization and Training of the Texas Militia, 1870-1897," *Texas Military History,* vol. 2, no. 2, May 1962, page 45.
55. Allan Robert Purcell, "The History of the Texas Militia," unpublished Ph.D. dissertation, University of Texas at Austin, 1981, page 309.
56. Weigley, pages 597-598.
57. Cantor, page 46.
58. William H. Riker, *Soldiers of the States: The Role of the National Guard in American Democracy:* Washington, D.C.: Public Affairs Press, 1957, page 44.

59. John K. Mahon, *History of the Militia and the National Guard*, New York: Macmillan Publishing Company, 1983, page 113.

60. *New York Times*, August 28, 1897.

61. *Army Navy Journal*, February 20, 1897, page 438.

62. Graham A. Cosmas, *An Army for Empire: The United States Army in the Spanish-American War*, Shippensburg, PA: White Mane Publishing Company, 1994, page 326.

63. Col. R. Ernest Dupuy, *The Compact History of the United States Army*, New York and London: Hawthorn Books, Inc., 1961, page 172.

64. Alan Robert Purcell, *The History of the Texas Militia: 1835-1903*, unpublished Ph.D. dissertation, University of Texas at Austin, 1981, page 284.

CHAPTER 3: GETTING READY FOR WAR

1. Letter of Private Leo Muller, *Yorktown News*, September 19, 1918.

2. Clarence C. Clendenen, *The United States and Pancho Villa*, Ithaca, New York: Cornell University Press for the American Historical Association, 1961, page 287.

3. Barbara W. Tuchman, *The Zimmermann Telegram*, New York: The Viking Press, 1958, pages 3-24.

4. Lonnie J. White, "Forming the 36th Division in World War I," *Military History of Texas and the Southwest*, vol. 17, no. 2, 1982, page 6.

5. John A. Logan, *The Volunteer Soldiers of America*, Chicago: R. S. Peale, 1887, page 105.

6. Brevet Major General Emory Upton, United States Army, *The Military Policy of the United States*, New York: Greenwood Press, Publishers, 1968 reprint of 1904 edition, page xi.

7. Emory Upton, *The Military Policy of the United States from 1775*, Washington, D.C.: Government Printing Office, 1904, page 323.

8. Russell F. Weigley, *History of the United States Army*, New York: The MacMillan Company, 1967, page 281.

9. Philip Jessup, *Elihu Root*, New York: Dodd, Mead & Company, 1938, vol. I, page 220.

10. Secretary of War, Annual Report, 1899, page 58.

11. William C. Sanger, "Report on the Reserve and Auxiliary Forces of England and the Militia of Switzerland": Prepared for President McKinley and Secretary of War Root, Washington, D.C.: Government Printing Office, 1903, page 9.

12. Louis Cantor, "The Creation of the Modern National Guard: The Dick Militia Act of 1903," unpublished Ph.D. dissertation, Department of History, Duke University, 1963, page 131.

13. Robert Bacon and James Brown Scott, collectors and editors, *The Military and Colonial Policy of the United States, Addresses and Reports by Elihu Root*, Cambridge: Harvard University Press, 1916, page 141.

14. MG Bruce Jacobs, "How the Dick Acts Happened," *National Guard*, September 1988.

15. The Statutes at Large of the United States of America, from December

1901 to March 1903. Edited, Printed and Published by Authority of Congress, volume XXXII–Part I, Washington, GPO: 1903, chapter 196, page 775.

16. John J. Pershing, *My Experiences in the World War*, New York: Frederick A. Stokes Company, 1931, volume I, pages 264-265.

17. Edward M. Coffman, *The War to End All Wars: The American Military Experience in World War I*, Madison: The University of Wisconsin Press, 1986, page 25.

18. Marvin G. Kreidberg and Merton G. Henry, *History of Military Mobilization in the United States*, Washington: Department of the Army, 1955, pages 306-207.

19. Bruce L. Brager, "Civil War Medicine," *National Guard Magazine*, volume XLIV, no. 4, April 1990, page 26.

20. Kriedberg and Henry, pages 287-289. Adams, page 239, cites similar figures broken down as to white and black troops in the Federal army.

21. Jim Dan Hill, *The Minute Man in Peace and War*, Harrisburg, PA: The Stackpole Company, 1964, page 264.

22. Meiron and Susie Harries, *The Last Days of Innocence*, New York: Random House, 1997, pages 123-124.

23. Memorandum, From: The Chief, Militia Bureau; To: The Adjutant General of Texas; Subject: Federal recognition, new units, July 31, 1917; John Hulen Papers, Texas State Archives.

24. Memorandum, From: The Chief, Militia Bureau; To: The Adjutants General of all States, Territory of Hawaii, District of Columbia, all Inspector-Instructors and Officers in Charge of Militia affairs, Department Headquarters; Subject: Organization and entry into Federal service of the National Guard; May 5, 1917, John Hulen Papers, Texas State Archives.

25. Brigadier General John H. Hulen, General Order #3, June 16, 1917, John Hulen Papers, Texas State Archives.

26. "Instruction for the use of those designated as officers of proposed organizations of the National Guard of Texas," By Command of Brig. Gen'l John A. Hulen, C. L. Test, Capt. Q. M. C. N. G. of Tex. Adjutant, undated by C. June 1917.

27. Ibid., page 1.

28. General Orders No. 2; Headquarters, Commanding General, National Guard of Texas; by command of Brigadier General Hulen; C. L. Test, Captain, Q. M. Corps, Adjutant; June 15, 1917; John A. Hulen Papers, Texas State Archives.

29. *Austin Statesman*, June 25, 1917.

30. White, page 12.

31. Will T. Little, L.G. Putnam, and R. J. Barker, *The Statutes of Oklahoma, 1890*, Guthrie: State Capitol Printing Company, 1891, page 692.

32. Kenny A. Franks, *Citizen Soldiers: Oklahoma's National Guard*, Norman: University of Oklahoma Press, 1984, page 3.

33. White, page 20.

34. Alexander White Spence, *The History of the Thirty-Sixth Division, U.S.A., 1917-1919*, unpublished typescript dated 1919. Spence is identified on the title page as "Formerly Captain of Infantry, 36th Division, Aide-de-Camp to

Major General Wm. R. Smith." Reading through the manuscript indicates that this is an authoritative primary history. National Archives staff agree with author evaluation that the presence of a 1947 declassification stamp shows that document was at one time classified, and can be considered accurate. Check through division files also shows memos from Spence to several 36th element commanders to ascertain further details about particular actions.

35. White, page 27.

36. Jim Dan Hill, page 273; and Capt. Ben H. Chastaine, *Story of the 36th*, Oklahoma City: Harlow Publishing Company, 1920, page 21.

37. Raymond S. McLain, Col. Inc., Chief of Staff, 45th Division; To, Col. Dallas J. Matthews, Headquarters, Thirty Sixth Division; September 15, 1933; 36th Division Association collection, Texas State Archives.

38. Spence, page 5.
39. White, page 43.
40. Quoted White, page 45.
41. Chastaine, page 17.
42. White, page 46.
43. Weigley, page 371.

44. Lonnie J. White, "Training the 36th Division in World War I," Military History of Texas and the Southwest, vol. 17, no. 3, 1982, page 55.

45. White, page 55.
46. Weigley, page 372.
47. Kreidberg and Henry, pages 294-295.
48. Ibid., pages 281-283.
49. Spence, page 26.
50. White, pages 69 and 70; Spence page 27.
51. White, Training, page 94.
52. Spence, page 30.

53. Letter from First Lieutenant E. C. Toy, 111th Field Signal Battalion, *Houston Chronicle,* December 27, 1918; firing incidents are also reported in Spence, pages 30-31.

54. Ibid.

55. Letter of Private Leo Muller, *Yorktown News,* September 19, 1918.

56. SWC: 46-1, Report; Baker Mission to England and France (Received: At War Department, July 26, 1917); From: Military Mission to England and France; To: The Chief of Staff.

57. Program of Training (first phase) for 36th Division, transmitted August 9, 1918.

58. Muller letter.

59. Frederick Palmer, John J. Pershing: General of the Army, Harrisburg, Pennsylvania: Military Service Publishing Company, 1948, page 307.

CHAPTER 4: COMBAT AT ST. ETIENNE

1. A. Page, 144th Infantry, letter undated, published in *Van Alstyne Leader,* date not give, Hornaday Collection, Texas State Archives, Box 3, page 309.

2. Hubert C. Johnson, *Breakthrough! Tactics, Technology and the Search for*

Victory on the Western Front in World War I, Novato, California: Presidio Press, 1994, page 51.

3. Frank Richards, *Old Soldiers Never Die*, London: Faber & Faber, 1933, page 35.

4. Rudolf Binding, *A Fatalist at War*, London: Allen & Unwin, 1929, page 21.

5. Johnson, page 59.

6. Bruce I. Gudmundsson, *Stormtroop Tactics: Innovation in the German Army, 1914-1918*, New York: Praeger, 1989, page 172.

7. Ernest D. Swinton, *Eyewitness*, London: Hoddor & Stoughton, 1932, page 115.

8. Marvin A. Kreidberg and Merton G. Henry, "History of Military Mobilization in the United States," Washington, D.C.: Department of the Army, 1955, page 292.

9. John J. Pershing, *My Experiences in the World War*, New York: Frederick A. Stokes Company, 1931, volume I, pages 264-265.

10. March 23, 1918, Telegram, From War Office, To: General Wagstaff [Chief, British Military Mission] Headquarters American Expeditionary Forces, France. "United States Army in the World War, Training and Use of American Units with British and French," Washington: History Division, Department of the Army, 1948, pages 71-72.

11. Russell L. Weigley, *History of the United States Army*, enlarged edition, Bloomington: Indiana University Press, 1984, page 383.

12. Ibid., pages 381-385.

13. Quoted in Pershing, volume I, pages 38-39.

14. Final Report of Gen. John J. Pershing, Commander-in-Chief, American Expeditionary Forces, Washington, D.C.: Government Printing Office, 1919, page 22.

15. Memorandum, June 14, 1918, French Group of Armies of the North, United States Army in the World War, Training and Use of American Troops with British and French, Washington, D.C.: History Division, Department of the Army, 1948, page 321.

16. Pershing, Volume II, page 189. Not published until 1948.

17. Training and Use of American Troops, pages 2-3.

18. Ibid., page 1.

19. American Battle Monuments Commission, "36th Division: Summary of Operations in the World War," Washington, D.C.: Government Printing Office, 1944, page 5. Files of the Center for Military History.

20. General Hunter Liggett, *AEF: Ten Years Ago in France*, New York: Dodd, Mead and Company, 1928, page 70.

21. American Battle Monuments Commission, page 5.

22. G-3 Report File, First Army, Messages Received, 107.04 Operations Report, V Army Corps, September 30, 1918.

23. Alexander White Spence, *The History of the Thirty-Sixth Division, U.S.A., 1917-1919*, Unpublished manuscript, 1919, pages 82-83.

24. Ibid., page 6.

25. Lonnie J. White, "The Combat History of the 36th Division in World War I," *Military History of Texas and the Southwest,* vol. 17, no. 4, 1982, page 124.

26. "Statement of Captain Stephen D. Ridings," Medical History of the 142nd Infantry, Historical File, 36th Division, RG 120, National Archives.

27. Chastaine, Capt. Ben H., *Story of the 36th,* Oklahoma City: Harlow Publishing Company, 1920, page 85.

28. Sgt. John A. Cegner, Co. G, 141st Infantry, letter of November 13, 1918, *Arlington Journal,* January 17, 1919.

29. White, page 129. An examination of what appears to be one of these maps, at the National Archives, confirms the comment.

30. Spence, page 78.

31. "Statement of Oscar F. Washam," Personnel War Experiences, 36th Division, RG120.

32. Letter by George McCall, Headquarters Company, 144th Infantry, of October 18, 1918, *Llano News,* December 5, 1918.

33. Letter of Olin N. Buchanan, Co. B, 144th Infantry, of October 18, 1918, *Lone Oaks News,* November 22, 1918.

34. Spence, page 83.

35. Chastaine, page 89.

36. Ibid., page 91.

37. John A. Lejeune, *The Reminiscences of a Marine,* Philadelphia: Dorrance and Company, 1930, pages 360-361.

38. Ibid.

39. White, page 132.

40. Memorandum, Headquarters 142nd Infantry, A. E. F., January 23, 1919, From: C.O. 142nd Infantry, To: The Commanding General 36th Division (Attention Captain Spence), Subject: Transmitting Messages in Choctaw.

41. Spence, page 117.

42. Lieutenant Colonel Ernst Otto, German Army, Retired, "The Battle at Blanc Mont," part III, United States Naval Institute Proceedings, vol. 56, no. 325, March 1930, pages 177-199, citation from page 184.

43. McCall letter.

44. White, page 135. Also Washam, Personnel War Experiences, Supplemental File, 36th Division, RG 120, National Archives.

45. Sergeant W. S. McBirnie, speech before Dallas Automobile Club, undated typescript circa 1919, Hornaday Collection.

46. Spence, page 119.

47. Chastaine, 105.

48. A. Page, 144th Infantry, letter undated, published in *Van Alstyne Leader,* date not given, Hornaday Collection, Texas State Archives, Box 3, page 309.

49. Headquarters, 142nd Infantry, A. E. F., APO #796, 21 December 1918, Memo To: Captain Spence, Division Headquarters, From: John K. Boyce, Captain, 142nd Infantry, Adjutant.

50. Cpl. Wayne Wheeler, Co G. 142nd Infantry, letter of December 13, 1918, Hereford Brand, February 6, 1919.

51. Spence, page 107.

52. Chastaine, page 121.

53. Headquarters Second Division (Regular), American Expeditionary Forces, 9 October, 1918, From: A. C. of S. G-3, To: C. G. 7st Brigade, Subject: Instructions for October 9th, 1918.

54. White, page 138.

55. Otto, Part IV, Naval Institute Proceedings, vol. 56, no. 326, April 1930, pages 304-316, citation from page 312.

CHAPTER 5: FOREST FARM

1. George McCall, Headquarters Company, 144th Infantry, letter of October 18, 1918, published in *Llano News,* December 15, 1918.

2. American Battle Monuments Commission, "36th Division: Summary of Operations in the World War," Washington: United States Government Printing Office, 1944, pages 9-10.

3. Lonnie J. White, "The Combat History of the 36th Division in World War I," *Military History of Texas and the Southwest,* vol. 17, 1982, no. 4 (entire special issue), page 149.

4. Capt. Ben H. Chastaine, *Story of the 36th,* Oklahoma City: Harlow Publishing Company, 1920, page 179.

5. Chastaine, page 179.

6. A. Page, 144th Infantry, letter, Hornaday Collection Box 3, Page 309, Texas State Archives.

7. White, page 152.

8. McCall, letter.

9. Chastaine, page 183.

10. Ibid., page 184.

11. Cited in White, page 152.

12. Intelligence Section, P.C. 144th Infantry, 25th October, 1918, "Report of Patrol," Harvey M. Radey, 2nd Lieut, 144th Infantry, Scout Officer.

13. Alexander White Spence, *The History of the Thirty-Sixth Division, U.S.A., 1917-1919,* unpublished manuscript, 1919. Official Division Records, National Archives.

14. Ibid., page 193.

15. White, page 155.

16. Ibid., page 158.

17. Barnes, *History of the 36th,* page 506, quoted in White.

18. Clyde S. Fosselman, 111th Engineers, letter of January 23, 1919, published in Navasota Examiner Review, Hornaday Collection, Box 1, Pages 269 and 270, Texas State Archives.

19. Headquarters, 71st Infantry Brigade, A. E. F. A. P. O. #796, 7 Nov 1918, From: G-3, 71st Infantry Brigade, To: G-3, 36th Division, Subject: Operations Report of Engagement 27 October 1918 in accordance with instructions of the Chief of Staff.

20. Spence, page 303. The section of the Spence authoritative history of 36th Division on Forest Farm was actually written by Capt. Frank A, Loftus, of the 141st Infantry.

21. Spence, page 304.
22. John J. Pershing, *My Experiences in the World War*, New York: Frederick A. Stokes Company, 1930, part II, pages 303, 317-341.
23. Chastaine, page 222.
24. Spence, page 310.
25. White, pages 161-162.
26. See Spence, pages 310-313, for an "on the scene" narrative of the discussions between Smith and Prax.
27. Conversation with National Archives staff expert, March 14, 1994.
28. White, page 164; and 1st Lieutenant John R. Eddy, Report on the American Indian Soldier, Historical Section, GHQ, AEF Records, RG 120, National Archives. See also White, "Indian Soldiers of the 36th Division," *Military History of Texas and the Southwest*, XV, pages 17-18; "Native Americans in World War I," short article in National Archives, Calendar of Events, August 1993; Spence 319-320; and "Headquarters 142nd Infantry, A. E. F., January 25, 1919, A.P.O. No. 796, From: C. O. 142nd Infantry, To: The Commanding General 36th Division (Attention Captain Spence), Subject: Transmitting Messages in Choctaw.
29. Chastaine, page 227.
30. 71st Brigade, Operations Report of Engagement of 27 October 1918.
31. Chastaine, page 229.
32. Walter B. Nichols, Company F, 141st Infantry, letter, published February 9, 1919, *Fort Worth Star-Telegram*, Hornaday Collection, Box 2, page 73, Texas State Archives.
33. 71st Infantry Brigade, Operations Report of Engagement 27 October 1918.
34. White, page 167.
35. Nichols letter, page 74.
36. Chastaine, page 238.
37. White, page 169.

CHAPTER 6: BETWEEN THE WARS

1. Bruce Sides, 111th Engineers Headquarters Detachment, letter of January 4, 1919, published in the *Brownsville Herald*, Hornaday Collection, Box 1, pages 145-146, Texas State Archives.
2. Ibid.
3. Lonnie J. White, "The Combat History of the 36th Division in World War I," *Military History of Texas and the Southwest*, vol. 17, no. 4, 1982, page 173.
4. Richard M. Burrage, "How the T-Patch Came to Be," *The Fighting 36th Historical Quarterly*, vol. X, no. 4, Winter 1990, pages 70-72.
5. Lonnie J. White, "The Homecoming of the 36th Division in World War I," *Military History of Texas and the Southwest*, vol. 18, no. 1, 1983, page 181.
6. George Marshall, quoted in Geoffrey Perret, *There's A War to be Won*, New York: Random House, page 481.
7. Robert H. Griffith, Jr., *Men Wanted for the U.S. Army*, Westport, Connecticut: Greenwood Press, 1982, page 21.

8. *Infantry Journal*, vol. 15, no. 7, January 1919, page 537.

9. Quoted in *The Literary Digest*, vol. 64, no. 7, February 14, 1920, pages 19-20.

10. Gene Smith, "The Seventeenth Largest Army," *American Heritage*, December 1992, pages 99-107, and other material cited elsewhere.

11. Ibid.

12. Charles W. Beacham, *Citizen Soldier*, unpublished manuscript, circa 1955, 36th Division Association materials, Texas State Archives, page 121.

13. Robert Bruce Sligh, *The National Guard and National Defense*, New York: Praeger, 1992, page 26.

14. Annual Report of the Adjutant General of Texas for the Period from January 1, 1921 to December 31, 1921, page 5.

15. Report of the Adjutant General for 1928 and 1929, page 150.

16. Annual Report of the Adjutant General for 1924, page 26.

17. Alan "Chum" Williamson, "Bring Your Bathing Suit and Fishing Tackle," *The Fighting 36th Historical Quarterly*, vol. V, no. 2, Summer 1985, pages 60-65, quote on page 62.

18. Report of the Adjutant General for 1923, page 7.

19. Williamson, page 61.

20. Harry Lynn Krenek, "A History of the Texas National Guard Between World War I and World War II," unpublished Ph.D. dissertation, Texas Tech University, 1979.

21. *Dallas Morning News*, September 16, 1919, page 1.

22. Ibid., September 18, 1919, page 1.

23. Harry Krenek, *The Power Vested*, Austin: Presidial Press, 1980, page 2.

24. "Guardsmen Rescue Galveston from Dock Strike Anarchy," *The National Guardsman and Citizen Soldier*, page 5.

25. James A. Clark, *The Tactful Texan: A Biography of Governor William Hobby*, New York: Random House, 1958, page 133.

26. Jacob F. Wolters, *Martial Law and its Administration*, Austin: Gammel's Book Store, Inc., 1930, page 78.

27. *Galveston Daily News*, June 20, 1920, page 23.

28. Krenek, *Power Vested*, page 19.

29. Ibid., pages 33-58.

30. *History and Pictorial Review, National Guard of the State of Texas*, Baton Rouge, Louisiana: Army and Navy Publishing Company, 1940, page xxxix.

31. Report of the Adjutant General for 1919 and 1920, page 51.

32. United States Army, Center for Military History, considers this revived 36th Division to have the lineage and honors of the 36th Division from World War I.

33. Ibid.

34. Jimmy M. Skaggs, "Lieutenant General John A. Hulen: Portrait of a Citizen-Soldier," *Texas Military History*, vol. 8, no. 3, 1970, page 135.

35. Annual Report of the Adjutant General of the State of Texas for the Year Ending August 31, 1924, pages 23-24.

36. Report of the Adjutant General of the State of Texas for the Year Ending August 31, 1928, page 26.

37. John K. Mahon, *History of the Militia and the National Guard*, New York: MacMillan Publishing Company, 1983, page 175.
38. Joseph G. Dawson, III, "Introduction," in Joseph G. Dawson III, editor, *The Texas Military Experience*, College Station: Texas A&M University Press, 1995, pages 4-13.
39. Report of the Adjutant General of the State of Texas from January 1, 1937 to December 31, 1938, pages 10-11.
40. Krenek, page 37.
41. Ibid., page 113. Also Arthur F. Raper, *The Tragedy of Lynching*, Chapel Hill, North Carolina: The University of North Carolina Press, 1933, pages 222-223.
42. *Sherman Democrat*, May 9, 1930.
43. Krenek, *The Power Vested*, page 117.
44. Raper, page 326.
45. Durward Pruden, "A Sociological Study of a Texas Lynching," MA Thesis, Southern Methodist University, 1935, pages 90-91.
46. Krenek, *Power Vested*, pages 124-126.
47. *Sherman Democrat*, May 16, 1930.
48. May 13, 1930 letter to F. A. Robinett from Dan Moody, Dan Moody papers, Texas State Archives.
49. *Fort Worth Press*, quoted Raper page 345.
50. Krenek, *Power Vested*, 171. "History of the Texas National Guard," unpublished manuscript prepared by the Office of the Historian and Records Manager, Texas National Guard, July 1992, Appendix #3, lists only one guard law enforcement deployment after 1931—September 1937 mob violence in Marshall, Texas.

CHAPTER 7: GETTING READY FOR WAR AGAIN

1. Bruce Kouser letters, undated but written winter or spring 1944, author's personal collection.
2. Quoted page 180, Richard M. Ketchum, *The Borrowed Years: 1938-1941*, New York: Random House, 1989.
3. Gene Smith, "The Seventeenth Largest Army," *American Heritage*, December 1992, page 99.
4. Secretary of War Harry H. Woodring to President Franklin R. Roosevelt, September 8, 1939.
5. Robert Dallek, *Franklin Roosevelt and American Foreign Policy, 1932-1945*, New York: 1979, page 203.
6. Charles M. Beacham, "Citizen Soldier: Prepare to Assault," unpublished manuscript, circa 1955, 36th Division Association Collection, Texas State Archives, page 44.
7. The Institute of Texan Cultures, Oral History Program, Interview with Brig. Gen. Armin Puck, August 23, 1983, page 6.
8. Ibid., page 7.
9. Kent Roberts Greenfield and Robert R. Palmer, "Organization of the Army Ground Forces: General Headquarters, United States Army, 1940-42," in *The United States Army in World War Two, The Army Ground Forces, The*

Organization of Ground Combat Troops, Kent Roberts Greenfield, Robert R. Palmer and Bell I. Wiley, Washington, D.C.: Historical Division, United States Army, 1947, page 36.

10. Leonard Wood, *Our Military History: Its Facts and Falacies,* Chicago: Reilly and Britton, 1915, page 115.

11. J. Garry Clifford, and Samuel R. Jr., Spencer, *The First Peacetime Draft,* Lawrence, University Press of Kansas, 1986, page 19.

12. Annual Report of the Secretary of War, 1939, page 4.

13. Ibid., 1938, page 40.

14. George Marshall interview, January 22, 1957, quoted in Clifford and Spencer, page 52.

15. Marshall to Congress Walter G. Andrews, May 26, 1940, Larry I. Bland, editor, *The Papers of George Catlett Marshall: We Cannot Delay,* July 1, 1939 to December 6, 1941, Baltimore: 1986, vol. II, page 217.

16. Ibid.

17. Samuel Rosenman, editor, *The Public Papers and Addresses of Franklin D. Roosevelt, New York: 1938-1950,* volume IX, page 252.

18. Kennett, pages 3-21.

19. Bruce Kouser letters, author's personal collection.

20. Hondon B. Hargrove, *Buffalo Soldiers in Italy: Black Americans in World War II,* Jefferson, North Carolina: McFarland and Co. Inc., Publishers, 1985, pages 104.

21. Kennett, pages 36-37.

22. Headquarters 36th Division, November 17, 1941, Release #122, 36th Division Association Collection, Texas State Archives.

23. Headquarters 36th Division, November 14, 1941, Release #108.

24. Ibid., November 24, 1941, Release #231.

25. Lee F. Allison to Wife, May 19, 1941, 36th Division Association Collection, Texas State Archives.

26. Walker, page 65.

27. Chaplain Herbert E. MacCombie, memoirs "Chaplains of the Thirty Sixth Division," Thirty-Sixth Division Association Collection, Texas State Archives, page 1.

28. "Address made to new men of the 36th Infantry Division, by the Division Commander, Major General Fred L. Walker, March 13, 1942."

29. Bell I. Wiley, "The Building and Training of Infantry Divisions," in Robert R. Palmer, Bell I. Wiley, and William R. Keast, *The Procurement and Training of Ground Forces, United States Army in World War II, The Army Ground Forces,* Washington, D.C.: Office of the Chief of Military History, 1948, pages 434-436.

30. Beacham, page 164.

31. John Sloan Brown, *Draftee Division, The 88th Infantry Division in World War II,* Lexington, The University Press of Kentucky, 1986, page 10.

32. Walker, page 66.

33. Robert Bruce Sligh, *The National Guard and National Defense: The Mobilization of the Guard in World War II,* New York: Praeger, 1992, page 147.

34. Morale is a major underlying theme in a recent book by Stephen E. Ambrose, *Citizen Soldiers*, New York: Simon & Schuster, 1997.

35. McNair to Marshall, October 14, 1941.

36. Hilton Howell Railey, "Morale of the United States Army: An Appraisal for the New York Times," unpublished manuscript, September 29, 1941, page 16.

37. Ibid., pages 18-19.

38. George Marshall to Claude Birkhead, July 30, 1941, Texas State Archives, 36th Division Association Collection.

39. Ibid.

40. Scott Summers, "36th May Be Unconventional But Sure Can Get Results," Express, dateline of August 25, 1941, Thirty-Sixth Division Collection, Texas State Archives.

41. Marshall letter.

42. Ibid.

43. Bruce Kouser letters, undated but written winter or spring 1944, author's personal collection.

44. General sources on combat replacements include Robert I. Palmer and William R. Keast, "The Provision of Enlisted Replacements," and William R. Keast, "The Training of Enlisted Replacements," in Palmer, Wiley and Keast, *The Procurement and Training of Ground Combat Troops*, pages 165 to 240, Brown, Kennett, Kouser letters and units histories consulted.

45. Francis C. Steckel, "Morale Problems in Combat: American Soldiers in Europe in World War II," *Army History*, Summer 1994, page 1.

46. Evan E. Voss, "Transition of a Green Recruit into a Combat Soldier," *The Fighting 36th Historical Quarterly*, vol. VII, no. 3, Fall 1987, pages 6-21, quote on page 8.

47. Ibid.

48. Ibid., page 2.

49. Michael D. Doubler, *Closing with the Enemy: How GIs Fought the War in Europe, 1944-1945*, Lawrence: University Press of Kansas, page 28.

50. Ambrose, *Citizen Soldiers*, page 286.

51. Ibid., page 287.

52. Associated Press, August 26, 1941, 36th Division Association Collection, Texas State Archives.

53. Quoted in Fred L. Walker, *From Texas to Rome*, Dallas: Taylor Publishing Company, 1969, page 1.

54. *Star Telegram*, January 1, 1942, *Dallas News*, January 16, 1942. Thirty-Sixth Division collection, Texas State Archives.

55. Charles M. Beacham, "Citizen Soldier: Prepare to Assault," unpublished manuscript, circa 195, Thirty-Sixth Division Association Collection, Texas State Archives, pages 193-194.

56. Walker, page 4.

57. Ibid.

58. *Dallas News*, October 3, 1941, Files of the Center for American History, University of Texas at Austin.

59. Lee J. Allison to wife, November 29, 1940, Thirty-Sixth Division Association Collection, Texas State Archives.

60. Lee Allison to wife, July 29, 1942, Thirty-Sixth Division Association Collection, Texas State Archives.

61. Palmer and Keast in Palmer, Wiley and Keast, cited above, pages 228-230.

62. Ibid., page 239.

63. Marvin A. Kreidberg and Merton C. Henry, *History of Military Mobilization in the United States Army, 1775-1945,* Washington, D.C.: U.S. Army Center of Military History, 1984, pages 647-648.

64. Eric Morris, *Circles of Hell,* New York: Crown Publishers, Inc. 1993, page 404.

CHAPTER 8: SALERNO

1. James T. Padgitt, "That Long Day at Salerno," *Texas Military History,* vol. 5, no. 2, Summer 1965, page 53.

2. Albert C. Wedemeyer, *Wedemeyer Reports,* New York: Hurt, 1958, page 230.

3. Winston S. Churchill, *The Second World War,* vol. 5, *Closing the Ring,* Boston: Houghton, Mifflin, 1951, page 131.

4. Samuel Eliot Morrison, *Strategy and Compromise,* Boston: Little Brown and Company, 1958, pages 45-47.

5. Quoted in Cornelius Ryan, *The Longest Day,* New York: Simon and Schuster, 1959, page 27.

6. Morrison, page 46.

7. Ulrich von Hassell, *The von Hassell Diaries,* Garden City, New York: Doubleday and Company, 1947, page 325.

8. Martin Blumenson, *Salerno to Cassino,* The Mediterranean Theater of Operations, United States Army in World War II, Washington, D.C.: Office of the Chief of Military History, United States Army, 1969, page 15.

9. Ralph S. Mavorgordato, "Hitler's Decision on the Defense of Italy," in Kent Roberts Greenfield, general editor, *Office of the Chief of Military History, Department of the Army, Command Decisions,* New York: Harcourt, Brace and World, 1959, pages 224-225; and Eric Morris, *Circles of Hell,* New York: Crown Publishers, 1993, pages 99-100.

10. Fred L. Walker, *From Texas to Rome,* Dallas: Taylor Publishing Company, 1969, page 271.

11. Fred L. Walker, Major General, USA, Retired, "Comments of the Commanding General, 36th Division Relating to Salerno, February 4, 1958, Thirty-Sixth Division Association Collection, Texas State Archives, page 1.

12. Morris, page 122.

13. Colonel Vincent Lockhart to author, undated memo (circa June 1994).

14. Ibid.

15. Ibid.

16. Blumenson, page 32.

17. Clark Diary, August 14, 1943, The Citadel Archives.

Endnotes 289

18. Geffrey Perret, *There's A War to Be Won*, New York: Random House, 1991, page 197.
19. A good description of the Italian surrender negotiations in found in Morris, pages 94-117.
20. Ralph Bennett, *Ultra and Mediterranean Strategy*, New York: William Morrow and Company, Inc., 1989, page 242.
21. Blumenson, page 65.
22. *Our Century: Masters of War*, "Battle for the Boot," Arts and Entertainment Cable Network, March 16, 1994.
23. Armand G. Jones, "Personal Account of the Italian Campaign and the Invasion of Southern France," page 2, 36th Division Association Collection.
24. Mark W. Clark, *Calculated Risk*, New York: Harper and Brothers, 1950, page 188. Also Operation "Avalanche," 21 September 1943, Annex #9 Division Artillery.
25. Walker, Comments of the Commanding General, pages 6-7.
26. Headquarters, One Hundred Forty Second Infantry, APO 36, U.S. Army, "Record of Events, 142d Inf., 36th Division, September 3d to September 20th, 1943," page 4.
27. LeRoy R. Houston, "Dead Men by Mass Production," *The Fighting 36th Historical Quarterly*, vol. X, no. 2, Summer 1990, pages 52-55, quote on page 52.
28. G-2 Periodic Reports, 36th Division, September 10, 1943, page 1.
29. Walker, Comments, page 8.
30. Ibid.
31. Record of Events, 142nd Infantry, page 5.
32. Ibid., page 4.
33. Blumenson, page 459.
34. Headquarters Fifth Army, General Orders Number 112.
35. Ibid., pages 4 and 5.
36. Robert L. Wagner, *The Texas Army*, Austin: State House Press, 1972 and 1991, page 14.
37. James T. Padgitt, "That Long Day at Salerno," *Texas Military History*, vol. 5, no. 2, Summer 1965, page 53.
38. Record of Events, 142nd Infantry, page 4.
39. Ralph Bennett, *Ultra and Mediterranean Strategy*, New York: William Morrow and Company, 1989, page 247.
40. Walker, *From Texas To Rome*, page 241.
41. Cited from German documents in Martin Blumenson, Salerno to Cassino, United States Army in World War II, The Mediterranean Theater of Operations, Washington, D.C.: Office of the Chief of Military History, United States Army, 1969, page 106.
42. "Personal Account by Armand G. Jones of the Italian Campaign," 36th Division Association Collection, Texas State Archives.
43. Blumenson, 106.
44. "Report of the Battle of Altavilla–143rd Infantry Regiment, 36th Infantry Division, Statement of Colonel William H. Martin, Regimental Commander, 143rd Infantry."

45. Headquarters, 36th Infantry Division, Operation "Avalanche," Annex #21.
46. Walker, "Comments of the Commanding General. . ." page 11.
47. Morris, page 167.
48. Clark, *Calculated Risk,* pages 198-202.
49. Bruce Jacobs, *Heroes of the Army: The Medal of Honor and Its Winners,* New York: W. W. Norton and Company, 1956, pages 107-108.
50. Ibid., 109-111.
51. Morris, page 172.
52. Blumenson, page 114.
53. Walker, September 12, 1943, page 244.
54. Dan S. Ray, "We Who Where There," part 2, *The Fighting 36th Historical Quarterly,* vol. X, no. 3, Fall 1990, page 46.
55. Headquarters, 36th Infantry Division, Operation "Avalanche," Annex #8, 143rd Infantry, page 13.
56. Walker, Comments, page 13.
57. Wagner, 218. Morris, 176-177.
58. Clark Diary, September 16, 1943, The Citadel Archives.
59. Walker, Comments, page 20.
60. Lucian K. Truscott, Jr., *Command Decisions,* New York: E. P. Dutton and Comapny, 194, page 253.
61. Bennett, page 249.
62. Wagner, page 53.
63. Blumenson, page 115.
64. Bernard L. Montgomery, *The Memoirs of Field Marshall Montgomery,* Cleveland, Ohio: World Publishing Company, 1954, page 179.
65. Lee Caraway Smith, *A River Swift and Deadly,* Austin, Texas: Eakin Press, 1989, page 91; Ronald Lewin, *Ultra Goes to War,* New York: McGraw Hill Book Company, 1978, pages 239, 282; and Bennett, pages 254-294, rate Clark as either someone who ignored Ultra reports or only grew to appreciate their value after Anzio.
66. Siegfried Westphal, *The German Army in the West,* London: Cassel and Company, 1954, page 154.
67. Harold Alexander, *The Alexander Memoirs: 1940-1945,* edited by John North, London: Cassel and Company, 1962, pages 115-116.
68. Admiral A. B. Cunningham, *A Sailor's Odyssey,* London: Hutchinson, 1951, page 570.
69. Headquarters, One Hundred Forty Second Infantry, APO 46, U.S. Army, "Record of Events, 142d INF., 36th DIVISION, September 3d to September 20th, 1943," page 13.

CHAPTER 9: SAN PIETRO
1. American Forces in Action Series, Fifth Army at the Winter Line, Historical Division, Washington, D.C.: Government Printing Office, 1945, page 7.
2. Diary of MG John Lucas, quoted page 46-47, Carlo D'Este, *Fatal Decision: Anzio and the Battle for Rome,* New York: HarperCollins, Publishers, 1991.

3. Siegfried Westphal, *The German Army in the West,* London, Cassell and Company, 1951, page 155.
4. American Forces in Action Series, Fifth Army at the Winter Line, Historical Division, Washington, D.C.: Government Printing Office, 1945, page 7.
5. Lee Fletcher, "San Pietro Memories," *The Fighting 36th Historical Quarterly,* vol. X, no 4., Winter 1990, pages 73, 75.
6. Al Dietrick, "A World War II Story," *The Fighting 36th Historical Quarterly,* vol. X, no. 2, Summer 1990, page 19.
7. Ibid.
8. Lieutenant-Colonel Robert D. Burhans, The First Special Service Force: A War History of the North Americans, 1942-1944, Washington: Infantry Journal Press, 1947, pages 1-19, 39-59; and Geoffrey Perret, *There's A War To be Won,* New York: Random House, 1991, pages 175-180.
9. Perret, page 178-179.
10. *The Devil's Brigade,* United Artists, 1968.
11. Colonel Vincent M. Lockhart, "Million Dollar Mountain at San Pietro," *The Fighting 36th Historical Quarterly,* vol. XIII, Fall 1993, page 69.
12. Martin Blumenson, Salerno to Cassino, United States Army in World War II, The Mediterranean Theater of Operations, Washington, D.C.: Office of the Chief of Military History, United States Army, 1969, page 259.
13. Interview, Dr. Sidney Mathews with Harold Alexander, cited Blumenson, page 265.
14. Fred L. Walker, *From Texas to Rome,* Dallas, Texas: Taylor Publishing Company, 1969, entry for November 18, 1943, page 278.
15. Ibid., pages 278-279.
16. 36th Div Annex 2 to FO 38, 30 November 1943.
17. 36th Division FO 38, Operation Raincoat, 30 November 1943.
18. Lockhart, page 69.
19. Burhans, pages 87-126.
20. Lockhart, page 70.
21. Walker, entry for December 1, 1943, page 282.
22. Lockhart, 78-79.
23. Robert L. Wagner, *The Texas Army,* Austin, Texas: State House Press, 1972 and 1991, page 74.
24. Wagner, 74, continues his practice of giving the home communities of mentioned members of the division.
25. Report of Operations in Italy–143rd Inf. Regiment, 36th Inf. Div., December 1943, page 3.
26. Blumenson, 252-253, 275; Eric Morris, *Circles of Hell,* New York: Crown Publishers, 1993, page 217.
27. Letter, Walker to Blumenson, July 1960, cited in Blumenson, page 275; Walker, entry of December 4, 1943, page 282 and entry of December 5, 1943, page 283.
28. Walker, entry for December 8, 1943, page 284.
29. Ibid., entry for December 13, 1943, page 285.
30. Ibid.

31. Lockhart, page 71.
32. Fletcher, page 78.

CHAPTER 10: THE RAPIDO RIVER

1. Captain Zerk O. Robertson, 143rd Infantry, L Company, quoted in Dallas, *Texas Daily Times Herald,* January 20, 1946.
2. Robert C. Patterson to A. J. May, Chairman, Committee on Military Affairs, House of Representatives, reprinted in Investigation of the Rapido River Crossing, page iv.
3. Martin Blumenson, *Bloody River: The Real Tragedy of the Rapido,* Boston: Houghton Mifflin Company, 1970, page 3.
4. Geoffrey Perret, *There's a War to be Won,* New York: Random House, 1991, page 213.
5. Quoted in Winston S. Churchill, *Closing the Ring,* vol. 5 of *The Second World War,* Boston: Houghton Mifflin Company, 1951, page 429.
6. Ralph Bennett, *Ultra and Mediterranean Strategy,* New York: William Morrow and Company, 1989, page 257.
7. Clark Diary, entries for December 12 and 18, 1943, Citadel Archives.
8. Clark Diary, December 25, 1943, Citadel Archives.
9. Ibid.
10. Lucas Manuscript Diary, page 303, quoted Perret page 213-214.
11. Clark Diary, January 2, 1944, Citadel Archives.
12. Clark to Alexander, January 2, 1944, quoted in 5th Army History, Part IV, Cassino to Anzio page 17.
13. Ibid., page 57.
14. Bennett, page 261-262.
15. Oran C. Stovall, "The Odyssey of a Texas Citizen Soldier," edited by Robert L. Wagner, *The Southwestern Historical Quarterly,* vol. LXIII, no. 1, July 1968, pages 60-87. Quotation on page 69.
16. The Rapido River Crossing, Hearings before the Committee on Military Affairs, House of Representatives, Seventy-Ninth Congress, Second Session, February 20 and March 18, 1946, page 25, selection from testimony of General Walker.
17. Clark Diary, January 23, 1944, Citadel Archives, and Martin Blumenson, Salerno to Cassino, United States Army in World War II, The Mediterranean Theater of Operations, Washington, D.C.: Office of the Chief of Military History, United States Army, 1969, page 321.
18. Interview by Dr. Philip A. Crowl (Department of the Army) with Lt. Gen. Geoffrey Keyes, USA (Ret.), Washington, D.C., 22 September 1955, page 1. Texas State Archives collections.
19. Letter James M. Wilson, Jr. (Truscott's aide for two years) to Carlo D'Este, March 2, 1989, quoted in Carlo D'Este, *Fatal Decision: Anzio and the Battle for Rome,* New York: Harper Collins, Publishers, 1991, page 80.
20. "An Interview with Genfldm Albert Kesselring: Rapido River Crossing," Ethnit 71, 6 May 1946, Historical Section, ETOUSA, page 3.

21. Interview by Philip A. Crowl with General F. von Senger u. Etterlin, 22 September, 1955, copy in Texas State Archives.
22. Crowl Interview with Keyes, page 1.
23. Clark Diary, January 7, 1944.
24. Fred L. Walker, *From Texas to Rome*, Dallas: Taylor Publishing Company, 1969 (Walker diary), entry for January 16, 1944, page 300.
25. Memorandum for General North, Subject: Report of Interview with Major General Fred L. Walker, 4 February 1946, William A. Walker, Colonel GSC, Deputy Chief, Current Group, OPD. Copy found in Citadel Archives, Mark Clark Papers.
26. Walker diary, entry for January 6, 1944, page 294.
27. Ibid., entry for January 8, 1944, page 295.
28. Vincent M. Lockhart to author, 9 November 1993.
29. Blumenson, page 327.
30. Crowl interview with Keyes, page 2.
31. "General Walker's Story of the Rapido Crossing," submitted to publication by Joe F. Presnall, written circa 1962 by Fred L. Walker, *The Fighting 36th Historical Quarterly*, vol. X, no. 2, Summer 1990, page 11. Text appears to be the same as Fred L. Walker, "My Story of the Rapido Crossing," *Army*, XIII, No. 2, September 1962, pages 52-60.
32. George E. Lynch to Robert L. Wagner, August 25, 1968, 36th Division Association Collection, Texas State Archives.
33. Author's own theory, supported by D'Este, page 79, Morris, 253, and Martin Blumenson, *Bloody River*, page 21.
34. Rick Atkinson, *Crusade*, Boston and New York: Houston Mifflin Company, 1993, page 71.
35. Walker Diary, entry for January 19, 1944, page 302.
36. Ibid., entry for January 20, 1944, pages 305-306.
37. Statement of Tech 5 Clayton H. Nelson, 19th Engineer Combat Group, Narrative of Operations, January 1944.
38. Ibid.
39. See Wagner, pages 224-225 for discussion of this point.
40. Clark diary for January 20, 1944, The Citadel Archives.
41. Blumenson, From Salerno to Cassino, page 321.
42. Bennett, page 263.
43. *Target: Pearl Harbor*, The Discovery Channel, Sunday, December 6, 1992.
44. Lee Caraway Smith, *A River Swift and Deadly*, Austin: Eakin Press, 1989, pages 91, 101-102; Bennett, pages 354-264; and Lewin, 239, 282, rate Clark as either someone who ignored Ultra intelligence or only grew to appreciate its value after Anzio.
45. Kesselring, Rapido River, page 3.
46. 141st Regiment, Narrative of Operations, 36th Division Report, Annex #6.
47. "Narrative of Operations of 141st Infantry in the Crossing of the RAPIDO River on January 20 to 23, 1944," 36th Division records, National Archives, page 3.

48. 141st Regiment Narrative of Operations, page 2.
49. Ibid., page 3.
50. HQ 2nd Battalion, 141st Infantry, APO #36, Memorandum: To General Walker, January 30, 1944.
51. 141st RCT, Narrative of Operations, page 3.
52. Quoted in Smith, page 36.
53. Ibid.
54. Kesselring, Rapido, Page 5.
55. 141st Narrative of Operations, page 3.
56. Robert L. Wagner, *The Texas Army*, Austin, Texas: State House Press, 1991, page 107.
57. Walker diary, entry of January 21, 1944, pages 307-308.
58. Ibid., entry of January 21, 1944, page 308.
59. Bill Hartung, "My Baptism of Combat - The Rapido River," *Fighting 36th Historical Quarterly*, vol. XIII, no. 3, Fall 1993, pages 40-42.
60. 141st Narrative of Operations, Rapido, page 4.
61. Ibid., pages 4 and 5.
62. *Army Times*, April 28, 1945.
63. Walker diary, entry of January 22, 1944, page 313.
64. Memorandum for General North, page 2.
65. Clark Diary, entry for January 23, 1944, Citadel Archives.
66. Walker diary, entry for January 23, 1944, page 316.
67. See Mark W. Clark, *Calculated Risk*, New York: Harper and Row, 1950, pages 281-282.
68. Walker diary, entry for January 29, 1944, page 322.
69. Ibid., entry for February 6, 1944, page 325.
70. Ibid., entry for March 2, 1944, page 336.
71. Col. Reid Beveridge to author, 14 May 1992.
72. Wagner, pages 227-228.
73. Telegram Clark to Eisenhower, circa January 1946, The Citadel Archives.
74. This writer's research, including into Clark's papers at The Citadel, in Charleston, South Carolina, has found no indication that Clark blamed the division itself.
75. Clark, pages 281-282.
76. Ibid.
77. Alfred H. Gruenther, Major General, U.S. Army, Deputy Commandant, to Colonel R. J. Wood, Operations Division, War Department General Staff, 11 January 1946, copy in Clark materials, The Citadel Archives.
78. Robert C. Paterson to A. J. May, Chairman, Committee on Military Affairs, House of Representatives, reprinted in Investigation of the Rapido River Crossing, page iv.
79. Anson Jones Papers, untitled draft narrative, September 9, 1943-May 8, 1945, 36th Division Collection, Texas State Archives.
80. Clark Diary, January 23, 1944, The Citadel Archives; *Calculated Risk*, pages 279-282.

81. Armin F. Puck interview, James B. Sweeney, interviewer, August 23, 1983, page 21, transcript in The University of Texas, Institute of Texan Cultures.
82. Walker, Story of the Crossing, page 7.
83. Transcription of Cable, Clark to Alfred Gruenther, circa January 1946, The Citadel Archives.
84. Wagner, pages 224-225.

CHAPTER 11: VELLETRI
1. Eric Morris, *Circles of Hell,* New York: Crown Publishers, Inc., 1993, page 337.
2. Eric Severeid, "Velletri," *The American Legion Magazine,* October 1944, page 51.
3. Fred L. Walker, *From Texas to Rome,* Dallas: Taylor Publishing Company, 1969, entry for January 23, 1944, page 317.
4. Lt. Col. Hal Reese, "Intermission at Cassino," 19 May 1944, 36th Division Association Collection, Texas State Archives.
5. Ibid.
6. James M. Estepp, "I Left My Friend on Mt. Artemisio," *The Fighting 36th Historical Quarterly,* vol. I, no. 4, Winter 1990, page 65. March 1944 eruption is also mentioned in Richard A. Huff, editor, *A Pictorial History of the 36th "Texas" Infantry Division,* Austin: The 36th Division Association, 1946.
7. Ralph Bennett, *Ultra and Mediterranean Strategy,* New York: William Morrow and Company, 1989, page 264.
8. Ernest F. Fisher, Jr., *Cassino to the Alps, United States Army in World War II, The Mediterranean Theater of Operations,* Washington, D.C.: Center of Military History, 1977 and 1989, page 23.
9. Ibid., pages 84-100.
10. Carlo D'Este, *Fatal Decisions: Anzio and the Battle for Rome,* New York: HarperCollins Publishers, 1989, page 348.
11. Interview with Mark Clark, quoted Morris, page 338 and Fisher, page 542.
12. This issue is discussed in Geoffrey Perret, *There's A War to Be Won,* New York: Random House, 1989, pages 225-226 and passim.
13. Mark W. Clark, *Calculated Risk,* New York: Harper and Row, 1950, page 352.
14. Ibid.
15. Sidney T. Mathews, "General Clark's Decision to Drive on Rome," in Kent Roberts Greenfield, general editor for the *Office of the Chief of Military History, Department of the Army, Command Decisions,* New York: Harcourt, Brace and Company, 1959, pages 273-284.
16. Nigel Nicolson, Alex: *The Life of Field Marshal Earl Alexander of Tunis,* New York: Antheneum, 1973, pages 252-253.
17. Heinz Greiner, *Kamp un Rome - Inferno Am Po,* Neckargemund, Germany: Kurt Volwinckel, 1968. Partial 1969 translation by Mrs. Marilyn Von Kohl, files of the 36th Division Association Collection, Texas State Archives, Austin, Texas.

18. Walker diary, entry for May 27, 1944, page 372.
19. Ibid., entry for May 28, 1944, page 372.
20. Ibid., entry for May 30, 1944, page 375.
21. Lucian K. Truscott, Jr., *Command Missions*, New York: E. P. Dutton and Company, Inc., 1954, page 377.
22. Robert L. Wagner, *The Texas Army*, Austin: The State House Press, 1991 and 1972, page 161.
23. *Fifth Army G-2 Journal*, December 1943.
24. Ibid., page 24, and Wagner, page 164.
25. Kenneth L. Dixon, "The Night of the Knife," *Argosy*, September 1964.
26. Letter to Authors, quoted in Robert H. Adelman and Colonel George Walton, *Rome Fell Today*, Boston: Little, Brown and Company, 1968, page 25.
27. "VELLETRI, as remembered by Sgt. Alden W. Williams, 111th. Engineers, May 30–June 1, 1944. Undated but circa mid-1960s, 36th Division Association Collection, Texas State Archives.
28. Operations in Italy, Regimental Commander's Comments, May 1944, Headquarters, One Hundred Forty-Second Infantry, U.S. Army, July 1, 1944, page 5.
29. Fisher, page 189.
30. 143rd Infantry, Operations Report, June 1944.
31. Truscott, page 377.
32. Questionnaire quoted in Adleman and Walton, page 27.
33. Estepp, page 67.
34. John Bob Parks, "The Road that Could Be Built," undated, circa 1968, 36th Division Association Collection, Texas State Archives.
35. Report of Operations in Italy, 141st Inf Regt 36th Inf Div. History of Operations of Hdqs. Co. 2nd BN, 141st Infantry June 1944, page 71.
36. Walker diary, entry for June 1, 1944, page 377.
37. *A Pictorial History of the 36th "Texas" Infantry Division, 1946.* Pages not numbered. (Excellent source of illustrations.)
38. Walker diary, entry for June 1, 1944, page 378.
39. Clark Diary, 31 May 1944, Citadel Archives.
40. Parks, "The Road That Could Not Be Built."
41. Clark Diary, 3 June 1944, Citadel Archives.
42. *88th Division G-3 Journal*, 4 June 1944.
43. Walker diary, entry for June 12, 1944, pages 392-396.
44. Ibid., entry for March 5, 1944, page 336.
45. Lieutenant Colonel Reuben D. Parker, "Infiltration as a Form of Maneuver," *Military Review*, December 1959, page 44.
46. Fisher, page 190.

CHAPTER 12: OPERATION ANVIL/DRAGOON

1. Clifford H. Peek, Jr., editor, *Five Years–Five Countries–Five Campaigns*, Munich: Germany, The 141st Infantry Regiment Association, c 1945, page 58.
2. Jeffrey J. Clark and Robert Ross Smith, *Riviera to the Rhine*, United

States Army in World War Two, The European Theater of Operations, Washington, D.C.: Center of Military History, United States Army, 1993, page 3.

3. Morris, pages 348-350.
4. Clarke and Smith, page 71.
5. Ibid., page 95.
6. "Operations in France for the Month of August 1944," Headquarters, One Hundred Forty-Second Infantry, APT #36, U.S. Army," After Action Report found in microfilm file entitled Historical Record of Operations in France for the Month of August 1944, dated 21 September 1944. Narrative written by Staff Sergeant Richard A. Huff, 142nd Infantry.
7. Clark and Smith, page 87.
8. Brigadier General Frederick B. Butler, "Task Force Butler," part I, pages 12-18; *Armored Cavalry Journal*, vol. LVII, January-February 1948, no. 1, page 15.
9. Ibid., page 19.
10. Russell F. Weigley, *Eisenhower's Lieutenants*, Bloomington: Indiana University Press, 1981, page 227.
11. Ibid., page 5.
12. Peek, page 58.
13. Ibid.
14. Ibid.
15. Ibid.
16. 2nd Battalion History (Daily Journal) 10 August - 16 September 1944, 2nd Battalion, 141st Infantry Regiment, 36th Division.
17. Lockhart, Vincent M, *T-Patch to Victory*, Canyon, Texas: Staked Plains Press, 1981, pages 1-17.
18. Peek, page 60.
19. 142nd RCT, "Operations in France for the Month of August 1944," cited above, page 19.
20. August After Action Report, 142nd Infantry. The diversion is described, in virtually identical terminology, in a memo, "Battle Experience, Coordination of Various Arms," 05 September 1945, To: Commanding General, Seventh Army, Attn. G-3, from R. T. Stack, Brigadier General, U.S. Army, Commanding.
21. Stack to Commanding General, 7th Army, September 5, 1945, page 2.
22. Lt. General L. K. Truscott, Jr. *Command Missions: A Personal Story*, New York: E. P. Dutton and Company, Inc., 1954, page 413.
23. George E. Lynch, Jr. (colonel commanding 142nd at time of invasion), as quoted in Lockhart, page 59.
24. Truscott, page 414.
25. Ibid., page 414.
26. Clarke and Smith, page 123.
27. Quoted Lockhart, pages 60 and 61.
28. Lockhart, pages 61 and 63.
29. 142nd RCT, "Operations," cited above, page 19.
30. Weigley, Russell F. Eisenhower's Lieutenants, Bloomington, Indiana: University of Indiana Press, 1981, pages 91-93; *Hastings, Max, Overlord*, New York: Simon and Schuster, 1984, pages 86-88.

31. Headquarters 36th Infantry Division, "Narrative, August 1944," page 4.
32. 142nd RCT, "Operations," above cited, page 19.
33. After Action Report.
34. Clarke and Smith, pages 134-135.
35. Ralph Bennett, *Ultra in the West*, London: Hutchinson, 1979, Page 159.
36. After Action Report, Company I, One Hundred Forty-Second Infantry, "Operations in France for the Month of August 1944."
37. After Action Report, Company A, One Hundred Forty-Second Infantry, "Operations in France for the Month of August 1944."

CHAPTER 13: THE BATTLE OF MONTELIMAR
1. Interview with Theodore H. Andrews, then lieutenant colonel commanding 3rd Battalion, 143rd Regiment, quoted in Lockhart, pages 45-46.
2. Brigadier General Frederick B. Butler, "Task Force Butler," pages 12-18, *Armored Cavalry Journal*, vol. LVII, January-February 1948, no. 1, page 13.
3. Jeffrey J. Clarke and Robert Ross Smith, *Riviera to the Rhine, United States Army in World War II, The European Theater of Operations*, Washington, D.C.: Center of Military History, United States Army, 1993, page 144.
4. Clarke and Smith, page 147.
5. Vincent M. Lockhart, *T-Patch to Victory*, Canyon, Texas: Staked Plains Press, 1981, pages 31-32.
6. 141st Infantry Regiment, Operations in France, August 1944, page 12.
7. Butler, "Task Force Butler," part II, *Armored Cavalry Journal*, March-April 1948, pages 30-38.
8. Ibid.
9. Quoted in Lockhart, page 63.
10. Clarke and Smith, page 151.
11. Lt. General L. K. Truscott, Jr., *Command Missions: A Personal Story*, New York: E. P. Dutton and Company, Inc., 1954, page 427.
12. Interview is quoted and described in Lockhart, pages 65-67.
13. Ibid., page 66.
14. Truscott, page 427.
15. Narrative, August 1944, Headquarters 36th Infantry Division, APO #36, U.S. Army, page 6.
16. 141st Infantry, Operations in France, August 1944, pages 12-13.
17. Clifford H. Peek, editor, *Five Years - Five Countries - Five Campaigns*, Munich, Germany, 141st Infantry Regiment Association, C 1945, page 64.
18. Interview with James H. Critchfield (then lieutenant colonel commanding 2nd Battalion), quoted and described in Lockhart, pages 33-34.
19. Quoted in Lockhart, page 34.
20. Clarke and Smith, page 157; Lockhart, page 40.
21. Interview with Theodore H. Andrews, then lieutenant colonel commanding 3rd Battalion, 143rd Regiment, quoted in Lockhart, pages 45-46.
22. Quoted in Lockhart pages 46-47.
23. Truscott, pages 430-431.
24. Ibid., page 430.

25. Martin Blumenson, *Liberation,* Alexandria, Virginia: Time-Life Books, 1978, pages 114-115.
26. Peek, page 65.
27. Quoted Clarke and Smith, page 169.
28. Ibid.
29. Peek, page 68.

CHAPTER 14: ROUTINE COMBAT
1. 36th Infantry Division AAR, October 1944, page 2.
2. Keith E. Bonn, *When the Odds Were Even: The Vosges Mountains Campaign,* October 1944–January 1945, Novato, California: Presidio, 1994, page 67.
3. The Seventh United States Army Report of Operations, France and Germany, 1944-1945, Heidelberg, Germany: Aloys Graf, 1946, volume I, pages 265-266.
4. Clifford H. Peek, Jr., editor, *Five Years - Five Countries - Five Campaigns,* Munich, Germany: 141st Regiment Association, C 1945, pages 66-67.
5. Vincent M. Lockhart, *T-Patch to Victory,* Canyon, Texas: Staked Plains Press, 1981, page 103.
6. 36th Infantry Division AAR, October 1944, page 2.
7. Lockhart, page 115.
8. Francis C. Steckel, "Morale Problems in Combat: American Soldiers in World War II," *Army History,* Summer 1994, pages 4-5.
9. Stephen E. Ambrose, *Citizen Soldiers,* New York: Simon and Schuster, 1997, page 287.
10. Jeffrey J. Clarke and Robert Ross Smith, *Riviera to the Rhine, United States Army in World War II, The European Theater of Operations,* Washington, D.C.: Center of Military History, United States Army, 1993, page 291.
11. Gen. Paul D. Adams oral history, quoted in Lockhart, pages 115-117.
12. Headquarters, One Hundred Forty-Second Infantry, "Operations in France for the month of September, 1944."
14. Ben F. Wilson, Jr., "As I Recollect," *The Fighting 36th Historical Quarterly,* vol. X, no. 4, Winter 1990, pages 59-61.
15. Pvt. Bruce Kouser (142nd Regiment) to Kouser family, October 16, 1944, author's personal materials; Lockhart, page 125, and personal family evidence. Some of Bruce Kouser's letters home were written on "American Red Cross" stationary.
16. Col. Benjamin F.Wilson, Jr., quoted in Lockhart, pages 125-126.
17. Bruce Kouser to Kouser family (this author's mother's family), October 16, 1944, author's personal materials.
18. Ibid., October 17, 1944, author's personal materials.
19. Headquarters, One Hundred Forty-Second Infantry, APO #36, U.S. Army, "Operations in France For the Month of October 1944."
20. Operations in France for the Month of October 1944, Headquarters, One Hundred Forty-Second Infantry, APO #36, U.S. Army, page 10.
21. Lee R. G. Ward, Lt. Col, AGD, Acting Chief of Branch (War

Department, The Adjutant General's Office, Demobilized Personnel Records Branch) to Mrs. Blanche Kouser, 17 March 1945, author's personal materials.

22. Raymond J. Schloder, Captain, 142d Infantry, Commanding (Company A), to Mrs. Blanche Kouser, 30 April 1945, author's personal materials.

CHAPTER 15: THE LOST BATTALION

1. "'Lost Battalion' Story," Richard A. Huff, editor, *A Pictorial History of the 36th Division*, Austin, Texas" 36th Division Association, C 1946, unpaginated.

2. Jeffrey J. Clarke and Robert Ross Smith, *Riviera to the Rhine, United States Army in World War Two, The European Theater of Operations*, Washington, D.C.: Center of Military History, 1993, page 322.

3. Richard Overy, *Why The Allies Won*, New York and London: W. W. North and Company, 1975, page 207. The entire chapter, pages 180-207, gives a good overview of the economic situation in Germany, the United States, and the Soviet Union.

4. Stephen Ambrose, *Citizen Soldiers*, New York: Simon and Schuster, 1997, page 113.

5. Keith E. Bonn, *When the Odds Were Even*, Novato, California: Presidio Press, 1994, page 87.

6. Ibid., page 89.

7. Vincent M. Lockhart, *T-Patch to Victory*, Canyon: Texas, Staked Plains Press, 1980, page 131.

8. Lockhart, page 147.

9. Clarke and Smith, page 330.

10. Lockhart, page 147.

11. *New York Times*, November 6, 1944, dated October 31 (delayed).

12. 36th Division G3 (operations) Journal, October 29, 1944.

13. 141st Regimental S-3 Operations report, entry for 1500 hours, October 27-28, 1944.

14. *New York Times*, November 6, 1944.

15. "'Lost Battalion' Story," Richard A. Huff, editor, *A Pictorial History of the 36th Division*.

16. Ibid.

17. Ibid.

18. Ibid.

19. *The Story of the 442nd Regimental Combat Team*.

20. Ibid.

21. Clarke and Smith, page 333.

22. "Engineers Build a Road," *A Pictorial History of the 36th Division*, unpaginated.

23. Ibid.

CHAPTER 16: WINTER WARFARE

1. Clifford Peek, editor, *Five Years–Five Countries–Five Campaigns*, Munich, Germany: 141st Infantry Regiment Association, 1945, page 89.

2. Jeffrey J. Clarke and Robert Ross Smith, *Riviera to the Rhine, United States Army in World War Two, The European Theater of Operations*, Washington, D.C.: Center of Military History, 1993, page 343.

3. Ernest F. Fisher, Jr., *Cassino to the Alps, United States Army in World War II, The Mediterranean Theater of Operations*, Washington, D.C.: Center of Military History, United States Army, 1989 and 1977, page 415.

4. Seventh Army Report of Operations, vol. II, Heidelberg: Aloys Graef, 1946, page 328.

5. Keith E. Bonn, *When the Odds Were Even*, Novato, California: Presidio Press, 1994, page 127.

6. Seventh Army Report of Operations, pages 431-432.

7. Ibid., pages 403-404.

8. Quoted in Vincent M. Lockhart, *T-Patch to Victory*, Canyon, Texas: Staked Plains Press, 1981, pages 174-177.

9. Ibid., page 177.

10. Incident described in Russell F. Weigley, *Eisenhower's Lieutenants*, Bloomington: Indiana University Press, 1981, page 407.

11. Clarke and Smith, page 439.

12. Ibid., 439-440.

13. See Eric Morris, *Circles of Hell*, New York: Crown Publishers, 1993, pages 205-211.

14. Forest C. Pogue, *The Supreme Command, United States Army in World War Two, The European Theater of Operations*, Washington, D.C.: Office of the Chief of Military History, Department of the Army, 1954, pages 397-404.

15. Martin Blumenson, editor, *The Patton Papers*, Boston: Houghton Mifflin, 1972-1974, page 583.

16. See Clarke and Smith, pages 440-445, for discussion of the decision.

17. Weigley, page 410.

18. Clarke and Smith, page 486.

19. Headquarters, One Hundred Forty-Second Infantry, APO #36, U.S. Army, "Operations in France for the Month of December 1944."

20. Jacob Devers Diary, December 19, 1944, quoted Clarke and Smith, page 491.

21. Bud Hutton and Andy Rooney, *The Story of Stars and Stripes*, New York: Farrar and Rinehart, 1946, pages 72-75. Note, however, that undated memo from Col. Vincent Lockhart, circa May 1994, retired staff officer from the 36th Division, states "I just flat don't believe the Stars and Stripes guys who claimed to be the ONLY Americans in Strasburg. I spent Christmas in a suburb of Strasburg and I was in the division rear echelon at the time!"

22. Clarke and Smith, page 505.

23. Clifford Peek, editor, *Five Years - Five Countries - Five Campaigns*, Munich, Germany: 141st Infantry Regiment Association, 1945, page 89.

24. *143rd Regiment Unit Journal*, "Operations in France," for January 1945, page 11.

25. Ibid.

26. Ibid.

27. "The Relief of Belsen, April 1945, Eyewitness Accounts," London: Imperial War Museum, 1991, page 9.

CHAPTER 17: THE END OF THE WAR

1. Clifford Peek, editor, *Five Years — Five Countries — Five Campaigns*, Munich: 141st Regiment Association, 1945, page 113.
2. Interview with Brig. Gen. Armin Puck, interviewer Col. James B. Sweeney, August 23, 1983, The University of Texas, Institute of Texan Cultures, page 46.
3. Figures cited in Russell F. Weigley, *Eisenhower's Lieutenants*, Bloomington: Indiana University Press, 1981, page 659.
4. Ibid.
5. Pages 12 and 13.
6. Ibid., page 13.
7. Charles B. MacDonald, *The Last Offensive, United States Army in World War II, The European Theater of Operations*, Washington, D.C.: Office of the Chief of Military History, 1973, page 257.
8. Weigley, pages 640-646.
9. H. Essame, *Patton: A Study in Command*, New York: Charles Scribner's Sons, 1974, page 244.
10. Headquarters, 36th Infantry Division, APO #30, U.S. Army, "Narrative March 1945," 30 March 1945.
11. R. K. Doughty, "Beyond the Dragon's Teeth," *The Fighting 36th Historical Quarterly*, vol. X, no. 2, Summer 1990, pages 42-43.
12. Vincent M. Lockhart, *T-Patch to Victory*, Canyon, Texas: Staked Plains Press, 1981, pages 266-270.
13. Ibid., pages 269-270.
14. Clifford Peek, editor, *Five Years - Five Countries - Five Campaigns*, Munich: 141st Regiment Association, 1945, page 115.
15. Ibid., page 113.
16. Lockhart, pages 275-276.
17. Conversation with research library staff at United States Holocaust Memorial Museum, March 18, 1994. Basic criteria for certification as a liberating division is that the division had to be the first into a camp, or arrive within forty-eight hours. The 36th Division being a supporting unit for the final advance into Bavaria does not detract from the horror of what they saw, or its effect on the men.
18. *1945: The Year of Liberation*, Washington, D.C.: United States Holocaust Memorial Museum, 1995, pages 286 and 287.
19. Paul Berben, *Dachau, 1933-1945: The Official History*, London: Comite Internationale de Dachau, 1968 and 1975, page 183.
20. Felix Sparks, "The Capture of Dachau," in *1945: The Year of Liberation*, page 128.
21. Doughty, pages 45-46.
22. Ibid., page 46.
23. Puck interview, page 54.

24. Headquarters, 36th Infantry Division, Operations in Germany and Austria, 1-10 May 1945, Division Narrative and Statistics, page 10.
25. "Casualties Exceeded 100 Pct. in 14 Divisions," *Philadelphia Inquirer*, April 14, 1946.
26. Cited in Lockhart, 307.
27. Ibid.
28. Ernest F. Fisher, Jr., *Cassino to the Alps, United States Army in World War II, The Mediterranean Theater of Operations*, Washington, D.C.: Center of Military History, United States Army, 1989 and 1977, pages 512-534.
29. Ibid., page 545.
30. Ibid., 532; and Report of Operations, the Seventh United States Army in France and Germany, 1944-1945, vol. III, Stuttgart, Germany: May 1946, page 847.

POSTLUDE AND MEANING
1. 1st Lt. Kevin McAndrews, "Keeping Your Guard Up," *National Guard*, March 1996, pages 24-26, quote on page 25.
2. Col. Harold B. Simpson, *Audie Murphy: American Soldier*, Hillsboro, Texas: Hill Junior College Press, 1975, page 339.
3. Col. John D. Stuckey and Col. Joseph H. Pistorius, "Mobilizing the Army National Guard and Army Reserve" Historical Perspective and the Vietnam War," Carlisle Barracks, Pennsylvania: U.S. Army War College, Strategic Studies Institute, 7 September 1984, page viii.
4. McAndrews, page 25.
5. Ibid., page 26.

Bibliography

Acheson, Dean, *Present at the Creation*, New York: Signit, 1970.
Adelman, Robert, and George Walton, *Rome Fell Today*, Boston: Little, Brown and Company, 1968.
Alexander, Bevin, *How Great Generals Win*, New York and London: W. W. Norton & Company, 1993.
Alexander, Harold, *The Alexander Memoirs: 1940–1945*, edited by John North, London: Cassel and Company, 1962.
Allen, William Lusk, *Anzio: Edge of Disaster*, New York: Elsevier-Dutton, 1978.
Ambrose, Stephen E., *Americans at War*, Jackson: University of Mississippi Press, 1997. Compilation of Ambrose's articles on various aspects of military history.
———, *Citizen Soldiers*, New York: Simon and Schuster, 1997. This excellent work focuses on the individual soldiers in the European campaign.
———, *Upton and the Army*, Baton Rouge: Louisiana State University Press, 1964.
American Battle Monuments Commission, *36th Division: Summary of Operations in the World War*, Washington, D.C.: United States Government Printing Office, 1944.
American Forces in Action Series, *Fifth Army at the Winter Line*, Historical Division, Washington, D.C.: Government Printing Office, 1945.
Annual Report of the Adjutant General of Texas, various years from 1865 to 1946.
Army Navy Journal, editorial, February 20, 1897, page 438.
Ash, Bernard, *The Lost Dictator: A Biography of Field Marshall Sir Henry Wilson*, London: Cassell, 1968.
Atkinson, Rick, *Crusade: The Untold Story of the Persian Gulf War*, Boston and New York: Houghton Mifflin, 1993.
Bacon, Robert, and James Brown Scott, collectors and editors, *The Military and Colonial Policy of the United States, Addresses and Reports by Elihu Root*, Cambridge: Harvard University Press, 1916.
Baldwin, Hanson W., *The Crucial Years: 1939-1941*, New York: Harper, 1976.
Baenziger, Ann Patton, "The Texas State Police During Reconstruction: A Reexamination," *Southwestern Historical Quarterly*, Volume LXXII, Number 4, April 1969.
Beacham, Charles W., *Citizen Soldier*, unpublished manuscript, circa 1955, 36th Division Association materials, Texas State Library and Archives.

Bennett, Ralph, *Ultra in the West*, New York: Charles Scribner's Sons, 1980.
Beunger, Walter L., *Secession and the Union in Texas*, Austin: University of Texas Press, 1984.
Binding, Rufolf, *A Fatalist at War*, London: Allen & Unwin, 1929.
Bland, Larry I., editor, *The Papers of George Catlett Marshall: We Cannot Delay, July 1, 1939–December 6, 1941*, Baltimore: Johns Hopkins University Press, 1986.
Blumenson, Martin, *Anzio: The Gamble That Failed*, Westport, Connecticut: Greenwood Press, 1978.
———, *The Battle of the Generals*, New York: William Morrow and Company, 1993.
———, *Bloody River: The Real Tragedy of the Rapido*, Boston: Houghton Mifflin, 1970.
———, editor, *The Patton Papers*, Boston: Houghton Mifflin, 1972-74.
———, *Salerno to Cassino*, Washington, D.C.: Office of the Chief of Military History, 1969. "Green Book" on first half of the World War II Italian campaign.
Bond, Brian, *The Pursuit of Victory: From Napolean to Saddam Hussein*, New York and Oxford: Oxford University Press, 1996.
Bonn, Keith E., *When the Odds Were Even: The Vosges Mountains Campaign, October 1944–January 1945*, Novato, California: Presidio Press, 1994.
Boritt, Gabor S., editor, *War Comes Again: Comparative Vistas on the Civil War and World War II*, with an introduction by David Eisenhower, New York and Oxford: Oxford University Press, 1995.
Bowden, J. J., *The Exodous of Federal Forces from Texas, 1861*, Austin, Texas: Eakin Press, 1986.
Brager, Bruce L., "Civil War Medicine," *National Guard*, Volume XLIV, Number 4, April 1964, pages 26-29.
Braim, Paul, *The Test of Battle*, Newark, Delaware: University of Delaware Press, 1987.
Brice, Donaly E., Supervisor, Reference Services, Texas State Library and Archives Commission, correspondence with author, 1994-1999.
Brown, John Sloan, *Draftee Division*, Lexington: University Press of Kentucky, 1986.
Burhans, Lt. Col. Robert E., *The First Special Service Force: A War History of the North Americans, 1942-1944*, Washington, D.C.: Infantry Journal Press, 1947.
Butler, Brig. Gen. Frederick R., "Task Force Butler," Part I, *Armored Cavalry Journal*, Volume LVII, January-February 1948, pages 12-18; Part II, March-April 1948, pages 30-38.
Campbell, Randolph B., *Grassroots Reconstruction in Texas, 1865-1880*, Baton Rouge: Lousiana State University Press, 1997.
Cantor, Lewis, *The Creation of the Modern National Guard: The Dick Militia Act of 1903*, unpublished Ph.D. dissertation, Department of History, Duke University, 1963.
Cantrell, Gregg, *Stephen F. Austin: Empresario of Texas*, New Haven, Connecticut, and London: Yale University Press, 1999.
———, "Whither Sam Houston?", *Southwestern Historical Quarterly*, Volume XCVII, No. 2, October 1993, pages 344-357.

Catton, Bruce, *Reflections on the Civil War*, edited by John Leekley, introduction by E. B. Long, Garden City, New York: Doubleday and Company, 1981. The thoughts of a great historian on the American Civil War. This war created the Reconstruction environment in Texas, which is where the story of the 36th Division begins.
Center for American History, University of Texas at Austin, collections not cited elsewhere.
Center for Military History, United States Army, Washington, D.C., general files and collections not cited elsewhere.
Chambers, John Whiteclay, *Draftees or Volunteers: A Documentary History of the Debate over Military Conscription in the United States, 1787-1973*, New York: Garland Pub., 1975.
Chastain, Capt. Ben, *Story of the 36th*, Oklahoma City: Harlow Publishing Company, 1920. This book focuses on Oklahoma troops in the World War I 36th Division.
Churchill, Winston S., *Closing the Ring*, Volume 5, *The Second World War*, Boston: Houghton Mifflin, 1951.
Clark, James A., *The Tactful Texan: A Biography of Governor William Hobby*, New York: Random House, 1958.
Clark, Mark, *Calculated Risk*, New York: Harper and Brothers, 1950.
———, papers and diary at the Library and Research Center, The Citadel, Charleston, South Carolina.
Clarke, Jeffrey J., and Robert Ross Smith, *Riviera to the Rhine*, The United States Army in World War II, The European Theater of Operations, Washington, D.C.: Center of Military History, United States Army, 1993.
Clarke, Mary Whatley, *David G. Burnet*, Austin: Pemberton Press, 1969.
Clendenen, Clarence C., *The United States and Pancho Villa*, Ithaca, New York: Cornell University Press for the American Historical Association, 1961.
Clifford, J. Garry, and Samuel R. Spencer, Jr., *The First Peacetime Draft*, Lawrence: University Press of Kansas, 1986.
Coffman, Edward M., *The War to End All Wars: The American Military Experience in World War I*, Madison: The University of Wisconsin Press, 1986.
The Confederate Research Center, Hill College, Hillsboro, Texas, collections.
Constitution of the United States, edition published by Commission on the Bicentennial of the United States Constitution, circa 1987.
Cooper, Jerry M, *The Military and National Guard in America Since Colonial Times: A Research Guide*, Westport, Connecticut: Greenwood Press, 1993.
Corum, James S., *The Roots of Blitzkrieg*, Lawrence: University Press of Kansas, 1992.
Cosmas, Graham A., *An Army for Empire: The United States Army in the Spanish-American War*, Shippensburg, Pennsylvania: White Mane Publishing Company, 1994.
Cunningham, Adm. A. B., *A Sailor's Odyssey*, London: Hutchinson, 1951.
Daily Democratic Statesman, January 16, 1874.
Daily State Journal, Febuary 10, 1871, August 13, 1871, April 9, 1872.
Dallas Morning News, September 16, 1919, September 18, 1919.
Dallek, Robert, *Franklin D. Roosevelt and American Foreign Policy, 1932-1945*, New York: Oxford University Press, 1979.

Davis, William C., *Three Roads to the Alamo*, New York: HarperCollins Publishers, 1998. Texas background is background to the 36th Division. This is an excellent lengthy study of the lives of three major players at the Alamo—William Travis, James Bowie and David Crockett.

Daughters of the Revolution in Texas Archives, Alamo, San Antonio, Texas, general collections.

Dawson, Joseph G., III, editor, *The Texas Military Experience: From the Texas Revolution through World War II*, College Station: Texas A&M University Press, 1995. Useful collection of essays on Texas military history.

Denman, Clarence P., "The Office of the Adjutant General in Texas: 1835-1881," *Southwestern Historical Quarterly*, XXVII.

Department of the Army, Lineage and Honors, 141st Infantry. Found in files of the Center for Military History.

D'Este, Carlo, *Fatal Decision: Anzio and the Battle for Rome*, New York: Harper Collins, 1991.

The Devil's Brigade, United Artists, 1968. (Movie)

Dexter, David, *Allegiance*, New York: Harcourt, Inc., 2001.

Dixon, Kenneth L., "The Night of the Knife," *Argosy*, September 1964.

Doubler, Michael D., *Closing with the Enemy: How the GIs fought the War in Europe, 1944-1945*, Lawrence: University Press of Kansas, 1994.

Dupuy, Col. R. Ernest, *The Compact History of the United States Army*, New York and London: Hawthorn Books, 1961.

Durant, Will and Ariel, *The Lessons of History*, New York: Simon and Schuster, 1968.

Eisenhower, Dwight David, *Crusade in Europe*, Garden City, New York: Doubleday, 1948.

Essame, H., *Patton: A Study in Command*, New York: Charles Scribner's Sons, 1974.

Faulk, J.J. *History of Henderson County, Texas*, Athens, Texas: Athens Review Publishing Company, 1929.

Fifth Army History, Historical Section, Headquarters Fifth Army, undated but circa 1945. Printed Florence, Italy: L'Impronta Press.

Fehrenbach, T. R., *Seven Keys to Texas*, El Paso: Texas Western Press, The University of Texas at El Paso, 1986.

The Fighting 36th Historical Quarterly, various issues cited in reference notes. A primary source collection of articles and reminiscences by members of the 36th Division.

Final Report of Gen. John J. Pershing, Commander in Chief, American Expeditionary Forces, Washington, D.C.: Government Printing Office, 1919.

Fisher, Ernest F., Jr., *Cassino to the Alps*, The United States Army in World War Two, The Mediterranean Theater of Operations, Washington, D.C.: Center of Military History, United States Army, 1989.

Fitzhugh, Lester N., "Writing Texas Military History," *Texas Military History*, Volume 1, Number 1, May 1961.

Flynn, George, *The Draft, 1940-1973*, Lawrence: University Press of Kansas, 1993.

Foner, Eric, *America's Reconstruction: People and Politics After the Civil War*, New York: HarperPerennial, 1995.

Franklin, John Hope, *The Militant South: 1800-1861*, Cambridge: Belknap Press, 1970.

Franks, Kenny A., *Citizen Soldiers: Oklahoma's National Guard*, Norman: University of Oklahoma Press, 1984.
Gammel, H.P.N., *The Laws of Texas, 1822-1897*, Austin: Gammel's Book Store, 1902.
Gartner, Scott Sigmund, *Strategic Assessment in War*, New Haven, Connecticut: Yale University Press, 1997.
Gibson, Arrell Morgan, *The History of Oklahoma*, Stillwater, Oklahoma: Forum Press, 1983.
Gilbert, Martin, *The First World War: A Complete History*, New York: Henry Holt and Company, 1994.
Galveston Daily News, June 20, 1920.
Graham, Dominick, and Shelford Bidwell, *Tug of War: The Battle for Italy, 1943-1945*, New York: St. Martin's Press, 1986.
Green, Michael, former research archivist, Texas State Archives, author's conversations with and correspondence with, 1992-1993.
Greenfield, Kent Roberts, general editor for the Office of the Chief of Military History, *Command Decisions*, New York: Harcourt, Brace and Company, 1959.
Greenfield, Kent Roberts, Robert R. Palmer, and Bell I. Wiley, *The Organization of Ground Combat Troops*, Washington, D.C.: Historical Division, United States Army, 1947. "Green Book."
Griffith, Robert H., Jr., *Men Wanted for the U.S. Army*, Westport, Connecticut: Greenwood Press, 1982.
Gundmundsson, Bruce I., *Stormtroop Tactics: Innovation in the German Army, 1914-1918*, New York: Praeger, 1989.
Hamilton, D. H. *History of Company M, First Texas Volunteer Infantry*, Waco, Texas: W. M. Morrison, 1962.
———, *History of Company M, First Texas Volunteer Infantry, Hood's Brigade, Longstreet's Corps, Army of the Confederate States of America*, no publisher or city listed, 1925. This is likely an earlier version, perhaps privately published, of D. H. Hamilton's 1962 book. This was located in the files of the research center at Hill College.
Handel, Michael, editor, *Intelligence and Military Operations*, London: Frank Cass, 1990.
Hardin, Stephen L., *Texian Iliad*, Austin: Universtiy of Texas Press, 1994.
Hargrove, Hondon B., *Buffalo Soldiers in Italy: Black Americans in World War II*, Jefferson, North Carolina: McFarland and Co. Inc., Publishers, 1985.
Harries, Meirion and Susie, *The Last Days of Innocence: America at War, 1917-1918*, New York: Random House, 1997.
Hassell, Ulrich von, *The Von Hassell Diaries*, Garden City, New York: Doubleday and Company, 1947.
Hastings, Max, *Overlord*, New York: Simon and Schuster, 1984.
Haythornthwaite, Philip J., *The World War One Sourcebook*, London: Arms and Armour Press, 1992.
Henderson, Harry McCorry, *History of the 141st Infantry, 36th Infantry Division, Texas National Guard*, San Antonio, Texas: Naylor Company, 1950. This is an interesting work but of uncertain accuracy.
Henderson, Wm. Darryl, *Cohesion: The Human Element in Combat*, introduction

by Charles C. Moskos, Washington, D.C.: National Defense University Press, 1985.

Hickey, Des, and Gus Smith, *Operation Avalanche: The Salerno Landings, 1943*, London: Heinemann, 1983.

Hill, Jim Dan, *The Minute Man in Peace and War*, Harrisburg: Pennsylvania: The Stackpole Company, 1964.

History and Pictorial Review, National Guard of the State of Texas, Baton Rouge: Louisiana: Army and Navy Publishing Company, 1940, page xxxix. Interesting for its reflection of 1940 attitudes as well as its history.

Houston Daily Post, September 27-30, 1887.

Houston Daily Telegraph, October 1, 1874.

Hornaday Collection, Texas State Library and Archives. This is a compilation, prepared at the time, of 6,000-8,000 typed pages, roughly 2,400 individual letters and accounts, of World War I narratives and letters to the editor from Texas soldiers. World War I letters from individual soldiers cited in the references are from this highly valuable collection of firsthand material on that war not feasible to collect individually.

Hulen, John, papers, Texas State Library and Archives.

Hutton, Bud, and Andy Rooney, *The Story of the Stars and Stripes*, New York: Farrar and Rinehart, 1946.

Huff, Richard A., editor, *A Pictorial History of the 36th "Texas" Infantry Division*, compiled by the 36th Division Pictorial History Team, Austin: 36th Division Association, undated but circa 1945-49.

Infantry Journal, "Universal Military Training: Legislation Imperative Before Demobilization" (Statement of H. H. Sheets, secretary, National Association for Universal Military Training), Volume 15, Number 7, January 1919, pages 537-540.

Jacobs, Bruce, *Heroes of the Army: The Medal of Honor and its Winners*, New York: W. W. Norton and Company, 1956.

———, "How the Dick Acts Happened," *National Guard*, September 1988.

Jessup, Philip, *Elihu Root*, New York: Dodd, Mead and Company, 1938.

Johnson, Hubert E., *Breakthru! Tactics, Technology and the Search for Victory on the Western Front in World War I*, Novato, California: Presido, 1994.

Anson Jones Papers, untitled draft narrative, September 9, 1943–May 8, 1945, 36th Division Collection, Texas State Archives.

Jones, Archer, *Civil War Command and Strategy*, New York: The Free Press, 1992.

Keast, William R., Robert R. Palmer, and Bell I. Wiley, *The Procurement and Training of Ground Combat Troops*, Washington, D.C.: Office of the Chief of Military History, 1948. Volume in the authoritative "Green Book" series.

Keegan, John, *A History of Warfare*, New York: Alfred A. Knopf, 1993.

———, *The Mask of Command*, New York: Penguin, 1988.

Kennett, Lee, *G.I.: The American Soldier in World War II*, New York: Charles Scribner's Sons, 1987.

Kesselring, Albert, "An Interview with Genfldm Albert Kesselring: Rapido River Crossing," Ethnit 71, May 6, 1946, Historical Section, ETOUSA.

———, *A Soldier's Record*, New York: William Morrow & Company, 1954.

Ketchum, Richard M., *The Borrowed Years: 1938-1941, America on the Way to War*, New York: Random House, 1989. Excellent and lengthy study of the United States in the three years before American entry into World War II.

Keyes, Geoffrey, Interview by Dr. Philip A. Crowl (Department of the Army) with Lt. Gen. Geoffrey Keyes, USA (Ret.), Washington, D.C., September 22, 1955, Texas State Archives collections.
Kingston, Donald M., *Forgotten Summers: The Story of the Citizen's Military Training Camps, 1921-1940*, San Francisco: Two Decades Publishers., 1995.
Klein, Maury, *Days of Defiance: Sumter, Secession and the Coming of the Civil War*, New York: Alfred A. Knopf, 1997.
Kouser, Bruce, letters, unpublished and undated but written 1943-1944, author's personal collection.
Kreidberg, Marvin A., and Merton C. Henry, *History of Military Mobilization in the United States Army, 1775-1945*, Washington, D.C.: U.S. Army Center of Military History, 1984.
Krenek, Harry Lynn, *A History of the Texas National Guard Between World War I and World War II*, unpublished Ph.D. dissertation, Texas Tech University, 1979.
———, *The Power Vested*, Austin: Presidia Press, 1980.
Lack, Paul, *The Texas Revolutionary Experience: A Political and Social History, 1835-1836*, College Station: Texas A&M University Press, 1992.
John A. Lejeune, *The Reminiscences of a Marine*, Philadelphia: Dorrance and Company, 1930.
Lewin, Ronald, *Ultra Goes to War*, New York: McGraw Hill Book Company, 1978. Every history book written since the Ultra secret was revealed has to be read in a different light.
Liggett, Gen. Hunter, *AEF: Ten Years Ago in France*, New York: Dodd, Mead and Company, 1928.
Linderman, Gerald F., *Embattled Courage*, New York: The Free Press, 1987.
Little, Will T., L. G. Putnam, and R. J. Barker, *The Statutes of Oklahoma, 1890*, Guthrie: State Capitol Printing Company, 1891.
Lockhart, Vincent M., memo to author, November 9, 1993.
———, *T-Patch to Victory*, Canyon, Texas: Staked Plains Press, 1981. The 36th Division in the Southern France campaign, August 1944 to end of war. This is a personal memoir as well as division history. Lockhart was also kind enough to review a draft of sections of this author's history.
———, undated memos to author, May 1994 and June 1994.
Logan, John A., *The Volunteer Soldiers of America*, Chicago: R. S. Peale, 1887.
Long, Jeff, *Duel of Eagles*, New York: William Morrow and Company, 1990.
Luck, Hans von, *Panzer Commander*, New York: Praeger, 1989.
MacDonald, Charles B., *The Last Offensive*, United States Army in World War II: The European Theater of Operations, Washington: Office of the Chief of Military History, 1973. Volume of the authoritative "Green Book" series of histories of World War II. Virtually a primary source, these works provided campaign overviews, allowing the author to concentrate on the detailed history of the 36th Division itself.
———, *The Mighty Endeavor*, New York: Oxford University Press, 1969.
Mahafer, Col. Harry J., *Brave Decisions*, Washington and London: Brassey's, 1995.
Mahon, John L., *History of the Militia and National Guard*, New York: Macmillian, 1983.
Mansoor, Peter R., *The GI Offensive in Europe: The Triumph of American Infantry*

Divisions, 1941-1945, Lawrence: University Press of Kansas, 1999. Study of how well the American divisions actually did in World War II. This is a much-needed antidote to modern military historians' claims that the United States army won only by use of materiel and artillery.

Marten, James Alan, *Texas Divided: Loyalty and Dissent in the Lone Star State, 1856-1874*, Lexington, Kentucky: University Press of Kentucky, 1990.

Mathews, Lloyd J., and Dale E. Brown, editors, *The Challenge of Military Leadership*, introduction by Lt. Gen. Walter F. Ulmer, Jr., Washington, D.C.: Pergamon-Brassey's International Defense Publishers, Inc., 1989.

McAndrews, 1st Lt. Kevin, "Keeping your Guard Up," *National Guard*, March 1996, pages 24-26.

Messenger, Charles, *The Blitzkrieg Story*, New York: Charles Scribner's Sons, 1976.

Millett, Allan R., and Williamson Murray, editors, *Military Effectiveness*, three volumes, Boston: Unwin Hyman, 1988. Useful for World War I and World War II.

Millis, Walter, *Arms and Men: A Study in American Military History*, New York: Putnam, 1956.

Montgomery, Bernard L., *The Memoirs of Field Marshall Montgomery*, Cleveland: World Publishing Company, 1954.

Moody, Dan, May 13, 1930, letter to F. F. Robinett, Dan Moody Papers, Texas State Archives.

Moore, James Orville, *The Men of the Bayou City Guards*, MA Thesis, unpublished, University of Houston–Clear Lake, School of Human Sciences and Humanities, December 1988.

Morgan, H. Wayne, *Oklahoma: A History*, New York: Norton; Nashville: American Assocation for State and Local History, 1984.

Morris, Eric, *Circles of Hell: The War in Italy, 1943-1945*, New York: Crown Publishers, 1993.

Morrison, Samuel Eliot, *Strategy and Compromise*, Boston: Little Brown and Company, 1958.

Nackman, Mark E., "The Making of the Texas Citizen Soldier," *Southwestern Historical Quarterly*, LXXVIII (1974).

National Archives and Records Administration, *Record Group 407*. The basic collection of official records on the 36th Division in World War II.

National Archives and Records Administration, *Record Group 120*. Records of 36th Division from World War I.

National Guard Association of the United States, library, general collections.

"Native Americans in World War I," National Archives, Calendar of Events, August 1993.

Nelson, Christian G., "Organization and Training of the Texas Militia, 1870-1897," *Texas Military History*, Volume 2, Number 2, May 1962.

———, "Rebirth, Growth and Expansion of the Texas Militia, 1868-1898," *Texas Military History*, Volume 2, Number 1, February 1962.

Neustadt, Richard E., and Ernest R. May, *Thinking in Time*, New York: The Free Press, 1986.

New York Times, November 6, 1944, August 28, 1897.

Nicolson, Nigel, *Alex: The Life of Field Marshall Earl Alexander of Tunis*, New York: Antheneum, 1973.

Otto, Lt. Col. Ernst, German Army, Retired, "The Battle at Blanc Mont," Part III, *United States Naval Institute Proceedings*, Volume 56, Number 325, March 1930, pages 177-199; Part IV, *Proceedings*, Volume 56, Number 326, April 1930, pages 304-316.

Olson, Bruce, "The Houston Light Guards: A Case Study of the Texas Militia, 1873-1903," MA thesis, presented to the faculty of the Graduate School of the University of Houston, in partial requirements for the Degree of Master of Arts, May 1985.

Our Century: Masters of War, "Battle for the Boot," Arts and Entertainment Cable Network, March 16, 1994.

Overy, Richard, Why the Allies Won, New York and London: W. W. Norton & Company, 1995.

Palmer, Frederick, *John J. Pershing: General of the Army*, Harrisburg, Pennsylvania: Military Service Publishing Company, 1948.

Parker, Lt. Col. Reuben D. Parker, "Infiltration as a Form of Maneuver," Military Review, December 1959. Analysis of the Velletri infiltration.

Peek, Clifford H., Jr., editor, *Five Years—Five Countries—Five Campaigns*, Munich, Germany: The 141st Infantry Regiment Association, undated (circa 1945), page 58.

Perret, Geoffrey, *There's a War to Be Won*, New York: Random House, 1991.

Pershing, John J., *My Experiences in the World War*, New York: Frederick A. Stokes Company, 1930.

Padgitt, James T., "That Long Day at Salerno," *Texas Military History*, Volume 5, Number 2, Summer 1965.

Pogue, Forest C., *The Supreme Command*, United States Army in World War II, The European Theater of Operations, Washington, D.C.: Office of the Chief of Military History, Department of the Army, 1954.

Potter, David M., *The Impending Crisis*, completed and edited by Don E. Fehrenbacher, New York: Harper and Row, 1976.

Pruden, Durward, *A Sociological Study of a Texas Lynching*, MA Thesis, Southern Methodist University, 1935.

Puck, Brig. Gen. Armin, interview with General Puck, interviewed Col. James B. Sweeney, August 23, 1983, transcript from the University of Texas, Institute of Texan Cultures, San Antonio.

Purcell, Alan Robert, *The History of the Texas Militia: 1835-1903*, unpublished Ph.D. dissertation, University of Texas at Austin, 1981.

Railey, Hilton Howell, "Morale of the United States Army: An Appraisal for the New York Times," unpublished manuscript, September 29, 1941.

Ramsdell, Charles William, *Reconstruction in Texas*, Gloucester, Massachusetts: P. Smith, 1964.

Raper, Arthur F., *The Tragedy of Lynching*, Chapel Hill: The University of North Carolina Press, 1933.

The Rapido River Crossing, Hearings before the Committee on Military Affairs, House of Representatives, Seventy-Ninth Congress, Second Session, February 20 and March 18, 1946.

Reagan, John, *Memoirs*, edited by Walter F. McCaleb, New York: Neale Publishing Company, 1906.

Reagan, John, papers, Center for American History, University of Texas, Austin.

Reese, Lt. Col. Hal, "Intermission at Cassino," May 19, 1944, 36th Division Association Collection, Texas State Library and Archives, Austin.

The Relief of Belsen, April 1945, Eyewitness Accounts, London: Imperial War Museum, 1991.

Richards, Frank, *Old Soldiers Never Die*, London: Faber & Faber, 1933.

Richter, William L., *The Army in Texas During Reconstruction, 1865-1870*, College Station: Texas A&M University Press, 1987.

———, *Overreached All Sides: The Freedman's Bureau Administrations in Texas, 1865-1868*, College Station: Texas A&M University Press, 1991.

Riker, William J., *Soldiers of the States: The Role of the National Guard in American Democracy*, Washington, D.C.: Public Affairs Press, 1957.

Roberts, Oran M., *Texas*, Volume XI of *Confederate Military History*, edited by Clement A. Evans, New York: Thomas Yoseloff, 1962 (reprint).

Robertson, James I., Jr. *Soldiers Blue and Gray*, Columbia, South Carolina: University of South Carolina Press, 1988. Valuable background on what soldiers of all wars have in common.

Ross, Steven T., *American War Plans: 1941-1945, The Test of Battle*, London, England, and Portland, Oregon: F. Cass, 1997.

Ryan, Cornelius, *The Longest Day*, New York: Simon and Schuster, 1959.

Samuel Rosenman, editor, *The Public Papers and Addresses of Franklin D. Roosevelt*, New York: Russell & Russell, 1969.

San Antonio Press, circa 1950 (clipping not dated), "Famed 'Washington Guards' Again Organized as honor unit for city," files of the Daughters of the Revolution in Texas archives, the Alamo.

Sanger, William C., *Report on the Reserve and Auxiliary Forces of England and the Militia of Switzerland: Prepared for President McKinley and Secretary of War Root*, Washington, D.C.: Government Printing Office, 1903.

Schwarzkopf, Gen. H. Norman, with Peter Petre, *It Doesn't Take A Hero: The Autobiography*, New York: Linda Gray, Bantam Books, 1992. Perspectives on a war in which the 36th Division's descendant stood ready to fight.

Scribner, Colonel (now B.G., retired, Texas State Guard) John, historian and records manager, Texas National Guard, author's conversations with and correspondence with, 1992-1998.

Segal, David R., *Recruiting for Uncle Sam: Citizenship and Military Manpower Policy*, Lawrence: University Press of Kansas, 1989.

von Senger, General F., interview by Philip A. Crowl with General F. von Senger u. Etterlin, September 22, 1955, copy in Texas State Library and Archives.

Seventh Army Report of Operations, Volume II, Heidelberg: Aloys Graef, 1946.

Severied, Eric, "Velletri," *The American Legion Magazine*, October 1944.

Shelley, George E., "The Semicolon Court of Texas," *Southwestern Historical Quarterly*, XLVIII, pages 449-468.

Sherman Democrat, May 16, 1930.

Shirer, William L., *The Rise and Fall of the Third Reich*, New York: Simon and Schuster, 1960.

Shook, Robert W., "The Federal Military in Texas, 1865-1870," *Texas Military History*, Volume VI, Number 1, Spring 1967.

Simpson, Harold, *Audie Murphy: American Soldier*, Hillsboro, Texas: Hill Junior College Press, 1975.

———, editor, Jerome B. Robertson, compiler, *Touched with Valor: The Civil War Papers and Casualty Reports of Hood's Texas Brigade*, Hillsboro, Texas: Hill Junior College Press, 1964.
———, *Hood's Texas Brigades: A Compendium*, Hillsboro: Hill Jr. College Press, 1977. Pages 10-250 list all members of Hood's Brigade.
———, *Hood's Texas Brigade: Lee's Grenadier Guard*, Waco, Texas: Texian Press, 1970. Best history available on Hood's Texas Brigade.
Singletary, Otis A., "The Texas Militia During Reconstruction," *Southwestern Historical Quarterly*, Volume LX, Number 1, July 1956, pages 23-35.
Skaggs, Jimmy M., "Lieutenant General John A. Hulen: Portrait of a Citizen-Soldier," *Texas Military History*, Volume 8, Number 3, 1970, page 135.
Sligh, Robert Bruce, *The National Guard and National Defense: The Mobilization of the Guard in World War II*, New York: Praeger, 1992.
Slonaker, *The Volunteer Army*, Carlisle Barracks, Pennsylvania: U.S. Army Military Research Collection, 1972.
Smallwood, James, *Time of Hope, Time of Despair: Black Texans During Reconstruction*, Port Washington, New York: National University Publications, 1981.
Smith, David Paul, *Frontier Defense in the Civil War: Texas' Rangers and Rebels*, College Station: Texas A&M University Press, 1992.
Smith, Gene, "Seventeenth Largest Army," *American Heritage*, December 1992, pages 99-107.
Smith, Lee Caraway, *A River Swift and Deadly*, Austin: Eakin Press, 1989. Highly useful book on the Rapido crossing.
Smythe, Donald, *Pershing: General of the Armies*, Bloomington: Indiana University Press, 1986.
Sneed, Edgar P., "A Historiography of Reconstruction in Texas: Some Myths and Problems," *Southwestern Historical Quarterly*, Volume LXXII, No. 4, April 1969.
"Souvenir: Houston Light Guard," commemorative brochure, undated but circa 1902, Texas State Archives.
Spence, Alexander White, *The History of the Thirty-Sixth Division, U.S.A., 1917-1919*, unpublished manuscript, official division records, National Archives.
Springer, Joseph A., *The Black Devil Brigade*, Pacifica, California: Pacifica Military History, 2001.
The Statutes at Large of the United States of America, from December 1901 to March 1903, edited, printed and published by Authority of Congress, Volume XXXII, Part I, Washington, GPO: 1903.
Steckel, Francis C., "Morale Problems in Combat: American Soldiers in World War II," *Army History*, Summer 1994, pages 4-5.
Stern, Sheldon, "Beware the Past as Present," *History Matters*, Volume 5, Number 9, May 1993.
Stuckey, Col. John D., and Col. Joseph H. Pistorius, "Mobilizing the Army National Guard and the Army Reserve: Historical Perspectives and the Vietnam War," Carlisle Barracks, Pennsylvania: U.S. Army War College, Strategic Studies Institute, September 7, 1984.
Swinton, Ernest D., *Eyewitness*, London: Hoddor & Stoughton, 1932.
Target: Pearl Harbor, The Discovery Channel, Sunday, December 6, 1992.

Texas. Constitutional Convention (1868-1869): *Journal of the Reconstruction Convention: Which Met at Austin, Texas,* Austin: Tracy, Siemering & Co, 1870.

Texas National Guard, Office of the Historian and Records Manager, *History of the Texas National Guard,* unpublished manuscript, July 1992.

Texas National Guard Museum collections, Fort Mabry, Texas.

Texas State Library and Archives, Austin, Texas, material and collections not cited elsewhere.

Thirty-Sixth Division Collection, Texas State Archives, Austin, Texas. Collection of primary source material on division.

Truscott, Lucian, Jr., *Command Missions: A Personal Story,* New York: E. P. Dutton, 1954.

Tuchman, Barbara W., *The Zimmermann Telegram,* New York: The Viking Press, 1958.

United States Army in the World War, Training and Use of American Units with British and French, Washington: History Division, Department of the Army, 1948.

United States Congress, House of Representatives, "Communication from Governor Pease of Texas Relative to the Troubles in that State," 40th Congress, 2nd Session, 1868, H Misc. Doc 127.

United States Congress, House Executive Documents, 40th Congress, 3rd Session, No 1.

United States Holocaust Memorial Museum, conversation of author with research staff.

Upton, Bvt. Maj. Gen. Emory, United States Army, *The Military Policy of the United States,* New York: Greenwood Press, 1968 reprint of 1904 edition.

Upton, Emory, *The Military Policy of the United States from 1775,* Washington, D.C.: Government Printing Office, 1904.

Wagner, Robert L., *The Texas Army: A History of the 36th Division in the Italian Campaign,* Austin: Statehouse Press, 1991 and 1972. This was the pathfinder for the Italian campaign history.

Walker, Fred L., *From Texas to Rome,* Dallas, Texas: Taylor Publishing Company, 1969. Diary of the commander of the 36th Division during the Italian campaign.

———, "General Walker's Story of the Rapido Crossing," submitted to publication by Joe F. Presnall, written circa 1962 by Fred L. Walker, *The Fighting 36th Historical Quarterly,* Volume X, Number 2, Summer 1990. Text appears to be the same as Fred L. Walker, "My Story of the Rapido Crossing," *Army,* Volume XIII, Number 2, September 1962, pages 52-60.

Walker, William A., Memorandum for General North, Subject: Report of Interview with Major General Fred L. Walker, 4 February 1946, William A. Walker, Colonel GSC, Deputy Chief, Current Group, OPD. Copy found in Citadel Archives, Mark Clark Papers.

Wallace, Ernest, *The Howling of the Coyotes: Reconstruction Efforts to Divide Texas,* College State: Texas A&M University Press, 1979.

Wedemeyer, Albert C., *Wedemeyer Reports,* New York: Hurt, 1958.

Weigley, Russell L., *Eisenhower's Lieutenants,* Bloomington: Indiana University Press, 1981.

———, *History of the United States Army*, enlarged edition, Bloomington: Indiana University Press, 1984.
Westphal, Siegfried, *The German Army in the West*, London: Cassel and Company, 1954.
White, Lonnie, "The Combat History of the 36th Division in World War I," *Military History of Texas and the Southwest*, Volume 17, Number 4, 1982, entire issue.
———, "Forming the 36th Division in World War I," *Military History of Texas and the Southwest*, Volume 17, Number 2, 1982, entire issue.
———, "The Homecoming of the 36th Division in World War I," *Military History of Texas and the Southwest*, Volume 18, Number 1, 1983, entire issue.
———, "Indian Soldiers of the 36th Division," *Military History of Texas and the Southwest*, Volume 15, Number 1, pages 7-20. (1981, but date not given on journal.)
———, *Panthers to Arrowheads: The Thirty-Sixth (Texas-Oklahoma) Division in World War I*, Austin: Presidial Press, 1984. Reprint of his valuable series of articles. Provided both direction and information.
———, "Training the 36th Division in World War I," *Military History of Texas and the Southwest*, Volume 17, Number 3, 1982, entire issue.
Williams, A. M., *Sam Houston and the War of Independence*, Boston: Houghton Mifflin Co., 1893.
Williams, John Hoyt, *Sam Houston: A Biography of the Father of Texas*, New York: Simon and Schuster, 1993.
Wilson, John, Center for Military History, Department of the Army, Washington, D.C., author's conversations with, 1992.
Wilt, Alan F., *The French Riviera Campaign of August 1944*, Carbondale: Southern Illinois University Press, 1981.
Winkler, Mrs. A. V. *The Confederate Capitol and Hood's Texas Brigade*, Austin: Von Boeckmann, 1894.
Wolters, Jacob F., *Martial Law and Its Administration*, Austin: Gammel's Book Store, Inc., 1930.
Wood, Leonard, *Our Military History: Its Facts and Falacies*, Chicago: Reilly and Britton, 1915.
Wood, W. D., *Reminiscences of Reconstruction in Texas and Reminiscences of Texas and Texans Fifty Years Ago*, (no city or publisher), 1902.
Wooster, Ralph, editor (and with an introduction by), *Lone Star Blue and Gray: Essays on Texas in the Civil War*, Austin: Texas State Historical Association, 1995.
Wooten, Dudley, editor, *A Comprehensive History of Texas, 1685-1897*, Austin: published by the Texas State Historical Association in cooperation with the Center for Studies in Texas History, the University of Texas at Austin.
Yale Law Library, *Legislative History: Draft and Selective Service Acts (1863-1967)*, New Haven, Connecticut: Yale Law Library, 1968.
Yockelson, Mitchell, reference archivist, National Archives and Records Administration, conversations with author, 1993.

Index

A—
Adams, Paul D., 173, 227
Adler, Julius, 101
African-Americans. *See* blacks
Aisne River, 70-71, 73
Alamo, viii, 4
Alban Hills, 180, 181
Alexander, Harold, 131, 143, 148, 150, 154, 160, 173, 178, 179, 181, 182
Alger, Russell A., 24-25, 30
Allex, France, 217
Alsace, France, 50
Altavilla, Italy, 133-134, 135-141
Amalfi, Italy, 127
Ambrose, Stephen, 117, 233
Ancon (flagship), 132
Antheor Cove, 198
anti-militarism, 80-81, 83
Anzio invasion, 142, 158-159, 160, 161, 166, 172, 173-174, 180-181
Appomattox Court House, 7-8, 10
Ardennes Canal, 70
Ardennes Forest, 250
Argens River, 198
Argonne Forest, 56, 73
Armies of Asia and Europe, The (book), 29
Army Appropriations Bill, 14, 83
Army Navy Journal, 25
Army War College, 31
Arrowhead, The (newspaper), 79

Attigny, France, 69
Austin, Stephen F., 3-4

B—
Badoglio, Pietro, 125, 153
Bains-Les-Bains, 227
Baker, Newton, 39, 52
Balck, Herman, 225
Bar-sur-Aube, France, 47
Barron, Gaines J., 135
Battipaglia, Italy, 124
Battle of Bladensburg, 1
Battle of Brownsville, 10
Battle of Flers, 52
Battle of Montelimar, 222
Battle of Monterrey, 5
Battle of the Marne, 50
Battle of the Bulge, 250, 255
Battle of Verdun, 51
Battle of Ypres, 121
Beacham, Charles M., 155
Belfort Gap, 229
Bernhardt line, 126
Biffontaine, France, 235
Birkhead, Claude V., 36, 112, 114, 117, 118, 175
black codes, 13
blacks, crime, 14; draft boards, 106; Houston Light Guard, 20; integration, 119; militiamen, 16, 18; rights, 13; state police, 16; universal mili-

tary training, 81; violence against, 13-14, 15, 16, 20; World War I, 40; World War II, 119
Blakley, George, 40, 43-44
Blanc Mont, 56, 59-60
Blaskowitz, Johannes, 197, 200, 206
blitzkrieg, 97, 102
Boone, Gordon, 86
bootleggers, 44
Borah, William E., 96
Bowden Woods, 254
Bradley, Omar, 232, 245, 246, 257
Brann, Donald W., 154
Brauschitz, Colonel, 263
Brazoria, Texas, 22
Brest, France, 46
British militia, 2
Brooks, Edward H., 231, 232, 234, 239, 242-243, 253, 255
Bruyeres, France, 233, 234
Burke, Edward R., 103
Burke-Wadsworth Bill, 102, 103, 108
Burnet, David G., 12-13
Bush, George W., 267
Butler, Frederick B., vii, 209-212, 214, 215-217

C—
Caeser line, 180, 182, 191
Calabria, Italy, 125-126
Camp Blanding, Florida, 105, 118
Camp Bowie, Texas, 39, 42, 43, 45
Camp Mabry, Texas, 26
Camp Mills, New York, 45
Cantigny, France, 55
Carey, Robert H., 130
Cassino, Italy, 162, 163, 178, 179
Chastaine, Ben, 68, 69, 74
Cheney, Dick, 266
Choctaws, 74
Churchill, Winston, 103, 121, 122, 139, 147, 159, 247, 258
Citizen Soldiers, 233
civil rights, 100
Civil War, 1, 3, 4, 5-6, 7-8, 9-11, 18, 29-30
Clark, Grenville, 100-101, 102

Clark, Mark W.: Anzio invasion, 159-160, 161; British 46 Division, 165-166; British 56 Division, 166, 177; Cassino, Italy, 178; 15th Army Group, 231; 5th Army, 124; intelligence reports, 142, 143, 166, 192; Italy campaigns, 195; naval support, 139; Rapido crossing, 162, 163, 164, 172-173, 174-175, 176, 177; Rome, Italy, 182, 184, 192; Salerno invasion, 127, 128, 131, 135; San Pietro, Italy, 144-145, 148-149; Sant Ambrosio, Italy, 165-166; Walker, Fred, 191, 193; Winter line, 148-149, 150
Coke, Richard, 17, 18
Colmar Pocket, 245-250, 253-256
Columbia, Texas, 22
Comanches, 4, 23
combat fatigue, 226, 227-228
Condilac Pass, vii, 218
conscription. *See* draft
Constitution, U.S., 2-3
Cope, W. D., 86, 87
Corpus Christi, Texas, 86
Crest, France, 210-211
Crisman, Captain, 189
Critchfield, James H., 216
Croix Haute Pass, 210
Cunningham, A. B, 139, 143

D—
Dachau, 261
Dahlquist, John E.: ability, 231-232; Colmar Pocket, 255; Goering, Hermann, 263; "Lost Battalion," 236, 239; Montelimar, France, 209, 210, 213, 214, 215-218, 220, 221; Operation Anvil/Dragoon, 203, 204; rest camps, 226-227; 36th Division, 196, 235, 255
Darby, William, 127
Davidson, James, 15
Davis, Edmund, 15, 17, 18, 19
Davis, Jefferson, 6, 9
Dawley, Ernest J., 124, 129, 135, 137, 138, 139, 140, 142

Index   321

de Gaulle, Charles, 251
de Lattre de Tassigny, Jean, 196-197, 205, 245, 248
de Zavala, Augustine, 40
de Zavala, Lorenzo, 40
Democratic Party, 17, 18
Devers, Jacob: Bradley, Omar, 246; Colmar Pocket, 245, 246-247, 248, 251, 255; Eisenhower, Dwight, 246-247, 250, 251; German surrender, 264; Goering, Hermann, 263; 6th Army Group, 197, 222, 245, 250; Strasbourg, France, 251
Dick Act, 19, 32-33
Dick, Charles, 32-33
Die Suedfront (newspaper), 185
Dietrich, Sepp, 263
Dixon, Kenneth, 185
Donovan, William J., 101
draft, 34, 36, 37, 82, 97, 99-100, 102-104, 105-106
draft boards, 105-106, 107
Drome River, 211
Dunlap, James E., 93

E—
Eagles, William W., 196
Eisenhower, Dwight D.: Badoglio, Pietro, 125; Battle of the Bulge, 250; "broadfront" strategy, 224, 247; Clark, Mark, 176; Colmar Pocket, 246-248; Devers, Jacob, 245-247; German surrender, 264; Italy, negotiations, 125; Normandy Invasion, 110; politics, 247; Rhine valley, 245, 246-247; Salerno invasion, 131; 7th Army, 222

F—
Falaise Pocket, 197, 206
Felber, Joseph G., 211
Ferguson, James E., 36
Foch, Marshall, 73
food safety, 25
Ford, Henry, 81
Forest Farm, 73-77
Forsythe, John D., 138, 140, 142

Fort Belvoir, Virginia, 113
Fort Benning, Georgia, 193
Fort Sam Houston, Texas, 89
Fort Sill, Oklahoma, 113
Fort Sumter, South Carolina, 7
Fort Warren, Massachusetts, 11-12
Fort Worth, Texas, 44
Fowler, Wick, 191
Franco-Prussian War, 30
Franklin, Benjamin, 4
Frazior, David M., 170
Frederick, Robert T., 147, 151, 192, 196
freedmen, 13-14
Freedmen's Bureau, 12, 13
freedom rides, 100
French Forces of the Interior (FFI), 196, 199-200, 209-210, 215
Frost, Texas, 86

G—
Galveston, Texas, 87-89
Garigliano River, 145
German Americans, 41, 45-46
Gillette, A. Ward, 244
Glendening, Parris, 267
Goering, Hermann, 183, 263
Gonsalves, George, 40
Gonzales, Manuel S., 130
Gouraud, General, 57, 60, 71
Graham, Samuel S., 130
Grane, France, 217
Granger, Gordon, 10
Grant, Ulysses S., 9
Gray, Allen Charles, 18
Gray, Edwin Fairfax, 18
Grayson County Courthouse, 92
Great Depression, 91, 97
Greble, Edward St. John, 37, 39-40, 42, 44, 45
Grenoble, France, 210, 211, 213, 214
Gruenther, Alfred, 162, 165, 192
Gulf of Gaeta, 123
Gulf of Naples, 123
Gulf of Salerno, 123
Gustav line, 158-159, 160, 161, 180, 181

H—
Haguenau Forest, 253
Hamilton, A. J., 10
Harmon, Ernest, 154, 162
Harmony, John W., 213, 215, 217-218
Harvard Club, 99, 100, 103
Hawkins, Burton, 20, 21
Hayes-Tilden presidential election, 19
Henderson, J. P., 5
Hershey, Lewis, 105, 107
Hess, Walter, 173
Hewitt, Henry Kent, 132, 135, 138, 197, 204
Hickey, Sheriff, 21, 22
Himmler, Heinrich, 248-249
Hindenburg line, 56-57
Hitler, Adolf: assassination, attempted, 197, 249; France, strategies, 206, 250-251; Goering, Hermann, 263; Himmler, Heinrich, 248-249; Italy, alliance, 125; Italy, strategies, 122, 126, 164; *Mein Kampf*, 261; Poland invasion, 96
Hitler line, 180
Hobby Loyalty Act, 87
Hobby, William D., 86
Hood's Texas Brigade, 7-8, 268
Houston Daily Telegraph, 18
Houston Light Guard, 18, 20, 21-22
Houston, Sam, 4, 5, 7, 10
Hube, Hans-Valentine, 145
Hughes, George, 92, 94
Hulen, John A., 35, 36, 37, 90
Hurricane Carl, 266
Hutchings, Henry, 35

I—
Indian attacks, 15
Indian "problem," 10-11
Indian troops, 74-75
Indians, 37, 38, 40, 41
isolationism, 96, 100, 102, 104
Italy, 121-143, 144-157, 158-177, 178-194

J—
Jackson, Andrew, 4

Japan, 264
Jenkins, Sam, 16
Jodl, Alfred, 264
Johnson, Andrew, 10, 15, 266-267

K—
Karankawas, 3
Kelly, Charles E., 135-136
Kesselring, Albert: Anzio invasion, 159; Italy, alliance, 125; Italy, strategies, 122, 125-126, 174, 180, 197; Rapido crossing, 161, 164, 168, 174; Rome, Italy, 192; skills, 159, 164, 179, 195, 257; Velletri, Italy, 183
Keyes, Geoffrey: Italy, strategies, 153; Rapido crossing, 161, 162, 163-164, 165, 171, 172-173, 176, 177; San Pietro, Italy, 149-150, 151, 152; II Corps, 149-150
Kneiss, Baptist, 218
Knox, Frank, 103
Korean War, 266
Kouser, Bruce, 96, 105, 114, 228, 229-230
Kreuger, Walter, 118
Ku Klux Klan, 13

L—
La Cosa Creek line, 139
La Cosa Creek, 138, 139
La Pieta, Italy, 167
Lange, Otto, 138, 139-140, 142
Le Bonhomme Pass, 242
Lee, Robert E., 10
Lejeune, John A., 57-58, 60, 61, 71
Lewis, Spencer S., 197, 204
Lincoln, Abraham, 6
Liri Valley, 148, 159, 160, 162, 166, 172, 175, 179
Logan, John, 28
Long Island, New York, 45
"Lost Battalion," 235-240
Lucas, John P., 142, 159, 174, 182
Ludendorff, Erich, 51, 55
Lusitania (ship), 100
Lynch, George E.: 142nd Regiment,

188; Operation Anvil/Dragoon, 204, 205; Ste. Marie Pass, 243-244, 245

M—
M1 rifle, 115
Mabry, Woodford H., 26
MacCombie, Herbert E., 191, 260
Machault, France, 69
Maginot line, 224, 248
Maine (battleship), 24
Maquis, 196
Marne River, 163
Marriott, Hugh F., 52
Marseille, France, 196, 203, 206
Marshall, George: draft, 101, 103; leadership, 110, 114; morale of troops, 79, 110; officers, 109- 110, 113; promoted, 97; Roosevelt, Franklin D., 102; training of troops, 110, 111-112
Martin, Captain, 7
Martin, Colonel, 138, 173
Matagorda County, Texas, 20-21
McCall, Thomas E., 171-172
McCreery, Richard, 138, 162, 165
McGee, L. E., 94
McKinley, William, 25, 31, 38
McNair, Leslie J., 99, 110, 113
McNelly, L. H., 16
McShane, LTC, 142
Mein Kampf, 261
Messina, Jerry, 20, 21
Meurthe River, 241-242, 243
Meuse River, 56
Mexican Americans, 40
Mexican revolution, 27
Mexican War, 4, 5, 23
Middleton, Troy, 138
Mignano Gap, 145, 149, 150
Military Policy of the United States, The (book), 29, 31
Military Training Camps Association (MTCA), 100-101
militia, viii, 1-5, 16, 17, 18, 19, 23, 24, 26, 32
Militia Act of 1903, 32-33

Monte Artemisio, 182, 183, 184, 186, 187, 189
Monte Camino, 149
Monte Cassino, 145, 166, 168, 181
Monte Castellone, 179
Monte la Difensa, 150-151
Monte la Remetanea, 151
Monte Lungo, 150, 153, 154-155
Monte Maggiore, 150, 151
Monte Rotondo, 157
Monte Sammucro, 147, 149, 152, 153, 154-157
Monte Soprano, 130-131
Monte Trocchio, 167
Montecorvino, Italy, 123
Montelimar Battle Square, 211
Montelimar, France, vii, 207, 208-221
Montgomery, Bernard: Colmar Pocket, 247-248; 8th Army, 132; Rhine crossing, 257-258; Rome, Italy, 148; Salerno, Italy, 131, 132, 142; uncooperativeness, 247
Moody, Dan, 94
morale, 227
Mortagne River, 234
Moselle River, 223
Muchert, Jules S., 45-46
Murphy, Audie, 266
Mussolini, Benito, 125

N—
Naples, Italy, 124, 144-145
National Association for Universal Military Training, 81
National Defense Acts, 36, 80, 81-83, 98
National Emergency Committee (NEC), 102
National Guard (*see also* Texas National Guard, 36th Division): development, 24; Dick Act, 32- 33; division designations, 34-35; equipment, 43; formation, 33; Mexican Revolution, 27; National Defense Act of 1920, 81-83; National Defense Act of 1933, 83; officers, 43; organization, 40; recruitment,

36-37; Root, Elihu, 32; structure, 34-35; training, 39-44; use, 25; World War I, 28, 39, 40-48; World War II, 97, 102-103
National Guard Association, 104, 109
Naulin, Stanislas, 60, 68, 71, 74
Nazi atrocities, 260-263
New London, Texas, 95
New York City, 45
1940 Draft Law, 105
Nisei Regiment, 233, 236
Normandy Invasion, 110, 192, 195, 196, 205
Norton, A. B., 5
Nuckols, John, 20, 21

O—
O'Daniel, John E., 138, 140, 141, 226, 196
O'Ryan, John F., 102-103
offensive warfare, 47-48, 54
Oklahoma, 37
Oklahoma militia. *See* Oklahoma National Guard
Oklahoma National Guard, 35, 37, 38-39, 40
Oklahoma Territorial Militia, 38
Omaha Beach, 205
Operation Anvil/Dragoon, 195-207
Operation Diadem, 180
Operation Dogface, 229, 231, 239, 242
Operation Northwind, 251-255
Operation Slapstick, 131
Outlaw, Ned, 17
Owens, Charles H., 236

P—
Padgitt, James T., 131
Palmer, John McAully, 81, 101
Panama Canal, 25
Panther Division. *See* 36th Division
Parks, John Bob, 188, 192
Patch, Alexander M., 196, 197, 204, 208, 209, 213, 222, 245, 255
Paterson, Robert C., 101, 175

Patton, George S., 117, 124, 222, 245-247, 248, 255, 258
Pearson Guards, 22
Pease, Elisha M., 13, 14, 15
Persano crossing, 137
Pershing, John J.: Burke-Wadsworth Bill, 104; European command, 53-54, 61, 77; Forest Farm, 73; Greble, Edward, 44; Marshall, George, 97; morale of troops, 79; offensive warfare, 47, 53- 54, 73; preparation of troops, 42, 55, 60; 36th Division, 48; universal military training, 81
Persian Gulf War, 265-266
Philippines, 104, 105
Philippines rebellion, 25, 31
Plattsburg Training Camp Movement, 100
Plombieres-Les-Bains, 227-228
Powell, Colin, 266
Powers, Lieutenant, 189
Prax, General, 72, 73-74
prohibition, 85
prostitutes, 44
Puck, Armin, 99, 256
Pullman Sleeping Car Company, 24
Puy St. Martin, 211-212

R—
race riots, 20-21, 22, 87, 92, 93-94
racism, 12, 13-14, 15, 20, 21, 81
railroad strikes, 24
Rambervillers Forest, 234
ranging companies, 3-4, 5
Rapido crossing, 145, 158-177, 268
Reagan, John H., 11-12
Reconstruction, 9-16, 23, 24
Reconstruction Acts, 13, 14, 23
Red Cross, 34, 86
Reese, Hal, 191
Reichardt, Frank A., 21, 22
rest camps, 227-228
return to normalcy, 80
Revolutionary War (American), 2
Reynolds, Joseph J., 14
Rhone River, vii, 206, 208, 210

Riviera campaign, 195, 196, 199, 212-213, 246, 260
Roberts, Oran M., 13, 19
Rodt, General, 162
Rommel, Erwin, 122, 125-126, 143
Roosevelt, Franklin D., 96, 98, 100, 101-102, 110
Root, Elihu, 30-32
Ross, Governor, 21, 22
Rough Riders, 38
Rutherford B. Hayes, 24

S—
Salerno, 121, 124, 126
San Jacinto, 4, 23
San Juan Hill, 38
San Pietro, Italy, 144, 145, 150-157
San Vittore, 154
Sanger, William, 31-32
Santiago, Cuba, 38
Sappington, H. O., 88
Schmalz, Wilhelm, 183
secession, 5, 12
Sedan, France, 56
Sele plain, 124
Sele River, 123
Sele-Calore corridor, 133-134, 136-138, 140
Selective Service Act of 1940, 99-100, 105
Selective Service System, 105
Selestat, France, 243, 249, 250
Sells, Cato, 41
Senate Joint Resolution 286, 104
Shapiro, Jerome N., 263
Shepard, Oliver, 20, 21
Sheridan, Philip, 13, 17, 30
Sherman, Texas, 92-95
Sicily, Italy, 122-123
Sides, Bruce, 78-79
Siegfried line, 254, 256, 257
Sitzkrieg (phony war), 96
Six-Division Plan, 54
Sladen, Fred W., 205, 214
slavery, 12, 13-14
Sloan, John, 113, 119
Smith, Edmund Kirby, 9

Smith, William R.: Aisne crossing, 70-71; armistice (World War I), 79-80; described, 45; Forest Farm, 67, 70-71, 73-75; promoted, 45; St. Etienne, France, 57; 36th Division, 45
Smuts, Jan, 121
Solofrone River, 128
Sorrento Peninsula, 124
Spanish-American War, 22, 24-25, 26, 30, 31, 32, 38, 80
St. Croix, France, 244-245
St. Etienne, France, 59-65
St. Raphael, 198, 201-205
St. Tropez, 198, 201
Stack, Robert I., 210, 214, 263
Stars and Stripes (newspaper), 79, 251
states' rights, 6
Ste. Marie Pass, 243-244
Ste. Marie-aux-Mines, France, 243-244, 245
Ste. Maxime, 198
Steele, William, 19
Sterling, Ross, 92
Stewart, Richard A., 152
Stimson, Henry L., 101, 103
Stonewall Brigade, 268
Stoval, Oran, 163, 183, 188
Straits of Medina, 122, 123
Strasbourg, France, 245, 251
Sunday, Zeb, 224
Swinton, E. D., 52

T—
Tallman, Clayton T., 134-135
tanks, 52, 154, 156, 188
Taranto, Italy, 131-132
Task Force Butler, vii, 200, 207, 209-210, 215, 217, 218
Taylor, Zachary, 5
Templer, Major General, 150, 154
Texas Highway Patrol, 95
Texas militia , 3-5, 16, 17, 18, 19, 23, 26
Texas National Guard: Dick Act, 33; disaster relief, 85-86, 95, 98; economics, 90-91, 91-92; federalization, 35, 83, 90, 98, 103, 104; fund-

ing, 98; Galveston, 87-89; inspection, 85; law enforcement, 87-89, 92-95, 98; mobilization, 104; National Defense Act of 1920, 83; officer recruitment in, 36; oil industry, 92; recruitment, 90-91, 98; reorganization (1920), 89-90; size, 97, 98; southern militarism, 91; supplies, 84; training, 83-84, 91, 99; uniforms, 99
Texas Rangers, 3-4, 23, 87, 92
Texas Revolution, 4
Texas State Guard, 15, 16
Texas State Police, 15, 16, 17
Texas Volunteer Guard (*see also* Texas National Guard): described, 22-23; formation, 18-20; race, 40- 41; segregation, 40-41
Texas War of Independence, 4
Thirteenth Amendment, 12
36th Division (*see also* National Guard, Texas National Guard, World War I, World War II): accidents, 43; American Indians, 41; Baptists, 108; Condilac Pass, vii-viii; creation, 39; equipment, 42; federal recognition, 90; federalization, 108; formation, viii, 35; France, 46-47; illness, 43-44; infantry designation, 89; officers, 43, 48; Oklahoma guard, 39, 40; organization, 39-40; patches, 79; race, 41; Task Force Butler, vii-viii; training, 39-44, 47-48, 79; World War I, 39, 40-48; World War II, vii-viii, 109
36th Division Association, 174
Throckmorton, James W., 12, 13
Toulon, France, 198, 203, 206
Truscott, Lucian K., Jr.: Dahlquist, John E., 220, 221; 5th Army, 231, 264; manpower, 140; Montelimar, France, 211, 212, 213, 214, 215, 216-217, 218, 220, 221; Operation Anvil/Dragoon, 204-205, 206, 209; Operation Dogface, 229, 231; Rapido crossing, 161; Rome, Italy, 222; strategies, 223; Task Force Butler, 200, 209, 211; 3rd Division, 140, 161; 35th Division, 184, 212, 213, 214, 232; Velletri, Italy, 182, 184, 187, 188
Tucker, Reuben H., 141

U—
U-boats, 53
Ultra, 125, 131, 142, 159, 166, 174, 176, 192, 199, 212, 251
Umberto, Prince, 154
United States Volunteer Cavalry, 38
universal military training, 81-82, 100
Upton, Emory, 28-30, 31-32
Utah Beach, 205

V—
Valentine tanks, 156
Velletri, Italy, 182, 183, 185, 186-191, 193-194, 269
Vietnam War, 266-267
Vietri Pass, 124
Villa, Francisco "Pancho," 27
Vincent, Stewart T., 213, 214
Volturno River, 145
von Bismarck, Otto, 30
von Griem, Ritter, 263
von Rundstedt, Gerd, 250-251, 263
von Senger u. Etterlin, Gen. F., 161-162, 163
von Steuban, Baron, 4
von Vietinghoff, Heinrich, 126, 136, 145, 264
von Wietersheim, Wend, 218
Vosges Mountains, 206, 208, 224-225, 231
Voss, Evan E., 116

W—
Wadsworth, James W., 102
Wadsworth, Sheriff, 20, 21
Walker, Charles, 173
Walker, Edwin A., 196
Walker, Fred L., Jr.: Rapido crossing, 162-165, 169-171, 172-173, 175, 176-177, 178-179, 181; Salerno in-

vasion, 123, 127-128, 129, 132, 137, 142; San Pietro, Italy, 150, 152, 153, 154; Texas National Guard, 266; 36th Division command, 108, 117-118; Velletri, Italy, 182-183, 184-185, 188, 190, 191, 269
War of 1812, 3
Washington, George, 2
weapons, 115
Weigley, Russell, 30
Wells, H. G., 52
Werner, Richard, 129, 173
West, OB, 197, 200
Whitworth, General, 61
Wiegley, Russell, 248
Wiese, Friedrich, 225
Wilbur, William H., 138, 140, 173
Wilkie, Wendell, 104
Wilson, Woodrow, 27, 34
Winter line, 126, 145, 148, 150, 152, 164, 174, 177
Wolters, Jacob F., 88, 90
Wood, Leonard, 100
Woodring, Harry H., 102, 103
World War I, 28, 33-35, 39-48; Armistice, 78-79, 80; Battle of the Marne, 50; Battle of Verdun, 51; casualties, 51, 66, 77, 97; Forest Farm, 67, 73-77; German advance, 50; Hindenburg line, 56-57; illness, 75, 79, 43-44; Indian code talkers, 74-75; Six-Division Plan, 54-55; St. Etienne, France, 59-65; tactical innovations, 50, 51, 52, 54; tanks, 52; training, 39-44, 47-48, 52, 53, 54, 55, 79; trench warfare, 49-50, 54
World War II: American neutrality, 96-97; beginnings, 96; casualties, 115-116, 130, 148, 155, 182, 187, 220, 230, 238, 256-257; conscription, 102, 103, 104, 105-106; discipline, 110; France, 195-206; German strategies, 102, 129-130, 140, 152-153, 185; isolationism, 96, 100, 102, 104; Italy, 121-143, 144-157; mines, 167-169; Montelimar, France, 208-221; morale, 110, 113, 135, 173, 254; officers, 109-110, 113, 118; replacement system, 117, 118-119, 120; Salerno, Italy, 121-143; San Pietro, Italy, 144-145, 150-157; soldiers' attitudes, 110-111; supplies, 146-147, 151, 220, 232, 237-238; tanks, 156-188; training, 109, 112, 113, 114, 115, 118-119, 120
Wyatt, Aaron W., 167

Y—
yellow fever, 26
Young, Ross, 244

Z—
Zimmermann telegram, 28